TEMPERATURE–SALINITY ANALYSIS OF WORLD OCEAN WATERS

Elsevier Oceanography Series, 11

TEMPERATURE – SALINITY ANALYSIS OF WORLD OCEAN WATERS

by

O.I. MAMAYEV

Translation from the Russian by *Robert J. Burton*

ELSEVIER SCIENTIFIC PUBLISHING COMPANY   *Amsterdam - Oxford - New York 1975*

ELSEVIER SCIENTIFIC PUBLISHING COMPANY
335 Jan van Galenstraat
P.O. Box 211, Amsterdam, The Netherlands

AMERICAN ELSEVIER PUBLISHING COMPANY, INC.
52 Vanderbilt Avenue
New York, New York 10017

Library of Congress Card Number: 74-10260

ISBN 0-444-41251-4

With 111 illustrations and 29 tables

Printed in The Netherlands

FOREWORD TO THE ENGLISH EDITION

In the preparation of the present edition the text of the monograph was critically reviewed, revised and expanded, mainly for the purpose of bringing it up to date and reflecting the results of new studies in the field of T-S analysis published since the Russian edition of the monograph, but also to remove errors and inaccuracies which had occurred in that edition.

The second and third chapters were the subject of the greatest revision, so much so that one may say that they have been completely rewritten. In addition, the following new sections have been added to these chapters: 7, 8, 10 and 11 (Chapter 2) and 17 and 22 (Chapter 3).

The revision of Chapters 2 and 3 reflects the interest in the equation of state and the thermodynamics of sea water displayed in the literature in recent years.

Chapter 5 has been enlarged by the addition of section 38, "The density flux function in oceanographic T-S analysis", which reflects the supplementary results obtained by Veronis in studying orthogonal functions on a T-S diagram (Section 33).

Thus the total number of sections has increased from 56 in the Russian edition to 63 in the present version.

Three tables (A8, A9 and A10) of numerical values of constants in the different modifications of the equation of state of sea water and one table (A11) of the polynomial expression for the density flux function have been added to the Appendix. Tables A4 and A5, which contained errors, have been corrected according to the corresponding tables of Fofonoff and Froese (1958) which are more precise.

The monograph has also been provided with a list of main symbols and a subject index.

For a number of reasons, the review of new research in the field of T-S analysis in the monograph does not go much beyond 1972 (with some exceptions). Therefore, the author has added in proof a selected, annotated list of new publications devoted to the development of T-S analysis (pp. 367–368).

O. Mamayev

CONTENTS

CHAPTER 1

INTRODUCTION

The present monograph is devoted to the theory and practice of *oceanographic T-S analysis*, the purpose of which is the identification and study of the water masses of the World Ocean, of their interaction and transformation, as well as of the processes of heat and salinity exchange occurring in them, by means of one of the most important characteristic diagrams of state in physical oceanography, the T-S diagram (T — temperature, S — salinity of sea water).

The history of T-S analysis goes back to the time when Forch et al. (1902) proposed an equation of state of sea water, and Helland-Hansen (1912) constructed a T-S diagram on which he plotted the real T-S relations of the waters of the ocean (1918). Helland-Hansen realized at once the advantages of the method of T-S curves he had proposed and the far-reaching possibilities it offered as compared with other methods of analyzing the waters of the ocean.

However, the exceptional theoretical and practical "capacity" of the T-S diagram was realized considerably later, since Defant and Wüst succeeded in 1929 in plotting the geometric images of primary water masses on a T-S diagram, when Defant demonstrated at the same time the applicability of equations of heat conduction and diffusion to the analysis of T-S relations, since Jacobsen in 1927 proposed a method of determining coefficients of exchange with the help of the T-S diagram and, finally, since Shtokman laid the foundations of the analytical theory of T-S curves in 1943 (Shtokman, 1943a, b).

In the course of approximately the last forty years, the development of thermohaline analysis has proceeded in two directions. The *first direction* has been concerned with the study of the water masses of the World Ocean, based on the surprising conservation of the images of the water masses on the T-S diagram; this trend, although it was to a certain extent empirical, has contributed very much to the understanding of the thermohaline structure of the ocean.

Here we should mention the work of Jacobsen and Helland-Hansen and Nansen (1929–1930) devoted to the waters of the North Atlantic, Thomsen's (1935) studies of the water masses of the southern parts of the Indian and Pacific Oceans, Wüst's (1935) work on the water masses of the stratosphere of the entire Atlantic Ocean, Iselin's (1936) study of the waters of the western

North Atlantic, Montgomery's (1938) work on the equatorial regions of the
Atlantic, Dobrovol'skii's research on the water masses of the northern part
of the Pacific Ocean (Dobrovol'skii, 1947, unpublished manuscript; also 1961,
1962), Timofeev's (1960) investigations of the water masses of the Arctic
Basin and, finally, the work of Sverdrup et al. (1942) on the types of water
masses of the entire World Ocean.

The authors of these fundamental works have revealed the thermohaline
structure of the waters of the World Ocean and have contributed many theoretical and methodological innovations to T-S analysis; among these are the
establishment of analytical relationships between the temperature and the
salinity of water masses (Helland-Hansen, Iselin), the use of nomograms and
triangles of mixing (Thomsen), the development of the "core method" (Wüst)
and of the method of isopycnic analysis of water masses (Parr and Montgomery), the discovery of water-mass structures connected not only with
thermohaline but also with circulation phenomena (Dobrovol'skii), as well as
other achievements.

The *second direction* in the development of T-S analysis has been characterized by the study of the analytical properties of T-S curves and of other
T-S relations of the real waters of the ocean, as well as (to a lesser extent) of
that background which the equation of state of sea water represents on a
characteristic thermohaline diagram. In 1943 Shtokman laid the groundwork
for the theory of analytical T-S curves; this theory was then developed by
Ivanov in 1944–1949. A further contribution to the theory of T-S relations
was made by Stommel, who examined certain questions of thermohaline convection as represented on a T-S diagram, as well as the causes of the formation
of the T-S curve as an image of water masses (1961, 1962). The equation of
state is introduced in Stommel's work, although in a simplified (linearized)
form. Zubov (1947, 1957b) and Zubov and Sabinin (1958), as well as Fofonoff (1961), developed the theory and practice of calculating contraction on
mixing — a most important factor in the interaction of water masses caused
by the non-linearity of the equation of state of sea water. Finally, the author
of the present work proved the theorem of the existence of a line integral in
the T-S plane (Mamayev, 1962b), thus making it possible to apply the principles of field theory to the study of the T-S relations of the ocean's water
masses.

It should be pointed out that the theory of T-S analysis and its methods
are practically inseparable and that therefore the division of T-S analysis into
two directions, or trends, is largely artificial: both groups of studies listed
were mutually complementary, and the development of one direction would
have been impossible without the other.

It is perfectly evident that T-S analysis is an integral part of physical
oceanography as a whole; it represents a kind of "theory of water masses",

and this theory is just as significant for the development of oceanography as a whole as, say, the theory of ocean currents. Incidentally, the interpenetration of the latter two theories has become quite apparent precisely in recent years, in connection with the development of the theory of abyssal circulation and the theory of the thermohaline circulation of the ocean as a whole.

In the past, comparatively little place was given to questions of T-S analysis in monographs and textbooks on physical oceanography; a start was made by Sverdrup et al. in the monograph *The Oceans,* where, for the first time in the literature, a comprehensive chapter on the water masses (and currents) of the whole World Ocean was included. Subsequently, considerably more attention began to be paid to questions of T-S analysis. In corroboration of this, the following examples may be quoted. In the two-volume monograph by Defant (1961), *Physical Oceanography*, there is a chapter entitled "The T-S relation and its connection with mixing processes and the main water masses". In the textbook by Lacombe (1965), *A Course in Physical Oceanography*, the chapter "T-S diagrams and the analysis of water masses" as well as a section under the same title in *Principles of Physical Oceanography* by Neumann and Pierson Jr. (1966) are devoted to the same question; finally, in the comparatively brief, but fully up-to-date course by McLellan (1965), *Elements of Physical Oceanography*, there is also a chapter on "Temperature—salinity relations". Thus, in all four of the books mentioned, there is an exposition of the fundamentals of T-S analysis, although very brief and given practically only in its qualitative form.

At the present time, the need has made itself felt to provide a sufficiently complete exposition of the fundamentals of T-S analysis; the monograph now offered to the reader accordingly represents the first attempt of such a kind. In writing it, the author was guided by the idea that the problems of T-S analysis itself should be preceded by a statement of questions connected with the equation of state of sea water and its thermodynamics, for the ocean, filled with sea water, represents a completely unique gigantic natural thermodynamic system: the picture of the vital activity of the ocean and such of its features as, say, the presence of a broad category of intermediate waters, characterized by extremes of temperature or salinity, are determined precisely by the properties of sea water as a thermodynamic system. Here attention should be drawn to the indissoluble link of T-S analysis as a branch of physical oceanography with thermodynamics as a whole; such problems of T-S analysis as mixing and contraction on mixing of water masses represent in fact problems of the thermodynamics of irreversible (non-equilibrium) processes. In Chapter 3 of the monograph, the fundamentals of the thermodynamics of sea water as a two-component (binary) system are set forth. The content of this chapter could have formed the subject of a separate study covering a wider range of questions than those which are considered in this book; the author realizes this, as well as the fact that some questions in the first two chapters, devoted to the equation of state and to thermodynamics, are set forth more fully than is required for what follows. This has been done in order to lay the foundation for the further development of thermohaline analysis within the framework of thermodynamics to a greater extent than exists at the present day.

Here it may be observed that questions of the thermodynamics of sea water are in point of fact only just beginning to be studied in any serious way. The groundwork was laid by Proudman, who had published the chapter "The Thermodynamics of the Ocean" in his classic course *Dynamic Oceanography* (1953); then the important works of Fofonoff should be noted, *Energy Transformations in the Sea* (1961) and his article "Physical properties of sea water" (1962), where the problems of the thermodynamics of sea water are set forth at some length, as well as the monograph of Horne (1969) *The Chemistry of the Sea*, which contains an extensive chapter on the thermodynamics of sea water.

Attention is also drawn to the need for studying the thermodynamics of sea water in the monographs of Eckart (1960), *Hydrodynamics of Oceans and Atmospheres*, and Phillips (1966), *The Dynamics of the Upper Ocean.*

Thus, the method of exposition is deductive and conforms to the following sequence: (1) the equation of state and the thermodynamics of sea water; (2) the properties of the characteristic *T-S* diagram, which represents the "thermodynamic background" for the superposition of the images of real water masses; (3) the analytical *T-S* relations, which model the natural waters of the ocean; (4) the practical methods of analysis resulting from the theory; (5) the water masses of the World Ocean as subjects investigated by the methods of *T-S* analysis. This sequence is sufficiently well reflected by the table of contents, and therefore we shall not elaborate on it any further.

The present monograph is closely related to another monograph of the author, *Oceanographic Analysis in the $\alpha-S-T-p$ System* (Mamayev, 1963), in which questions of the study of the stratified waters of the ocean are considered, going beyond the direct scope of *T-S* analysis, as well as practical methods for the principal oceanographic calculations.

The Appendix contains several tables which do not appear in the *Oceanological Tables* of Zubov (1957a), but which may prove necessary in work on *T-S* diagrams.

The numbering of the sections, figures and tables is continuous; formulae are numbered within each section. The tables in the text are numbered in Roman; tables in the Appendix are numbered A1−A12.

ACKNOWLEDGEMENT

At different times the author has had the opportunity of receiving advice from, and consulting with, different scientists on the questions dealt with in the present work. The most valuable advice was provided by Prof. S.S. Voit, Prof. A.D. Dobrovol'skii, Prof. A.I. Duvanin, Prof. V.G. Kort, Dr. K.D. Sabinin, Academician V.V. Shuleikin and the scientific editor of the monograph, Dr. V.V. Panov, as well as the comments of the reviewers of the manuscript of the monograph, Dr. G.I. Baranov and Dr. A.O. Shpeicher. The author expresses his particular gratitude to the memory of the late Professor V.B. Shtokman for his favorable substantive evaluation of the present work.

Yu.K. Gordienko, A.I. Koshitz and A.V. Shumilov extended great technical assistance to the author.

LIST OF MAIN SYMBOLS *

a coefficient of thermal (volumetric) expansion
c (1) speed of sound; (2) specific heat capacity
c_p specific heat at constant pressure
c_v specific heat at constant volume
Cl chlorinity
d dynamic height
D dynamic depth
E (1) Young's modulus (modulus of elongation); (2) vertical stability
F force
f specific free energy
g (1) gravity acceleration; (2) specific thermodynamic potential (Gibbs' potential)
h (1) height; (2) specific enthalpy
I integral along T-S curve
J (1) mechanical equivalent of heat; (2) Jacobian
k (1) coefficient of isothermic compressibility; (2) coefficient of turbulent mixing
k_S coefficient of turbulent salinity diffusion
k_T coefficient of turbulent heat conduction
l length
M, m mass
m_i mass of component i in mixture
n_i number of moles of substance i per unit volume
p pressure

---

* This list does not include generally accepted mathematical symbols, or symbols which appear in the text only once.

p_a atmospheric pressure
R (1) universal gas constant; (2) electrical conductivity
S salinity
s specific gravity of sea water
T temperature
T absolute temperature
t time
U internal energy
V volume
v specific volume or conventional specific volume of sea water (depend-
          ing on the context)
v_T conventional specific volume of sea water at atmospheric pressure
$\bar{v}$ partial molar specific volume
u,v,w velocity components
x number of particles in a one-component system
x_i mass fraction of substance i
x,y,z Cartesian coordinates
α (1) specific volume of sea water; (2) coefficient of thermal (volumetric)
          expansion
β coefficient of saline "contraction"
Γ adiabatic temperature gradient
γ (1) thermal coefficient of pressure (elasticity); (2) Poisson's coefficient,
          $\gamma = c_p/c_v$; (3) "density flux" function (orthogonal function in relation
          to ρ)
Δ anomaly of specific volume of sea water
Δ_{ST} anomaly of specific volume (thermosteric anomaly)
δ (1) thermohaline coefficient; (2) specific volume anomaly; (3) specific
          volume correction (general expression)
ϵ (1) anomaly of density (general expression); (2) specific internal energy
ζ coordinate (in section 37)
η (1) specific entropy; (2) coordinate (in section 37)
θ (1) potential temperature; (2) temperature of maximum density of sea
          water
μ (1) mean coefficient of compressibility of sea water; (2) chemical poten-
          tial; (3) angle of slope of tangent to T-S curve
ρ density
ρ_θ potential density
$\Sigma(\text{\textperthousand})$ actual sum of ions in sea water
Σ_0 conventional specific gravity of pure water at $0°C$
Σ_T conventional specific gravity of pure water at $T°C$
σ conventional specific gravity of sea water
σ_0 conventioanl specific gravity of sea water at $0°C$

σ_T conventional specific gravity, as well as conventional density, of sea water at $T°C$

σ_θ conventional potential density of sea water

τ freezing point of sea water

Φ probability integral

φ velocity potential

ψ stream function

CHAPTER 2

THE EQUATION OF STATE OF SEA WATER

1. THE PARAMETERS OF STATE OF SEA WATER

Sea water, like any single-phase thermodynamic system, is characterized by certain physical properties which represent the quantitative features of the system. These quantities are called the *parameters*, or the characteristics of state of the system. For sea water the parameters of state are mass, volume, pressure, temperature and salinity. The first three of these parameters are mechanical; temperature is a thermodynamic parameter, while salinity and the electrical conductivity associated with it are physico-chemical parameters.

Instead of mass and volume, it is more convenient to use specific parameters: specific gravity and density, as well as a quantity which is inverse to density — specific volume.

In the following subsections each of the parameters of state of sea water is considered separately.

Temperature

In oceanography, temperature is measured in degrees centigrade, or Celsius (°C); standard accuracy in determining temperature is 0.01°C. We shall distinguish between temperature in situ, designated by the letter T, and potential temperature θ. Temperature in situ is the temperature measured in a corresponding point in the sea by thermometer; potential temperature is the temperature of a particle adiabatically reduced to normal (atmosphere) pressure, i.e., that temperature which would be observed in this point if there were no adiabatic processes (see Section 27). In some cases it is necessary to express temperature in degrees of the absolute Kelvin scale (°K). The degrees of the Kelvin scale are equal in magnitude to degrees centigrade (1°C = 1°K), but are calculated from absolute zero, equal to 273.16°K below 0° centigrade. Temperature on the Kelvin scale is designated by the letter T in boldface.

ity and electrical conductivity

Forch et al. (1902, p. 116) give the following definition of salinity, which has been adopted in oceanography: "Salinity is defined as the weight in grams of the dissolved solid material contained in 1 kg of sea water when all the bromine and iodine have been replaced by an equivalent quantity of chlorine, all the carbonates converted into oxides and all organic matter oxidized".

It is clear from this definition that salinity S is measured in grams per kilogram, or parts per thousand, per mille (‰).

Thanks to the quasi-constancy of the saline composition of the World Ocean, established as early as 1859 by Forchhammer, salinity is determined from one of the saline components, chlorine, by the chlorine titration method (standard method).

The relationship between salinity and chlorinity for the waters of the World Ocean has the form:

$$S(‰) = 0.030 + 1.8050 \, Cl \qquad [1.1]$$

This formula (Knudsen, 1901) is valid for the range of salinities of the World Ocean from 2.69 to 40.18‰.

Making the concept of chlorinity more precise, Jacobsen and Knudsen (1940) proposed another definition, understanding by chlorinity "the number of grams of atomic weight silver necessary to precipitate the halogens in 0.3285234 kg of sea water".

Lyman and Fleming (1940) (see also Bruevich, 1961), derived the following formula for the relationship between the actual sum of ions $\Sigma(‰)$ and chlorinity:

$$\Sigma(‰) = 0.073 + 1.8110 \, Cl \qquad [1.2]$$

Subsequently, Lyman (1959) elaborated on this formula, obtaining it in the following form:

$$\Sigma(‰) = 0.069 + 1.8112 \, Cl \qquad [1.3]$$

Comparing formula [1.1] with formulae [1.2] and [1.3], we see that salinity is not the actual sum of ions. As Defant (1961) points out, salinity, obtained according to Knudsen's formula, is approximately 0.14‰ less than the actual salt content of sea water. However, in practice, Knudsen's formula [1.1] is used as reflecting the conventional value for salinity according to the definition of Knudsen et al. quoted above. The conversion of the conventional value "salinity" into the actual sum of ions and vice versa is carried out in the way demonstrated by Bruevich (1961).

In 1962, the Unesco Joint Panel on the Equation of State of Sea Water

(UNESCO, 1962; Ivanov-Frantskevich, 1963), instead of formula [1.1], proposed the formula:

$$S(\permil) = 1.80655 \, Cl(\permil) \qquad\qquad\qquad [1.4]$$

Formula [1.4] is obtained on the condition that it should not have a free term, and at the same time, should correspond to formula [1.1] with a salinity of 35‰; it is, indeed, from this condition that the chlorine coefficient of 1.80655 is calculated. The difference in the determination of salinity S according to formula [1.1] and salinity S' according to formula [1.4] is equal to:

$$\Delta S = S - S' = 0.03(1 - S/35) \qquad\qquad\qquad [1.5]$$

For various salinities this difference amounts to:

$S(\permil)$	$\Delta S(\permil)$
0	+0.03
10	+0.021
20	+0.013
30	+0.004
35	0.000
40	−0.004

In the range of salinities $32 \leqslant S \leqslant 38\permil$ the discrepancy amounts to only ± 0.0026‰, which is less than the error which occurs in determining salinity by modern methods. In addition, formula [1.4], which does not contain a free term, makes the concept of the chlorine coefficient S/Cl and of salinity itself more definite *.

The relationship between the salinity and the chlorinity of the waters of landlocked seas (the Caspian and the Aral), as well as of Mediterranean seas such as the Baltic, the Black Sea, the Sea of Azov and others, is substantially different from [1.1] and [1.4]. The relations between salinity and chlorinity for these seas are given in the *Oceanological Tables* (Zubov, 1957a, p. 330) and the *Oceanological Tables for the Caspian, Aral and Azov Seas* (1964).

In connection with the introduction of new instruments to determine salinity based on the electrical conductivity of sea water — salinometers (Schleicher and Bradshaw, 1956; Cox et al., 1968), it has become necessary to find a new exact relation between salinity and conductivity. This work was

---

* Formula [1.1], on the other hand, leads to the relation $S/Cl = 1.8050 + 0.030 \, (Cl)^{-1}$, which does not satisfy this condition. This question is considered in detail in Carritt's report (Unesco, 1962).

also performed by the Unesco Joint Panel; as a result the following relation
was obtained:

$$S(\text{‰}) = -0.08996 + 28.29720\,R + 12.80832\,R^2 -$$

$$10.67869\,R^3 + 5.98624\,R^4 - 1.32311\,R^5 \qquad\qquad [1.6]$$

In this formula $R = R_{\text{sample}}/R_{35\text{‰};15°C}$ is relative conductivity. The error
in the determination of conductivity according to formula [1.6], as compared with titration, salinity being defined according to formula [1.4],
amounts to approximately 0.008‰. Tables for the conversion of conductivity into salinity have been prepared according to formula [1.6] (International Oceanographic Tables, 1966). In 1967 these Tables were confirmed
by an international agreement (on this question see Wooster et al., 1969, and
also Lyman, 1969).

Formula [1.6] defines the relationship between conductivity and salinity
at atmospheric pressure and does not take account of the effect of pressure
on conductivity. We will not dwell on this question here, referring the reader
to a sufficiently authoritative primary source (Bradshaw and Schleicher,
1965).

The accuracy of determination of salinity by titration amounts to 0.02‰,
with the introduction of salinometers the error in determining salinity varies
between 0.02 and 0.005‰ (Alekin, 1966). The records of the oceanographical
survey of the Atlantic Ocean, undertaken during the International Geophysical Year (1957–1958), published in the *Atlas of the Atlantic Ocean*
(Fuglister, 1960) represent the first data in which the values of salinities at
oceanographic stations are given with an accuracy to the third decimal place.
It is clear that such an increase in accuracy already calls for a revision in the
near future of the very concept of the constancy of the saline composition
of the World Ocean. The relations between salinity and chlorinity may
therefore prove different not only for different seas, but also for different
zones of the ocean itself.

Pressure

The unit of pressure in the CGS system is the dyn cm^{-2}. Since this unit is
very small, a unit introduced by V. Bjerknes is used, which is 10^6 larger and
called the bar:

1 bar $= 10^6$ dyn/cm^2 $= 10^6$ g cm^{-1} sec^{-2}

The unit of pressure in the MTS system is the centibar, 1 cbar $= 10^{-2}$ bar
$= 10^4$ dyn/cm^2. In oceanography the most widely used unit of pressure is the
decibar:

1 dbar $= 10^{-1}$ bar $= 10^5$ dyn/cm^2

The relation between pressure expressed in bars (10^6 g cm^{-1} sec^{-2}), and pressure in kg/cm^2 (in the technical system) is expressed by the formula:

1 kg/cm^2 = 980.665 $\cdot$ 10^3 g cm^{-1} sec^{-2}

Here g_n = 980.665 cm/sec^2 is the normal acceleration of gravity.

Let us consider what the height Δh of a unit water column of density ρ should be in order to exert on its lower base pressure Δp, equal to 1 dbar = 10 cbar:

$$\Delta p = g\rho\,\Delta h = 10$$

whence:

$$h = \frac{p}{g\rho} = \frac{10}{9.8 \times 1.03} = 0.9907 \text{ m} \approx 1 \text{ m}$$

(here $\rho = 1.03$ is the mean density of sea water). Because of the fact that a change in pressure of 1 bar corresponds to a change in depth of approximately 1 m, in oceanography depth is sometimes measured in decibars, particularly in the dynamic method (Zubov and Mamayev, 1956). However, in a number of cases, for instance in calculating the speed of sound in the sea (Section 19), such a substitution is incorrect, and the exact relationship between pressure and depth must be known. This relationship is determined by the compressibility of sea water and is extremely complex; however, non-linear effects are determined by the vertically integrated equation of hydrostatics [1.34]:

$$p(z) = p_a + \int_0^z g(z)\,\rho(z)\,\mathrm{d}z \qquad\qquad [1.7]$$

where p_a is atmospheric pressure and $\rho(z)$ the real density in situ (see below, the Section "Density and specific volume").

For a standard ocean, i.e., an ocean which would exist if one and the same salinity (35‰) and temperature (0°C) were observed at all its depths, relationship (1.7) can be approximated with a sufficiently high degree of accuracy by the formula:

$$p = 1.033 + 0.1028126\,z + 2.38\cdot 10^{-7}\,z^2 \qquad\qquad [1.8]$$

where p is expressed in kg/cm^2 and z in meters. Formula [1.8] was derived by Polosin (1967) by the method of the least squares on the basis of the relations between p, ρ and z, given in the table of parameters of the standard ocean (Appendix, Table A1), taken from Defant's monograph (1961, p. 305).

Below we will consider the state of sea water, in particular at atmospheric pressure. This pressure is conventionally taken in oceanography as "zero" pressure (the pressure on the surface of the sea, i.e., at "zero depth"), and the state of the sea water, subjected to the pressure of overlying layers (at depth), is defined precisely with reference to "zero" pressure. Designating "zero pressure" by the number "0", we must remember, however, that in so doing we do not take account of the pressure of the atmospheric column, which is equal to approximately 10 decibars, or more exactly 1.01325 bars, or 1.03323 kg/cm² ("normal", or "physical atmosphere") (Dorsey, 1968). In other words, the conventional "zero" pressure is equal to full pressure minus 1 standard atmosphere. In those cases were p represents full (absolute) pressure, as, for example, occurs in Eckart's equation [6.1]–[6.3] and if a comparison must be made with "zero" pressure, it is necessary to subtract from the former 10.3323 decibars (Sturges, 1970).

Density, specific gravity and specific volume

In the present section we consider ways of expressing these quantities; the physical relationships will be taken up in Section 2 and further on.

The density of water ρ (dimension g/cm³ in the CGS system) represents the mass, contained in a unit of volume. The specific gravity s represents the relation of the density of the given sample of water to the density of a standard sample; thus, specific gravity is a dimensionless quantity (it is used mainly in instrument determinations). The specific volume α, or v, (dimension cm³/g) is a quantity which is the inverse of density:

$$\alpha\rho = 1 *$$ [1.9]

Density and specific volume at constant (atmospheric) pressure are functions of temperature and salinity:

_____

* It should be pointed out that the liter (and, correspondingly, the milliliter) are also used as units of volume in oceanography. In accordance with the "old" definition of the liter (as the volume of 1 kg of pure water at a temperature of 4°C at atmospheric pressure), the specific volume of pure water at 4°C equals 1.000.000 ml/g = 1.000027 cm³/g, and its density is 0.999973 g/cm³ (Dorsey, 1968, p.203). All data on the density and specific volume of sea water, based on the classical determinations of Knudsen (1901; Bjerknes and Sandström, 1910; Sverdrup, 1933; Zubov, 1957a and other sources – see Sections 2, 3 and others) are in fact expressed in milliliters, inasmuch as the difference noted above was not known at that time. For the conversion of specific volume expressed in ml/g into specific volume expressed in cm³/g, it must be multiplied by 1.000027 (Pollak, 1961; Fofonoff, 1962). This correction does not apply to the anomalies of specific volume (see below).

According to the new definition of the liter as the volume of 1 cubic decimeter, adopted in 1964 by the General Conference on Weights and Measures, the concepts of the cubic centimeter and the millimeter become identical.

$$\rho = \rho(T,S) \tag{1.10}$$

$$\alpha = \alpha(T,S) \tag{1.11}$$

In the general case ("in situ"), they are functions of temperature, salinity and pressure:

$$\rho = \rho(T,S,p) \tag{1.12}$$

$$\alpha = \alpha(T,S,p) \tag{1.13}$$

Density defined as a function of potential temperature and salinity is called potential density ρ_θ:

$$\rho_\theta = \rho(\theta,S) \tag{1.14}$$

The standard precision of determination of specific gravity, density and specific volume in oceanography amounts to 10^{-5}; in some cases, for example in the preparation of tables, accuracy is brought up to 10^{-6}.

Let us consider ways of expressing the quantities mentioned above:

(A) Specific gravity

Oceanographers have adopted the following expressions of specific gravity as a function of salinity, $s_0 = s_0(S)$ and a function of temperature and salinity, $s_T = s_T(T,S)$, at atmospheric pressure (Knudsen, 1901) *:

$$s_0 = \frac{\text{density of sea water at } 0°\text{C}}{\text{density of pure water at } 4°\text{C}} \tag{1.15}$$

$$s_T = \frac{\text{density of sea water at } T°\text{C}}{\text{density of pure water at } 4°\text{C}} \tag{1.16}$$

The expressions of conventional specific gravity (or the anomalies of specific gravity) are used correspondingly; they were introduced by Knudsen to shorten the notation (number of signs):

$$\sigma_0 = (s_0 - 1) \cdot 10^3 \tag{1.17}$$

$$\sigma_T = (s_T - 1) \cdot 10^3 \tag{1.18}$$

(B) Density $\rho = \rho(T,S)$ and specific volume $\alpha = \alpha(T,S)$ under constant pressure

Instead of density ρ, in order to reduce the number of signs the quantity

---

* The seldom used expression of specific gravity $s_{17.5}$, determined at a temperature of samples of sea and pure water equal to 17.5°C, and its anomaly $\rho_{17.5}$, are not considered in the present work.

of *conventional density* is used:

$$\sigma_T = (\rho - 1) \cdot 10^3 \qquad [1.19]$$

which, as we see, is designated in the same way as the quantity of conventional specific gravity [1.18]; however, unlike the former, it is a dimensional quantity. Attention is drawn in the work of Cox et al. (1968) to a certain confusion prevalent in oceanography with respect to the use of these quantities.

The quantity:

$$\sigma_\theta = (\rho_\theta - 1) \cdot 10^3 \qquad [1.20]$$

is potential conventional density.

Instead of specific volume α, also in order to reduce the number of signs, conventional quantities are used. The most common are the following quantities:

(1) Conventional specific volume v_T.

$$v_T = (\alpha_T - 0.9) \cdot 10^3 \qquad [1.21]$$

or:

$$v_T = (\alpha_T - 0.97) \cdot 10^3 \qquad [1.22]$$

The expression of conventional specific volume was introduced by Sund (1926), although it appears for the first time without a special letter designated in the work of Hesselberg and Sverdrup (1915a); on the proposal of Zubov (1929), it is widely used in Soviet oceanographic practice.

(2) Anomaly of specific volume Δ_{ST}.

$$\Delta_{ST} = \alpha_{ST0} - \alpha_{35,0,0} \qquad [1.23]$$

where α_{ST0} is the specific volume as a function of temperature and salinity (at zero pressure) *, $\alpha_{35,0,0} = 0.972643$ — the specific volume with temperature of $0°C$ and salinity of 35‰.

The expression of "surface" anomaly of specific volume Δ_{ST} was proposed by Sverdrup (1933) and is used in western oceanographic literature. As proposed by Montgomery and Wooster (1954), the quantity Δ_{ST} is called at the present time the "thermosteric anomaly" — sometimes it is designated as δ_T, to distinguish it from the "total" anomaly δ, formula [1.32]. In oceano-

---

* With the symbols α, σ, v etc. it would be more correct to insert indices defining to which quantities they are related: instead of σ_T to write σ_{ST0}, to designate the quantity $\rho = \rho(S, T, p)$ as ρ_{STp} etc. We see if only from the example of σ_T and v_T that the insertion of these indices is not strictly observed.

graphic practice the quantity $10^5 \Delta_{ST}$ is used; in this case its dimension is centiliters per ton.

(3) Conventional specific volume (or anomaly) Δ.

$$\Delta = 10^4 (1 - \alpha) \qquad [1.24]$$

This expression is rarely used; in particular, we find it in Dorsey's tables (1968) of the properties of sea water, in Eckart's (1958) work, etc.

(4) The relation between the quantities σ_T, v_T *and* Δ_{ST}. On the basis of expressions [1.9], [1.19] and [1.21] it is easy to obtain formulae interrelating the quantities σ_T and v_T:

$$v_T = \frac{10^6}{\sigma_T + 10^3} - 0.9 \cdot 10^3 \qquad [1.25]$$

$$\sigma_T = \frac{10^3}{10^{-3} v_T + 0.9} - 10^3 \qquad [1.26]$$

The relationship between Δ_{ST} and σ_T is obtained in the following way (Sverdrup, 1933). Since:

$$\rho_{ST0} = 1 + 10^{-3} \sigma_T$$

$$\alpha_{ST0} = \frac{1}{\rho_{ST0}} = 1 - \frac{10^{-3} \sigma_T}{1 + 10^{-3} \sigma_T}$$

and:

$$\alpha_{ST0} = \alpha_{35,0,0} + \Delta_{ST} = 0.97264 + \Delta_{ST}$$

then:

$$\Delta_{ST} = 0.02736 - \frac{10^{-3} \sigma_T}{1 - 10^{-3} \sigma_T} \qquad [1.27]$$

The relationship between Δ_{ST} and v_T is obtained as follows. Since:

$$\alpha_{ST0} = 0.97264 + \Delta_{ST}$$

$$v_T = (\alpha_{ST0} - 0.9) \cdot 10^3$$

then:

$$v_T = 72.64 + 10^3 \Delta_{ST} \qquad [1.28]$$

and

$$\Delta_{ST} = 10^{-3} (v_T - 72.64) \qquad [1.29]$$

As may be seen from the formulae given above, the quantities σ_T, v_T and Δ_{ST} which are important in oceanographic analysis, are equally inter-related. Therefore, the various sources of actual data give — as a function of temperature and salinity — any one of these three quantities, which makes it possible, where necessary, to carry out the corresponding conversions. The conversion of σ_T into v_T and the conversion of v_T into σ_T is carried out according to tables 12 and 13 of Zubov's *Oceanological Tables* (1957a). The conversion of σ_T into Δ_{ST} is performed according to Table A3, taken from Sverdrup et al. (1942). The tables for the conversion of Δ_{ST} into v_T and vice versa are not given, since it is easily carried out according to formulae [1.28] and [1.29].

(C) Density $\rho = \rho(S, T, p)$ and specific volume $\alpha = \alpha(S, T, p)$ in situ
The specific volume in situ is equal to:

$$v_{STp} = (\alpha_{STp} - 0.9) \cdot 10^3 \tag{1.30}$$

or, otherwise:

$$v_{STp} = (\alpha_{STp} - 0.97) \cdot 10^3$$

The quantity v_{STp} is used in Soviet oceanographic practice. The corresponding quantity σ_{STp} equals:

$$\sigma_{STp} = (\rho_{STp} - 1) \cdot 10^3 \tag{1.31}$$

The quantity of the anomaly of specific volume in situ (Bjerknes and Sandström, 1912, p. 27) equals:

$$\delta = \alpha_{STp} - \alpha_{35,0,0} \tag{1.32}$$

This anomaly is determined with respect to specific volumes on the level of p dbars in the *standard ocean*; the table of specific volumes in the *standard ocean* is given in the Appendix (Table A1). Anomalies of specific volume, for purposes of convenience in using them, are normally expressed in units of $10^5 \delta$.

The relationship between v_{STp} and $10^5 \delta$ is considered below in Section 23.

We have considered above the determinations of density and specific volume, the various ways of designating these quantities and the relationship between them. *The quantitative connection* between S, T, p, ρ and α, expressed by the equation of state of sea water, will be considered below (Sections 3 and 4).

Geopotential field and dynamic depth

Before proceeding to the study of the equation of state of sea water, attention should be drawn to the fact that we are considering sea water not in the laboratory, but in natural conditions, i.e., in the ocean. In other words, we are applying the equation of state to volumes of water considered in situ, i.e., in their natural conditions. Accordingly, the influence of one of the parameters of state, namely pressure, on the volume of water considered in situ manifests itself in the form of pressure of a water column situated above the volume considered, and all the other parameters of state are functions of the depth (or of the pressure): $S = S(z)$, $T = T(z)$, $\rho = \rho(z)$, $\alpha = \alpha(z)$.

This fact enabled us above to compare pressure with depth and to determine that a change in pressure of approximately 1 dbar corresponds to a change in depth of 1 m. However, the pressure of two water columns, equal in geometrical height, may be different if the densities (specific volumes) of these water columns are different. Therefore, it appears necessary to compare pressure not only with depth, as we have done above, but also with specific volume (density). For this purpose, as we shall now see, we must consider these relationships in the field of gravity.

Elementary work dD, performed on a unit of mass along a perpendicular along the distance dz equals:

$$dD = g\,dz \qquad\qquad\qquad [1.33]$$

where g is the acceleration of gravity and z the depth (height). The elementary increment of pressure corresponding to the same vertical distance dz, as follows from formula [1.7], equals:

$$dp = g\rho\,dz = \frac{g\,dz}{\alpha} \qquad\qquad\qquad [1.34]$$

(this expression is the equation of hydrostatics)

Comparing the last two formulae, we come to the following important relationship (Sverdrup et al., 1942, p. 405):

$$dp = \rho\,dD\,, \quad \text{or} \quad dD = \alpha\,dp \qquad\qquad [1.35]$$

This relationship associates pressure, density (specific volume) and work along a perpendicular in the field of gravity.

The unit of this work performed over a distance of $1/9.8$ m (with $dz = 1/9.8$ m the right part of formula [1.33] equals unity), in the MTS system is the *kilojoule* (kJ). Inasmuch as this work is performed along a perpendicular, its quantity may serve as a measure of the vertical distance; therefore, the *dynamic decimeter* is used in oceanography instead of the kilojoule, since the work of one kilojoule is performed on a vertical distance equal to approximately one geometric decimeter. Correspondingly, a dynamic meter is equal to 10 kJ. The distance along the perpendicular measured in the sea in units of work of gravity is called the *dynamic*, or *geopotential depth* (height); it may be obtained by integrating expression [1.33]:

$$D = \int_{z_a}^{z_b} g \, dz \tag{1.36}$$

or, inasmuch as with a high degree of accuracy g = constant:

$$D = g(z_b - z_a) \tag{1.37}$$

On the other hand, expression [1.35] shows that the dynamic depth can be associated with units of pressure; the expression for the dynamic depth between depth p_a dbar and p_b dbar can be obtained by integrating the right-hand expression (1.35):

$$D = \int_{p_a}^{p_b} \alpha(p) \, dp \tag{1.38}$$

Correspondingly, *dynamic height* is determined by the expression:

$$d = \int_{p_b}^{p_a} \alpha(p) \, dp = -D \tag{1.39}$$

Up to now our discussion has referred to the perpendicular alone; extending it to some volume of the sea, we must introduce the concept of *isobaric surfaces*, i.e., surfaces of equal pressure (p = const.), and of *isopotential* (geopotential) *surfaces*, i.e., surfaces of equal potential of gravity (D = const.). Since we assume g = const., the family of isopotential surfaces is fixed; all these surfaces are equidistant from each other on different verticals in the sea (for example, by one dynamic meter). On the other hand, it follows from expression [1.38] that because of the difference in specific volumes in different parts of the sea (we are not concerned with the causes of this difference here), the distances between these same isobaric surfaces will generally be different on different verticals. The fixed family of isopotential surfaces serves as a system for reading vertical distances between isobaric surfaces on different verticals in the sea; these distances, or dynamic depths (heights), are calculated according to formula [1.38].

The study of the relief of isobaric surfaces, and its expression in dynamic meters is of decisive importance in the *dynamic method of calculating sea currents* (Zubov and Mamayev, 1956).

Internal energy

The parameters of state of sea water which have been considered — pressure p, temperature T (absolute temperature T), specific volume α, salinity S — are associated with a fundamental thermodynamic function — specific internal energy ϵ (erg/g) * by the following relations (Eckart, 1962):

$$p = -\frac{\partial \epsilon}{\partial \alpha} \; ; \quad T = \frac{\partial \epsilon}{\partial \eta} \; ; \quad \mu = \frac{\partial \epsilon}{\partial S} \qquad\qquad [1.40]$$

Here η is specific entropy (erg $g^{-1}\,{}^{\circ}K^{-1}$), μ is the specific chemical potential of sea water.

We shall return to the elucidation of formulae [1.40] in Chapter 3.

2. THE EQUATION OF STATE OF SEA WATER

The relationship which links the parameters of state: density (or specific volume), temperature, salinity and pressure, is called the *equation of state of sea water*. In its general form the equation of state can be written as follows:

$$f(\rho, S, T, p) = 0 \qquad\qquad [2.1]$$

or:

$$f(\alpha, S, T, p) = 0 \qquad\qquad [2.2]$$

The expressions noted represent the *thermal* equation of state **. Writing [2.2] in the following form, more widely used in oceanography:

$$\alpha = \alpha(S, T, p) \; , \qquad\qquad [2.3]$$

one can construct the following expression of the total differential of specific volume α as a function of the three variables S, T and p:

$$d\alpha = \left(\frac{\partial \alpha}{\partial S}\right)_{T,p} dS + \left(\frac{\partial \alpha}{\partial T}\right)_{S,p} dT + \left(\frac{\partial \alpha}{\partial p}\right)_{S,T} dp \qquad\qquad [2.4]$$

Let us assume that $d\alpha = 0$; then, having divided [2.4] by dT, we will obtain the following relationship, true if volume remains unchanged (for the isostere):

$$\left(\frac{\partial \alpha}{\partial S}\right)_{T,p} \left(\frac{\partial S}{\partial T}\right)_{p} + \left(\frac{\partial \alpha}{\partial T}\right)_{S,p} + \left(\frac{\partial \alpha}{\partial p}\right)_{S,T} \left(\frac{\partial p}{\partial T}\right)_{S} = 0 \qquad\qquad [2.5]$$

or:

---

* $\epsilon = U/M$, where U is the internal energy and M the mass.

** The caloric equation of state $f(\alpha, \epsilon, p, S) = 0$ is not considered in the present work.

$$\frac{\left(\dfrac{\partial \alpha}{\partial S}\right)_{T,p} \left(\dfrac{\partial S}{\partial T}\right)_{p}}{\left(\dfrac{\partial \alpha}{\partial T}\right)_{S,p}} + \frac{\left(\dfrac{\partial \alpha}{\partial p}\right)_{S,T} \left(\dfrac{\partial p}{\partial T}\right)_{S}}{\left(\dfrac{\partial \alpha}{\partial T}\right)_{S,p}} = -1 \qquad [2.6]$$

Let us introduce the well-known determinations of coefficients (Bazarov, 1961; Reid, 1959) —

(1) the coefficient of thermal (volumetric) expansion at constant pressure (and constant salinity):

$$a = \frac{1}{\alpha_0} \left(\frac{\partial \alpha}{\partial T}\right)_{S,p} \qquad [2.7]$$

(2) the coefficient of saline "contraction":

$$\beta = -\frac{1}{\alpha_0} \left(\frac{\partial \alpha}{\partial S}\right)_{T,p} \qquad [2.8]$$

(3) the coefficient of isothermal (and isohaline) compressibility:

$$k = -\frac{1}{\alpha_0} \left(\frac{\partial \alpha}{\partial p}\right)_{S,T} \qquad [2.9]$$

(4) the thermal coefficient of pressure (elasticity):

$$\gamma = \frac{1}{p_0} \left(\frac{\partial p}{\partial T}\right)_{S,\alpha} \qquad [2.10]$$

(5) a new determination of (5) the *coefficient of thermohalinity*:

$$\delta = \frac{1}{S_0} \left(\frac{\partial S}{\partial T}\right)_{p} \qquad [2.11]$$

where α_0, p_0 and S_0 are some constant values of the corresponding parameters.

Then, substituting the values of these coefficients in [2.6], we will obtain:

$$a = \beta \delta S_0 + \gamma k p_0 \qquad [2.12]$$

This relation represents an extension to the case of a more complex system — sea water — of the well-known thermodynamic relation:

$$a = p_0 \gamma k \tag{2.13}$$

valid for a simple system, the thermal equation of state of which is:

$$f(\alpha, T, p) = 0 \tag{2.14}$$

(Bazarov, 1961, pp. 29–30); this is easily seen by means of similar substitutions. With p = const. we obtain from [2.12] a relation:

$$a = \beta \delta S_0 \tag{2.15}$$

valid for sea water at constant (atmospheric) pressure (Mamayev, 1968):

$$f(\alpha, S, T) = 0 \tag{2.16}$$

Relation [2.12] is interesting from the point of view of the further study of the equation of state of sea water (in particular, by the method of thermodynamic potentials). We see that it breaks down into three parts: the first term [2.12] describes the thermal expansion of sea water at constant pressure, the second term — the thermal expansion of fresh water, and their sum — the thermal expansion of sea water in situ. The coefficient of thermohalinity which has been introduced will be considered in greater detail in Sections 14 and 25.

The expression of the total differential [2.4] can also be written in the following form (Reid, 1959):

$$\frac{d\alpha}{\alpha_0} = a \, dT - \beta \, dS - k \, dp \tag{2.17}$$

Equation [2.17] (as well as [2.4]) represents the equation of state of sea water in differential form.

Thus, the density of water in the ocean depends upon three quantities:

(a) temperature T, because of thermal expansion of water, expressed by the coefficient a;

(b) salinity S, because of the fact that the salts dissolved in water change its specific gravity or because of the "saline contraction" of water, expressed by the coefficient β;

(c) pressure p, because of the property of compressibility of sea water under the pressure of the overlying layers of water, determined by the coefficient of compressibility k.

According to one of the hypotheses, water is a complex mixture of 8-, 4-, 2-molecular associations and of free molecules (Eucken, 1948). Due to its complex composition, water possesses strongly anomalous properties as compared with other liquids (high heat capacity, non-coincidence of the temperature of maximum density and the temperature of freezing, etc.). The presence of dissolved salts reinforces the anomalous properties of water, while the

existence of pressure still further complicates the possibility of a quantitative
description of its state.

Due to the complexity of the composition of water and its anomalous
properties, the equation of state of water is very complex, and the equation
of state of sea water even more so. Let us imagine a series of any equations of
state whatsoever of various substances arranged according to their degree of
complexity. This series naturally begins with the simplest – the equation of
state of an ideal gas – Clapeyron's equation:

$$pv = nRT \qquad\qquad\qquad [2.18]$$

where p is pressure, v specific volume, T (absolute) temperature, $R =$
$2.87 \cdot 10^6$ cm^2 sec^{-2} ($°$K)$^{-1}$ – the universal gas constant and n the number
of moles of gas.

The equation of state of sea water, as we shall see below, stands very far in
this imaginary series from Clapeyron's equation, although the latter does have
a place in the study of the thermodynamic properties of sea water (Section 15).

Approaching the question of the analytical and numerical form of the
equation of state of sea water, let us note that it may be obtained in various
forms and, accordingly, by various ways, based at the present time on an
empirical (or at best, a semi-empirical) approach. The main forms of expression of the equation of state, considered later in greater detail, are as follows:

(1) The equation of state can be expressed in the form of a power-series
polynomial for the parameters of state (Section 8), or by other empirical
formulae, close to the polynomial (Sections 3, 4).

(2) With a greater or less degree of approximation it can be expressed by
equations having a thermodynamic basis, for example: (a) by Tumlirz' equation (Section 6); (b) by Tait–Gibson's equation (Section 7); (c) by Van der
Waals' equation:

$$\left(p + \frac{an^2}{v^2}\right)(v - nb) = nRT \qquad\qquad\qquad [2.19]$$

(The symbols are the same as in formula [2.18]; a and b are functions of
temperature and salinity *.)

(3) The equation of state can be approximated by the equations of physical
chemistry which determine the density of solutions (Section 17).

Several versions of the equation of state will be considered below (refer-
ences to corresponding paragraphs have been given above). We shall start by
considering the empirical formulae of Knudsen (Section 3) and Ekman
(Section 4), relating to the first of the categories enumerated above. The

---

* The approximation of the equation of state of sea water by Van der Waals' equation has not been
carried out, to our knowledge. Such a possibility is noted in the work of Holser and Kennedy (1958).

equation of state in the Knudsen–Ekman form is a classical one generally accepted in oceanography as the relation $\alpha = \alpha(S, T, p)$ and its derivatives. All the tables and calculations in the present monograph, connected with the numerical representations of the equation of state, are based on the Knudsen–Ekman formulae. We shall first consider the *equation of state of sea water at constant (atmospheric) pressure*, $\alpha = \alpha(S, T)$ and then *the equation of state in situ*, $\alpha = \alpha(S, T, p)$.

The other representations of the equation of state to which reference was made above, and which will also be considered, may in some cases prove more convenient in solving problems of the thermodynamic and thermohaline analysis of ocean waters.

3. THE EQUATION OF STATE OF SEA WATER AT ATMOSPHERIC PRESSURE

The equation of state of sea water at atmospheric pressure is determined by a series of empirical formulae (Knudsen, 1901; Forch et al. 1902). The *first group* of formulae determines the relationship between the conventional specific gravity σ_0 and salinity S and chlorinity Cl of sea water:

$$\sigma_0 = \sigma_0(S), \qquad S = S(Cl) \qquad\qquad [3.1]$$

These empirical formulae are as follows:

$$S = 0.030 + 1.8050 \, Cl$$

$$\sigma_0 = -0.069 + 1.4708 \, Cl - 0.001570 \, Cl^2 + 0.00000398 \, Cl^3 \, * \qquad [3.2]$$

The *second group* of empirical formulae serves to determine the relationship between density σ_T and conventional specific gravity σ_0 and temperature T:

$$\sigma_T = \sigma_T(\sigma_0, T) \qquad\qquad [3.3]$$

This relationship is determined in the following way:

$$\sigma_T = \Sigma_T + (\sigma_0 + 0.1324)[1 - A_T + B_T(\sigma_0 - 0.1324)] \qquad [3.4]$$

where:

$$\Sigma_T = -\frac{(T - 3.98)^2}{503.570} \frac{T + 283°}{T + 67.26°}$$

---

* This formula is correct within the limits of chlorinity Cl from 1.47362 to 22.2306°/oo (Defant, 1961, p.41).

$$A_T = T(4.7867 - 0.098185\,T + 0.0010843\,T^2) \cdot 10^{-3}$$
$$B_T = T(18.030 - 0.8164\,T + 0.01667\,T^2) \cdot 10^{-6}$$

[3.5]

here $\Sigma_0 = -0.1324$ — the conventional specific gravity of pure water at $0°C$, Σ_T is the conventional density of pure water at temperature $T°C$ with respect to density at $4°C$, A_T and B_T are temperature coefficients. The formula for the determination of Σ_T was obtained by Thiesen (1897).

Formula [3.4] follows from the representation of relation [3.3] in the form of the following quadratic formula:

$$\sigma_T = A + B\sigma_0 + C\sigma_0^2 ,$$

[3.6]

amounting in the case of pure water to the form:

$$\Sigma_T = A + B\Sigma_0 + C\Sigma_0^2$$

[3.7]

(Fofonoff, 1962). In these formulae A, B and C are functions of temperature. Indeed, subtracting [3.7] from [3.6], we obtain formula [3.4] provided that:

$$B = 1 - A_T$$
$$C = B_T$$

[3.8]

Fofonoff points out that in some cases formula [3.6] proves to be preferable to formula [3.4] for calculations. Coefficient A in this formula is determined in the following way:

$$A = \frac{4.53168\,T - 0.545939\,T^2 - 1.98248 \cdot 10^{-3}\,T^3 - 1.438 \cdot 10^{-7}\,T^4}{T + 67.26}$$

[3.9]

Finally, relation [3.3] can be represented by the following formula, which is convenient for calculations (Knudsen, 1901):

$$\sigma_T = \sigma_0 - D$$

[3.10]

where:

$$D = \sigma_0 - \sigma_T = -\Sigma_T - 0.1324 + (\sigma_0 + 0.1324)[A_T - B_T(\sigma_0 - 0.1324)]$$ [3.11]

Formulae [3.2], [3.4] and [3.5], linking temperature, salinity and density of sea water, were obtained on the basis of laboratory investigations carried out by Knudsen, Forch, Jacobsen and Sörensen. For their experiments these investigators used 24 samples of sea water from the surface of the sea which had been collected in those parts of the ocean most studied at that time, namely, in the seas of Northwestern Europe (the Norwegian Sea, the Kattegat, the Baltic Sea and others). In particular, to obtain relation [1.1] only nine water samples collected in the regions mentioned (one of the samples was from the Red Sea) were used.

Basing himself on the hypothesis of the constancy of the saline composition of the world ocean, Knudsen extended his results to the entire world ocean and prepared, according to the formulae quoted, his *Hydrographic Tables* (Knudsen, 1901), which are the basis for all calculations of the density of the waters of the World Ocean at constant pressure.

These tables include:

(1) a table of correspondence of the quantities Cl, S, σ_0 and also $\rho_{17.5}$ with an accuracy of 0.01 for each of the values. The table is calculated according to formulae [3.2];

(2) a table of the quantities Σ_T, A_T and B_T for the entire range of temperatures of the World Ocean (from -2 to $33°C$) at intervals of $0.1°$, with an accuracy as follows: Σ_T – to 10^{-4}, A_T – to 10^{-6}, B_T – to 10^{-7};

(3) a table of the values of D for the calculation of σ_T according to formula [3.4] (with an accuracy of 10^{-3});

(4) a table of the values of D for the calculation of σ_0 according to formula [3.10] (with an accuracy of 10^{-3}).

The calculation of the density and specific volume of sea water according to Knudsen's tables and all sources based on Knudsen's tables is considered in detail in a work by the author (Mamayev, 1963).

As has already been pointed out, the standard accuracy of calculation of σ_T and v_T amounts to 10^{-2}. Knudsen's formulae also make it possible to obtain σ_T and v_T to the third decimal place, but this is already influenced by the actual inconstancy of the saline composition, which should never be lost sight of. For seas separated from the World Ocean, Knudsen's formulae are, strictly speaking, inapplicable.

The difference from the equation of state for the World Ocean, as determined by Knudsen's formulae [3.1], [3.2] and [3.3], was studied more or less in detail for the Caspian Sea. As a result, other expressions were obtained for the determination of the quantities σ_0, $\rho_{17.5}$, A_T and B_T, differing from the corresponding Knudsen formulae and given in the *Oceanological Tables for the Caspian Sea* (1949).

4. THE EQUATION OF STATE OF SEA WATER IN SITU

In order to obtain the equation of state of sea water in situ, the information on the relationship between specific volume and temperature and salinity must be completed by data on its relationship to pressure.

The relation between specific volume in situ α_{STp} and the specific volume on the surface of the sea α_{ST0} can be expressed by the following obvious formula:

$$\alpha_{STp} = \alpha_{ST0}(1 - \mu p) \tag{4.1}$$

where μ is the *mean* coefficient of the compressibility of sea water in a layer of p bar (decibar) thickness; the dimension of the coefficient, as follows from formula [4.1], is inverse bars (decibars).

The mean coefficient of compressibility μ for 1 bar between 0 and p bar depths is determined by the empirical formula of Ekman (1908):

$$10^8 \mu = \frac{4886}{1 + 0.000183\,p} - (227 + 28.33\,T - 0.551\,T^2 + 0.004\,T^3) +$$

$$+ p \cdot 10^{-3}(105.5 + 9.50\,T - 0.158\,T^2) - 1.5\,p^2 T \cdot 10^{-7} -$$

$$- \frac{\sigma_0 - 28}{10}\,[147.3 - 2.72\,T + 0.04\,T^2 - p \cdot 10^3 \times$$

$$\times\,(32.4 - 0.87\,T + 0.02\,T^2)] +$$

$$+ \left(\frac{\sigma_0 - 28}{10}\right)^2 [4.5 - 0.1\,T - p \cdot 10^{-3}(1.8 - 0.06\,T)] \tag{4.2}$$

Expressing depth not in bars, but in *decibars*, i.e. assuming in formula [4.1] quantity p to be ten times greater (referring to one and the same depth) we must take quantity μ, naturally, as ten times smaller. If pressure p in decibars is introduced in formula [4.2], its form will change affecting the multipliers of μ; modified formula [4.2] is given in Zubov's *Oceanological Tables* (1957a, p. 332). Thus, quantities $10^8 \mu$, $10^9 \mu$, $10^{10} \mu$ will be identical if we are referring to the mean compressibility for 1 bar, 1 dbar and 1 cbar respectively *.

Ekman's formula [4.2] is based on the results of the precise experiments of Amagat (1893) in determining the compressibility of pure water and the results of Ekman's experiments on sea water. For his experiments Ekman used only one sample of water, one half of which was diluted to a salinity of 31.13‰ while the other was evaporated to obtain a salinity of 38.53‰. Ekman carried out 36 measurements at pressures from 200 to 600 atm. and at temperatures from 0 to 20°C. Ekman's experiments were exact, but incomplete, and, in spite of some recent studies of the compressibility of sea water (Newton and Kennedy, 1965; Wilson and Bradley, 1968; and others), to which reference will be made below, Ekman's result, expressed by formula [4.2], is the sole source for the calculation of specific volume in situ. At the same time, it is supposed that it is also the source of fundamental errors both

---

* There is a misprint in the heading of the table: one should read $10^9 \mu$ instead of $10^8 \mu$ since, according to the explanation of the table (p. 332), the mean compressibility for one *decibar* is intended.

in the calculation of specific volumes and in the calculation of the speed of sound (Section 19) etc.

Knowing the mean coefficient of compressibility μ, determined by formula [4.2], we can define from formula [4.1] any specific volume in situ, α_{STp}. A table of mean coefficients of compressibility $10^9\mu$ is given in the *Oceanological Tables* (Zubov, 1957a, table 14) *. More convenient tables for this determination within the ranges of temperatures from -2 to $30°$, of salinities from 30 to 40‰ and of pressures up to 10,000 dbar have been compiled by Schumacher (1924). Schumacher breaks Ekman's formula down into four addends and gives tables for each of them **.

$$10^9\mu = \frac{4886}{1 + 1.83 \ 10^{-5}p} - (227 + 28.33\,T - 0.551\,T^2 + 0.004\,T^3) + \qquad \text{(a)}$$

$$+ 10^{-4}p(105.5 + 9.50\,T - 0.158\,T^2) - 1.5 \cdot 10^{-3}\,Tp^2 - \qquad \text{(b)}$$

$$- 10^{-1}(\sigma_0 - 28)\,[(147.3 - 2.72\,T + 0.04\,T^2) -$$

$$- 10^{-4}p(32.4 - 0.87\,T + 0.02\,T^2)] + \qquad \text{(c)}$$

$$+ 10^{-2}(\sigma_0 - 28)^2\,[4.5 - 0.1\,T - 10^{-4}p(1.8 - 0.06\,T)] = \qquad \text{(d)}$$

$$= a + b + c + d$$

The value of the last term (d) may be neglected for ocean waters; its value becomes important only for strongly desalinated waters.

This, however, is not the procedure for calculations of specific volume in situ manually, without the use of a computer. The method indicated of determining α_{STp} according to formula [4.1] represents a preliminary stage, and these calculations are made for some base values of specific volume. For the calculations themselves special tables have been compiled by Hesselberg and Sverdrup (1915a) and Bjerknes and Sandström (1910, 1912); the latter are the most convenient and are based on the expansion of the function $\alpha = \alpha(S, T, p)$ into Taylor's series (Section 23).

5. CHARACTERISTIC DIAGRAMS. THE *T-S* DIAGRAM

Let us turn once again to the equation of state of sea water at atmospheric pressure:

$$\alpha = \alpha(S, T) \qquad\qquad\qquad [5.1]$$

---

* See p. 28 (footnote).
** Schumacher's tables are also given in Landolt–Börnstein (1952, p. 432).

and consider the question of its graphic representation. Equation [5.1] shows that setting two parameters of state arbitrarily, we can determine the third:

$$\alpha = \alpha(S, T) ; \quad T = T(\alpha, S) ; \quad S = S(\alpha, T)$$

Let us consider the graphs of these three equations, constructed according to Knudsen's formulae, using instead of specific volume α conventional specific volume v_T.

The graph $v_T = v_T(S, T)$, represented in Fig. 1, is called a *T-S diagram* (thermohalogram); Fig. 2 represents the graph $S = S(v_T, T)$, or, a v_T-T diagram (volumethermogram); finally, Fig. 3 shows graph $T = T(v_T, S)$, or a v_T-S diagram (volumehalogram).

All three diagrams (Figs. 1–3) with the parameter v_T are isosteric (we encounter the term "isosteric T-S diagram" for the first time in the work of Zubov and Sabinin (1958). Using the parameter σ_T instead of v_T we will obtain the corresponding isopycnal diagrams. The diagrams considered are *characteristic diagrams of state* of sea water at atmospheric pressure; the term "characteristic diagram" was apparently proposed by Montgomery (1950).

Considering the T-S diagram (Fig. 1), we can immediately observe its very important property consisting in the presence of extremes of isosteres (isopycnals); the line of these extremes (the broken line in Fig. 1) bears the not very appropriate name of the line of the temperatures θ of maximum density (or least specific volume). This line coincides with isoline $\partial S/\partial T = 0$,

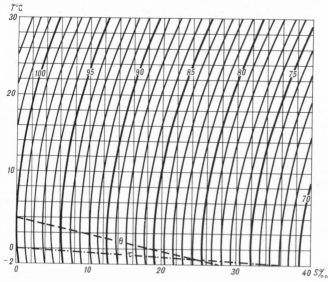

Fig. 1. Isosteric T-S diagram of sea water. The broken line is the line of temperature of maximum density θ, the dash-dotted line is the line of freezing point τ.

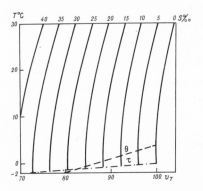

Fig. 2. v_T-T diagram of sea water.

where $\partial S/\partial T$ is the cotangent of slope of the angle of the isosteres (isopycnals) to the abscissa axis of the T-S diagram, or the thermohaline derivative (see Section 25). The temperature of maximum density θ, which decreases with an increase in salinity, is determined according to the formula (Zubov, 1938):

$$\theta = 3.95° - 0.2S - 0.011S^2 + 0.00002S^3 \qquad [5.2]$$

Helland-Hansen (1912) and Defant (1961), instead of [5.2], give the following approximate formula:

$$\theta = 3.95° - 0.266\,\sigma_0 \qquad [5.3]$$

The most exact formula for θ can be obtained starting with the concept of the thermohaline derivative (Section 25).

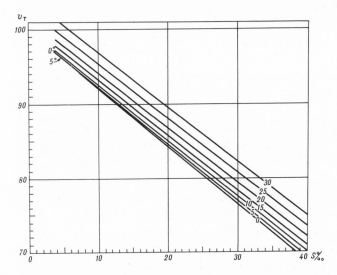

Fig. 3. v_T-S diagram of sea water.

The line of temperatures τ of freezing of sea water (dash-dotted line in Fig. 1) is also plotted on the T-S diagram, as determined by the empirical formula of Hansen (1904):

$$\tau = -0.0086 - 0.064633\,\sigma_0 - 0.0001055\,\sigma_0^2 \qquad\qquad [5.4]$$

The last relation can also be expressed through salinity (Zubov, 1938):

$$\tau = -0.003 - 0.0527\,S - 0.00004\,S^2 - 0.0000004\,S^3 \qquad\qquad [5.5]$$

Subsequently Thompson (1932) proposed the formula:

$$\tau = -0.0966\,Cl - 0.0000052\,Cl^3 \qquad\qquad [5.6]$$

which is correct for values of Cl up to 40‰.

It is apparent from Fig. 1 that for low salinities the temperature of maximum density is higher than the temperature of freezing; these temperatures become equal at $S_{\theta=\tau} = 24.695$‰; at the same time $\theta = \tau = -1.332°$; $\sigma_{\theta=\tau} = 19.825$.

The property mentioned of the T-S diagram — the presence of the extremes of isosteres (isopycnals), as well as the non-linearity of the relationship between density and temperature and salinity, is the consequence of the anomalous nature of the properties of sea water; this question will be considered in more detail in Section 21. In particular, the relationship considered between the temperature of maximum density θ and the temperature of freezing τ leads to the differing nature of the processes of mixing and to other differences between *sea waters* ($S > 24.695$‰) and *brackish waters* ($S < 24.695$‰).

Characteristic diagrams have a twofold application in oceanography:

(1) *The calculation of density (specific volume)*. For this purpose both the T-S diagram and the v_T-S diagram are convenient. Graphs for the computation of density (specific volume) are constructed on a large scale using *auxiliary tables* of the functions $S = S(\rho, T)$ or $S = S(\alpha, T)$ specially calculated for this purpose (Mamayev, 1954; Zubov and Mamayev, 1956; Burkov et al., 1957; Mamayev, 1963). In particular, the author (Mamayev, 1954) calculated an auxiliary table for the construction of isosteric T-S and v_T-S diagrams in the range $-2 < T < 30°C$ and $5 < S < 40$‰, i.e., for the entire range of temperatures and salinities of the World Ocean. The methods for calculating the table are considered in the article quoted; the table itself is given in the Appendix (Table A2) together with a brief description of the methods. The theoretical basis for the method of calculating auxiliary tables is considered in Section 29.

(2) *The analysis of the T-S relations of sea waters*. The T-S diagram is the basis for plotting the T-S relations of the water masses of the World Ocean. At the same time, the field of the T-S diagram can be used to plot, in addition, not only the isolines of density or of specific volume, but also other

functions of the equation of state of sea water — the isolines of the first derivatives of density by temperature and salinity (Section 24), the isolines of the
speed of propagation of sound (Section 19) and others. The study of the real
T-S correlations of the waters of the ocean and of their changes in time and
space in the field of particular functions of the equation of state constitutes,
as has already been said in the introduction, the subject of T-S analysis and
hence of the entire present work; the present paragraph represents the first
graphic introduction to this problem.

What has been said above accordingly justifies the application of the term
"characteristic diagram" to the T-S diagram, emphazising its role as a tool
(or means) for the analysis of the natural waters of the World Ocean. It is
well-known that in dynamic meterology a whole series of characteristic
diagrams is used for the analysis of the state of air masses; of them the
closest to the T-S diagram is probably Taylor's diagram, on which the relationship between temperature and vapor pressure is established (Montgomery,
1950); so far as dynamic oceanography is concerned, the only characteristic
diagram in its arsenal (if we are not to go beyond the limits of the present
context), is the T-S diagram, or the diagrams derived from it, considered
above.

The diagram of the state of sea water in situ gives us the possibility of constructing a *volumetric diagram of state*; for this it is necessary to plot the
temperature on axis y, the salinity on axis x and the pressure on z. Then the
planes parallel to plane $x0y$ in which the T-S diagram for the atmospheric
pressure is shown (Fig. 1), will represent T-S diagrams at corresponding
values of pressure. For example, Fig. 4 represents a T-S diagram for plane
$p = 2,000$ dbar, the values of specific volume $v_{S,T,2000}$, the field of the isolines
of which has been plotted on the figure, calculated with the help of the
Oceanological Tables (Zubov, 1957a) *. Comparing Figs. 1 and 4, it is easy to
see the differences between the fields of the isosteres, determined by pressure.
The T-S diagram shown in Fig. 4 can be used for the analysis of water masses
in the plane $p = 2,000$ dbar.

In addition to those considered above, there exist T-S diagrams with functional scales. As an example one may cite the diagram proposed by Weyl
(1970). On this diagram the isopycnals represent equidistant straight lines
inclined at an angle of 45° to the abscissa axis, the isotherms are parallel to
the abscissa axis and the isohalines come together slightly at the top (thus,
the system of coordinates is not rectangular). Weyl's T-S diagram is useful
for a comparative graphic evaluation of the influence of temperature and
salinity on density.

---

* Pingree (1972) points out that T-S diagrams for deep isobaric surfaces can be constructed approximately by displacing the isolines of σ_T or σ_θ by $-2.2°/1,000$ dbar.

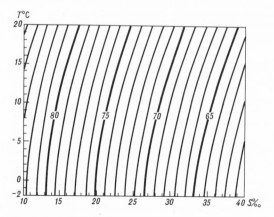

Fig. 4. Isosteric T-S diagram of sea water at $p = 2000$ dbar.

6. THE EQUATION OF STATE IN TUMLIRZ' FORM

The equation of state of sea water, expressed by Knudsen's empirical formulae, is not, generally speaking, an equation of state in the strict sense of the word; it is practically impossible to use it to describe an arbitrary thermodynamic process in which all the parameters of state are changing simultaneously. Therefore, it is necessary to seek new forms of the equation of state which may be equally acceptable both from oceanographic and thermodynamic points of view.

Eckart (1958) was the first to obtain a simplified form of the equation of state both for pure and for sea water, distinguished by quite a good degree of accuracy and satisfying the conditions noted above. For this purpose, Eckart analyzed the initial experimental data of Amagat, of Ekman and the later data of Kennedy (1957), as well as making use of Bjerknes' and Sandström's (1910) tables, which include Knudsen's data. Eckart showed that the equation of state could be expressed by Tumlirz' equation (1909):

$$(p + p_0)(\alpha - \alpha_0) = \lambda \qquad\qquad [6.1]$$

where p is pressure, α specific volume, p_0, α_0 and λ the functions of temperature (for pure water), temperature and salinity (for sea water).

Testing out Tumlirz' equation in relation to the experimental data (we do not consider the details of his methods, referring those interested to the primary source), Eckart came to the following results. Quantity α_0 proved constant both for pure and for sea water and equal to $\alpha_0 = 0.6980$. For pure water ($S = 0$) and sea water p_0 and λ are expressed by the following relations:

$$p_0 = 5890 + 38T - 0.375T^2 + 3S \qquad\qquad [6.2]$$

$$\lambda = 1779.5 + 11.25T - 0.0745T^2 - (3.80 + 0.01T)S \qquad\qquad [6.3]$$

i.e., the relationship between p_0 and λ and salinity turns out to be linear.

In Eckart's formulae [6.1]–[6.3] specific volume is expressed in ml/g
(see footnote on p.14), while p represents *total* pressure (i.e., taking
account of the air column) in atmospheres (1 atm. = 1.01325 bar).

Eckart's simplified equation of state was derived by him for the range of
temperatures $0 \leqslant T \leqslant 40°C$, of salinities from 0 to 40‰ and of pressures
from 0 to 1,000 atm. (10,000 dbar). With respect to the precision of the
equation, Eckart points out that errors in the determination of specific
volume amount to: random errors – not less than $\pm 2 \cdot 10^{-4}$ ml/g, systematic
errors – not less than $\pm 2 \cdot 10^{-7}$ ml/g; the latter lead to an uncertainty of the
order of $0.5 - 1.0\%$ in the determination of isothermal compressibility.
Eckart also points to the need for a more precise definition of the relationship between p_0 and λ_0 and salinity. We may note that equation of state in
Knudsen's form and in Eckart's form should be considered as equally valid,
since both are based on an independent study of the initial data; one can
speak only of a possibly greater accuracy of Knudsen's equation. Fig. 5,
taken from Fofonoff's (1962) work, gives an idea of the differences in the
determination of specific volume according to Knudsen and Eckart. These
variations do not exceed $5 \cdot 10^{-5}$ ml/g at a temperature of $18°C$ and they rise
sharply in the neighborhood of the high salinities and temperatures, reaching
$30 \cdot 10^{-5}$ ml/g at $T = 30°$ and $S = 35‰$.

These differences, as Fofonoff (1962) points out, are caused mainly by
inexact knowledge of the coefficients of thermal expansion of sea and fresh
water. As regards the coefficients of saline "contraction", like the coefficient

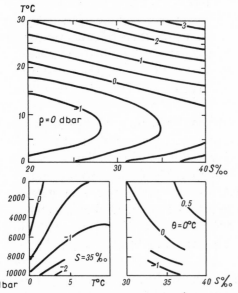

Fig. 5. Differences of specific volumes $10^4(\alpha_K - \alpha_E)$ (where α_K is given according to Knudsen, α_E
according to Eckart) as functions of temperature, salinity and pressure (in dbars), (Fofonoff, 1962.)

of isothermal compressibility it differs by not more than 1% in Knudsen's and Eckart's determinations.

Tumlirz' equation was chosen by Wilson and Bradley (1968) for an approximation of the results of their very extensive investigations of the equation of state of sea water. They performed 795 measurements of specific volume (cm^3/g) with values of temperature close to 0, 5, 10, ... 40°C, of salinity close to 0, 10, 20, 30, 35 and 40‰ and in the range of pressures between 1,013 and 965.258 bar (15 values). Samples for this purpose were taken from the Atlantic Ocean (in the Key West—Bermuda area), while the salinity of the samples was varied by dilution with distilled water or by evaporation in order to reach values close to those indicated above. The specific volume of pure water was also measured under similar conditions.

The experimental results of all 795 measurements are presented by these authors with an accuracy of up to 10^{-5} in the form of tables which served as a basis for the calculation of the parameters of Tumlirz' equation [6.1]. The coefficients in formulae [6.1]–[6.3], according to Wilson and Bradley (1968), differ somewhat from Eckart's coefficients; their values are given in Table A8. Wilson and Bradley's equation is valid within the ranges: $0 < S < 40$‰, $0 < T < 40$°C and $1 < p < 965$ bar.

The difference in specific volumes according to Wilson and Bradley and according to Knudsen—Ekman is small, and right up to 20°C does not exceed $2 \cdot 10^{-4}$ cm^3/g. For sea water of 35‰ salinity the greatest differences (Wilson and Bradley's data minus the classical data) are observed at a temperature of 40°C; $4 \cdot 10^{-4}$ at $p = 1$ bar and $7 \cdot 10^{-4}$ at $p = 1,000$ bar.

Wilson and Bradley (1968) consider the main shortcoming of their equation of state to be the fact that, according to it, the thermal expansion of pure water approaches zero at 3°C rather than at 4°C, and that the specific volume of pure water at 0°C proves equal to 1.0000 cm^3/g. They attribute this to the small quantity of measurements carried out in the low temperatures.

Fisher et al. (1970), analyzing anew the approximation by Wilson and Bradley of their experimental data with the help of Tumlirz' equation, found that the quantities p_0, α_0 and λ, according to Wilson and Bradley, are not precise enough. As a result of their second approximation of Wilson and Bradley's data to Tumlirz' equation, with the introduction of other data, in particular of the results of direct measurement of the thermal expansion of sea water under pressure (Bradshaw and Schleicher, 1970), Fisher et al. obtained other, more unwieldy expressions for the quantities p_0, α_0 and λ, which are also given in Table A8.

The difference in the values of specific volume (Fisher's data minus the classical) for water of 35‰ salinity do not exceed 10^{-4} in the range $0 < T < 30$°C and $1 < p < 800$ bar; the maximum density of pure water is

observed at 4.00°C, and, what is most interesting, the intersection of the
lines of maximum density temperatures and freezing temperatures on the
T-S diagram is observed at lower salinity – approximately 22‰ instead of
24.7‰.

7. THE EQUATION OF STATE IN TAIT–GIBSON'S FORM

Interesting work in investigating the influence of pressure on the specific
volume of pure water and sea water has been done by Li (1967), who
studied the question of approximating the experimental data of Amagat
(1893) and Ekman (1908) on the compressibility of pure water and sea
water respectively, by expressions simpler than those proposed in the past by
these authors, namely, by the old, simple equation of Tait (1888) for pure
water and by Tait–Gibson's equation for sea water (Gibson was the first to
study the applicability of Tait's equation to solutions). To verify his results,
Li also made use of more modern data (Kennedy et al., 1958; Wilson, 1959;
Newton and Kennedy, 1965, and others).

The equation of state of pure water in Tait's form is as follows:

$$v_{T,p} = v_{T,1} - C \log \frac{B+p}{B+1} \qquad [7.1]$$

with $1 \leqslant T \leqslant 45°C$

$1 \leqslant p \leqslant 1,000$ bar

Tait–Gibson's equation for sea water has the form:

$$v_{S,T,p} = v_{S,T,1} - (1 - S \cdot 10^{-3}) C \log \frac{B^*+p}{B^*+1} \qquad [7.2]$$

with $0 \leqslant T \leqslant 20°C$

$1 \leqslant p \leqslant 1,000$ bar

$30 \leqslant S \leqslant 40$‰

In formulae [7.1] and [7.2], according to Li's results:

$$C = 0.315 \cdot v_{T,1} \qquad [7.3]$$

$$B = 2668.0 + 19.867T - 0.311T^2 + 1.778 \cdot 10^{-3}T^3 \qquad [7.4]$$

$$B^* = (2670.8 + 6.89656S) + (19.39 - 0.0703178S)T - 0.223T^2 \qquad [7.5]$$

The symbols: $v_{T,1}$ and $v_{T,p}$ are the specific volumes (ml/g) of pure water
at atmospheric pressure (1 bar) and at pressure p bar, respectively; $v_{S,T,1}$ and
$v_{S,T,p}$ are the same for sea water, respectively.

It should be borne in mind that formula [7.2], in addition to $v_{S,T,1}$, also includes the specific volume of pure water at atmospheric pressure, $v_{T,1}$. Formulae [7.1] and [7.2] do not include any new relationships for the determination of these quantities and therefore, in calculating with these formulae, it is necessary to use known values of specific volume of pure and sea water at atmospheric pressure. Thus, Li's investigation concerns only the compressibility per se of water.

Formulae [7.1] and [7.2] yield a very good result. Thus, for standard sea water ($S = 35‰$, $T = 0°C$), the difference between the results of calculation according to formula [7.2] and according to the Knudsen–Ekman formulae does not exceed $\pm 10^{-5}$ in the range of pressure from 1 to 99 bar. At the same time, formula [7.2] is considerably simpler than Ekman's formulae [4.1] and [4.2].

8. THE EQUATION OF STATE IN POLYNOMIAL FORM

As has already been noted, the equation of state can be represented in particular, in the form of a regular polynomial by increasing powers of the parameters of state. Below are given the existing formulae for the equation of state of sea water at atmospheric pressure and in situ; these formulae may prove highly useful (and more convenient) for various types of electronic computations than others.

The equation of state at atmospheric pressure (Cox et al., 1970) was obtained by these authors in the following form (σ is specific gravity):

$$\sigma(T,S) = \sum_{i,j} a_{i,j} T^i S^j \qquad (0 \leqslant i,j \leqslant 3)$$

$$= a_{0,0} + a_{1,0}T + a_{0,1}S + a_{2,0}T^2 + a_{1,1}ST + a_{0,2}S^2 +$$

$$+ a_{3,0}T^3 + a_{2,1}ST^2 + a_{1,2}S^2T + a_{0,3}S^3 \qquad [8.1]$$

with $9 \leqslant S \leqslant 41‰$

$0 \leqslant T \leqslant 25°C$

The numerical values of coefficients $a_{i,j}$ of polynomial [8.1] are given in Table A9.

The data which served to obtain equation [8.1] were derived by Cox et al. directly (by the flotation method). As a reference standard of pure water a specially prepared specimen was taken (Cox et al., 1968), obtained by triple evaporation of a deep-water sample of the Mediterranean (it was established that this distillation did not lead to isotopic fractionation – Ménaché, 1971).

In the range of salinities of 15–40‰, the data of Cox et al. agree with Knudsen's data on the average within limits of $\Delta\sigma = 0.006$ (Knudsen's data are systematically lower as compared with Cox's data); the maximum divergence amounts to about 0.021. At lower salinities and temperatures, Cox' data exceed those of Knudsen to a considerably greater degree (reaching a maximum difference of $\Delta\sigma = 0.078$ at $T = 0°C$ and $S = 5‰$.

The equation of state of sea water in situ has been obtained by Crease (1962) in the following form (α is specific volume):

$$\alpha(p,T,S) = \sum_{i,j,k} a_{i,j,k}\, p^i T^j (S-35)^k , \qquad 0 \leqslant i,j,k \leqslant 4 \qquad [8.2]$$

Formula [8.2] is valid within the following ranges:
  (1) $p < 100$ kg/cm^2, $5 < S < 37‰$, freezing point $< T < 30°C$
  (2) $p < 500$ kg/cm^2, $33 < S < 37‰$, freezing point $< T < 30°C$
  (3) $p < 1,000$ kg/cm^2, $33 < S < 37‰$, freezing point $< T < 6°C$

The numerical values of coefficients $a_{i,j,k}$ of the polynomial [8.2] are also given in Table A9.

Equation [8.2] was obtained indirectly — from data on the speed of sound in the sea (Wilson, 1960b) by integration of relationship [19.11] and expressing it for $\partial\alpha/\partial p$. Crease's work was designed to show that the indirect determination of the density of sea water through the speed of sound can prove more exact than the direct.

We may mention that the equation of state of pure water in polynomial form can be found in the work of Kell and Whalley (1965). This equation (it is not given here) is valid in the ranges $0 < T < 150°C$ and $p < 1,000$ bar.

9. SIMPLIFIED EQUATION OF STATE OF SEA WATER

The unwieldiness of the equation of state of sea water, expressed by Knudsen's and Ekman's formulae (or by other formulae), sometimes constitutes an obstacle to its immediate application in various theoretical investigations directed towards elucidating the fundamental characteristics of oceanic thermohaline circulation or of the analysis of the dynamics of water masses.

For this reason the need often arises for an approximation of the equation of state by relationships as simple as possible. Two cases occur in this connection: the case of a linear approximation, when the equation of state is simplified to such an extent that it expresses nothing more than the linear relationship between density and temperature and salinity (or even only temperature), and the case of a non-linear approximation, when the equation of state is simplified as much as possible but retains its fundamental non-linear features (curvature of the isopycnals on the T-S diagram).

The first version is often applied in analytical models when the linearization of equations is necessary, and when, at first, it is enough to confine oneself solely to the very fact of the simplest relationship between density,

temperature and salinity. A non-linear approximation, of course, has a considerably larger field of application. Let us consider these two versions.

Linear approximation

The so-called Boussinesq approximation:

$$\rho = \rho_0(1 - \alpha T) \tag{9.1}$$

in which the relationship between density and salinity is neglected, is used in
many studies of thermohaline circulation; as an example we may refer to one
of the basic works — Robinson and Stommel (1959), as well as to others.
    A linear equation of state of the form:

$$\rho = \rho_0(1 - \alpha T + \beta S) \tag{9.2}$$

is introduced in the works of Lineikin (1962), Stommel (1961, 1962b) and
others (see, e.g., Section 46 of the present monograph).
    In formulae [9.1] and [9.2] ρ_0 is some constant value of density, α and β
are constant coefficients of heat expansion and saline "contraction" of sea
water, respectively. In the majority of cases it is taken that $\alpha = -2 \cdot 10^{-4}$ and
$\beta = 8 \cdot 10^{-4}$.
    Considering linear formula [9.2], closer to reality as compared with [9.1],
we must note that it still does not reflect the most important property of
sea water, which manifests itself in the existence of a curvature of the isopycnals on the T-S diagram, whereas it is precisely this property which is
responsible for many of the particularities of the thermodynamics of sea
water.
    Therefore, it is essential to obtain an approximate equation of state of sea
water, which, on the one hand, would be very simple by comparison with
Knudsen's formulae and, accordingly, might be freely used in various theoretical constructions, while reflecting, on the other hand, the fundamental
non-linear features of the composition of sea water. It is clear that the
existence of Eckart's equation of state does not do away with the need to
obtain even simpler relationships.

Non-linear approximation

The first simple relationship in such form is, apparently, Dorsey's (1968)
formula for the calculation of the specific volume of sea water:

$$\Delta_{S,T,d} = \Delta_{35,0,d} - (6.48 + 0.00375\,d)T -$$

$$- 0.46\,T^2 + (\delta - 0.283\,T - 0.005\,T^2)(S - 35) \tag{9.3}$$

where $\Delta = 10^5(1-\alpha)$, α is the specific volume, d is the depth in fathoms (1 fathom = 182.88 cm), S = the salinity, T = the temperature, δ = the tabulated function of d (otherwise, a different expression of d, see Table I).

This equation is suitable for salinities from 31 to 37‰, temperatures from 0 to 22° and d from 0 to 4,000 fathoms. Within these limits it yields specific volume α with an accuracy of up to $2 \cdot 10^{-5}$. Equation [9.3] was derived by Dorsey from the table of specific volumes given in Heck and Service's tables (1924). Like formula [9.3] itself, Heck and Service's tables are based on the tables of Bjerknes and Sandström (1910).

Dorsey also gives a table of the relationship between p (the pressure in bars) and d (the depth in fathoms).

Dorsey's formula is inconvenient for calculations and, in addition, can be simplified; therefore, it is desirable to calculate anew a relationship between temperature, salinity and density which is distinguished, on the one hand, by the greatest simplicity, and on the other, allows for the fact of curvature of the isopycnals in the field of the T-S diagram, which is not reflected in linear formula [9.2]. At first, we will confine ourselves to obtaining a formula for the determination of the density of sea water at constant (atmospheric) pressure, bearing in mind its possible application to that range of questions where sea water can be considered as an incompressible liquid.

The consideration of the T-S diagram or of the v_T-T diagram (Figs. 1 and 3) shows that the isopycnals have a parabolic form; on the other hand, the distances between neighboring isopycnals decrease with an increase in salinity, and this increase is at first sight linear in nature; this makes it possible to neglect the quadratic term in the second parenthesis of Dorsey's formula [9.3] and to approximate the desired relationship at constant pressure by the following formula [1]:

$$\rho - \rho_0 = a + b(T-T_0) + c(T-T_0)^2 + [f+g(T-T_0)](S-S_0) \qquad [9.4]$$

TABLE I

The relationship between δ and d in Dorsey's (1968) formula

d (fathoms)	100	300	500	700	900
0	76.3	75.3	75.0	74.3	73.8
1,000	73.3	72.8	72.3	71.8	71.3
2,000	70.8	70.3	69.8	69.5	69.0
3,000	68.5	68.2	67.8	67.2	66.8
4,000	66.2	66.0	65.8	65.2	–

[1] Maurer–Schumacher's formula for the determination of the speed of propagation of sound in the sea has the same form (see, for example, Mamayev, 1963).

where a, b, c, f and g are empirical coefficients subject to determination, ρ_0, T_0 and S_0 — some constant values of density, temperature and salinity. We will assume below, as is usual, that $T_0 = 0°C$, $S_0 = 35‰$, $\rho(0.35) = 1.02813$.

The values of the constant coefficients a, b and c were calculated by the method of the least squares with a fixed value of salinity $S = 35‰$ (in this case the last term in formula [9.4] equals zero). For calculations, from the exact table 10 of the *Oceanological Tables* of Zubov (1957a), calculated according to Knudsen's formulae, values of density were taken with $S = 35‰$ for temperatures of 0, 5, 10, ..., 30°C, seven values in all.

Coefficients f and g, with:

$$f + gT = \partial\rho/\partial S \qquad\qquad\qquad [9.5]$$

were calculated by the method of least squares from the table of gradients $\partial\rho/\partial S$ (Zubov, 1957a, table 24). The values of the gradients $\partial\rho/\partial S$ were averaged every 5‰ between 0 and 40‰ for the same values of temperature $T = 0, 5, 10, ..., 30°C$, seven values of gradients in all.

As a result of calculations, the following formula was obtained (Mamayev, 1964a), which we accordingly propose as a simplified equation of state of sea water at atmospheric pressure *:

$$\sigma_T = 28.152 - 0.0735T - 0.00469T^2 + (0.802 - 0.002T)(S - 35) \qquad [9.6]$$

where $\sigma_T = (\rho - 1) \cdot 10^3$, conventional density.

Calculations according to formula [9.6] show that the error in the determination of density ρ throughout the entire range of temperatures $(0 \leqslant T \leqslant 30°)$ and salinities $(0 \leqslant S \leqslant 40‰)$ of the World Ocean as compared with Knudsen's formulae on the average amounts to $0.5 \cdot 10^{-4} \div 10^{-4}$ g/cm³; it increases with a decrease in temperature and salinity and reaches the value $2 \cdot 10^{-4}$ when $T = 0°$ and $S = 0‰$. In the range of oceanic salinities $(32 \leqslant S \leqslant 37‰)$ and the range of temperatures $(0 \leqslant T \leqslant 30°)$ the error does not exceed so small a quantity as $0.3 \cdot 10^{-4}$. In particular with $T = 0°$ and $S = 35‰$ the error is equal to $2.6 \cdot 10^{-5}$.

Table II gives the values of the differences of densities $\Delta\sigma_T = (\sigma_T)_K - (\sigma_T)_M$, where $(\sigma_T)_K$ is conventional density according to Knudsen's formulae (Zubov, 1957a, table 11), $(\sigma_T)_M$ — conventional density according to formula [9.6], while Fig. 6 represents the same result graphically. From a comparison of Fig. 5 and Fig. 6 it follows that formula [9.6] is no less exact than Eckart's equation [6.1].

---

* In the author's work (Mamayev, 1964a), in formula [9.6] the first term is written erroneously; instead of 28.14 it should read 28.152. In addition, it is also stated erroneously in the same place that coefficients f and g were calculated from the values of gradients $\partial\rho/\partial S$, taken with $S = 35°/oo$. In this latter case, instead of the coefficients appearing in formula [9.6], the very close values $f = 0.801$ and $g = -0.0019$ are obtained.

TABLE II

Values $\Delta\sigma_T = (\sigma_T)_K - (\sigma_T)_M$

S(°/oo)	0	5	10	15	20
T(°C)					
−2	−	0.031	0.083	0.120	0.147
0	−0.214	−0.122	−0.088	−0.068	−0.057
5	+0.045	+0.099	+0.100	+0.088	+0.071
10	0.149	0.183	0.164	0.136	0.103
15	0.152	0.178	0.153	0.120	0.082
20	0.094	0.122	0.101	0.072	0.040
25	0.008	0.044	0.034	0.017	−0.001
30	−0.083	−0.033	0.029	−0.028	−0.028

S(°/oo)	25	30	35	40
T(°C)				
−2	0.169	0.192	0.218	0.255
0	−0.049	−0.041	−0.026	0.001
5	0.052	+0.036	+0.030	0.036
10	0.070	0.041	0.023	0.020
15	0.046	0.016	−0.004	−0.008
20	0.010	−0.013	−0.025	−0.021
25	−0.017	−0.025	−0.021	0.001
30	−0.023	−0.090	0.020	0.068

Let us note that the precision of equation [9.6] can be enhanced if instead of the quadratic relationship of density to temperature at constant salinity ($S = 35‰$) we take a cubic relationship. Calculations by the method of the

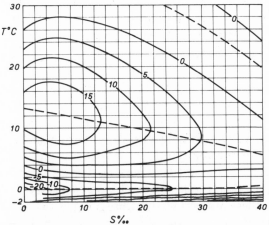

Fig. 6. The difference of densities $10^2 \Delta[(\sigma_T)_K - (\sigma_T)_M]$, where $(\sigma_T)_K$ is the density according to Knudsen, $(\sigma_T)_M$ according to formula [9.6].

least squares with $S = 35\%$ and $T = 0, 5, 10, ..., 30°$ lead to the following formula:

$$\sigma_T(T, 35) = 28.123 - 0.0576T - 0.00613T^2 + 0.000032T^3 \qquad [9.7]$$

which is accurate in density to 10^{-5}.

Let us note that without any particular loss of accuracy the last term in formula [9.6] can be simplified at low temperatures and written in the form $0.8(S - 35)$.

In conclusion, let us consider the question of an approximate calculation of the influence of pressure on density. Defant (1961, p.305) gives the following simplified formulae for the determination of specific volume and density in situ in a standard ocean:

$$10^5 \alpha_{35,0,p} = 97,264 - 0.44p$$

$$10^5 \rho_{35,0,p} = 102,813 + 0.46p$$

where p is pressure in decibars. From the last formula follows:

$$\sigma_T(35, 0, p) = 28.13 + 0.0046p \qquad [9.8]$$

This formula makes it possible approximately to calculate the influence of pressure in equation [7.6]; for this it is necessary to introduce into it the term $+0.0046p$. It should be noted, however, that Defant's relations given above are inexact; thus, when $p = 5,000$ dbar the error in the determination of density in a standard ocean reaches 0.5 units of σ_T and can increase if temperature and salinity deviate from their values in a standard ocean.

A comparative analysis of the exactness of the simplified equations of state enumerated above was carried out by Vasiliev (1968), who came to the conclusion that the most exact were equations [6.1] and [9.6], and that the second, thanks to its greater simplicity, was to be preferred. Equation [9.6], in particular, has been used for the construction of theoretical models of ocean circulation (see, for example, Sarkisian, 1966).

10. APPROXIMATION OF THE KNUDSEN–EKMAN EQUATION OF STATE

Bryan and Cox (1972), and Friedrich and Levitus (1972), published at the same time their independent results of the approximation of the classical Knudsen–Ekman equation of state by polynomials of varying degree, more convenient for numerical calculations than the Knudsen–Ekman formulae themselves. The immediate goal of these authors was to obtain equations for the determination of density in situ at levels varying in depth for use in the numerical calculations of models of ocean circulation.

However, their results are of great value for T-S analysis, in particular for the construction of T-S diagrams for deep levels (of the type shown in Fig. 4), and therefore it is highly desirable to give some account of them.

The content of this section could have become part of Section 8 (the re-presentation of the equation of state in polynomial form) or of Section 9 since, for example, Bryan and Cox give a set of equations of state even simpler in form than equation [9.6]. However, it appears separately, since, in the first place, the authors mentioned above are not seeking an alternative, thermodynamically-grounded equation of state of the type formulated by Tumlirz, Gibson-Tait or others; in the second place, they are not treating the results of independent experiments or indirect calculations; in the third place, the simplified equations form an integral part of their sets of formulae.

Bryan and Cox approximate the Knudsen–Ekman equation both by a third degree polynomial:

$$\sigma_T(T,S,Z) - \sigma_0(T_0,S_0,Z) = x_1\delta T + x_2\delta S + x_3(\delta T)^2 + x_4(\delta S)^2 + x_5\delta T\delta S +$$

$$+ x_6(\delta T)^3 + x_7(\delta S)^2\delta T + x_8(\delta T)^2\delta S + x_9(\delta S)^3 \qquad [10.1]$$

and by the simplest non-linear formula:

$$\sigma_T(T,S,Z) - \sigma_0(T_0,S_0,Z) = x_1\delta T + x_2\delta S + x_3(\delta T)^2 \qquad [10.2]$$

In these formulae $Z = 0, 250, 500, \ldots 6,000$ m — the depth of the level for which the equation of state in situ is calculated (25 levels at intervals of 250 m), $\delta T = (T - T_0)$, $\delta S = (S - S_0)$, where T_0 and S_0 are the mean values of temperature and salinity at corresponding levels for the entire World Ocean. Thus, to each of the formulae [10.1] and [10.2] correspond 25 approximations of the Knudsen–Ekman equation of state (according to the number of levels Z), which determine the conventional density in situ; the fact that deviations are determined with respect to values of T_0 and S_0 different in each case reduces the range of temperatures and salinities for which an approximation is necessary and, accordingly, reduces the error of the latter.

Table A10 gives the values of $\sigma_{T,0}$, T_0 and S_0, as well as coefficients x_1, x_2 and x_3 of the polynomial [10.2]. For the coefficients of the polynomial [10.1] the reader is referred to the primary source (Bryan and Cox, 1972). Fig. 7 shows the density deviations of Bryan and Cox from the Knudsen–Ekman equation of state. It may be seen from the figure that exactness increases substantially with depth.

Friedrich and Levitus (1972) took as a basis equation [9.6]:

$$\sigma_T(T,S) = C_1 + C_2T + C_3S + C_4T^2 + C_5TS \qquad [10.3]$$

and, in particular, recalculated the coefficients of equation [9.6], obtained by the author of the present work, for a narrower range, namely:

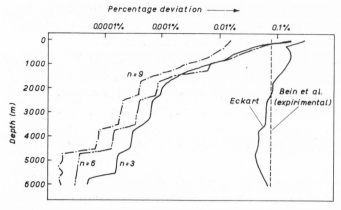

Fig. 7. The mean deviation (%) in the determination of density σ_T by Bryan and Cox' (1972) polynomials from the Knudsen–Ekman formulae in the oceanic range of temperatures and salinities (this range varies from $-2 < T < 29°$ and $28.5 < S < 36.7°/oo$ for the surface of the sea to $0 < T < 2°$ and $34.7 < S < 34.8°/oo$ for a depth of 6000 m – for details see the original). Also shown are the mean deviations of Eckart's (1958) equation and of the determinations of density by Bein et al., (1935). (From Bryan and Cox, 1972.)

$-2° < T < 30°C$

$30 < S < 38‰$ [10.4]

(the result for the low temperature range proves better as compared with equation [9.6]).

Striving for greater exactness, they also considered an approximation of the Knudsen–Ekman equation of state by an incomplete third degree polynomial:

$$\sigma_T(T, S) = C_1 + C_2 T + C_3 S + C_4 T^2 + C_5 ST + C_6 T^3 + C_7 ST^2 \qquad [10.5]$$

Their subsequent methods are close to those of Bryan and Cox (calculation of the pressure for discrete levels at intervals of 1 km) with the difference that the coefficients of polynomials [10.3] and [10.5] occur at continuous quadratic functions of depth, $C(Z) = \alpha + \beta Z + \gamma Z^2$, which makes it possible to carry out interpolation for the intermediate depths (the curves of $C(Z)$ are given in the work of these authors).

In order to obtain greater exactness, Friedrich and Levitus (1972) use equation [10.5] for the upper layer of the ocean, $0 < Z < 2$ km, and the range of temperatures and salinities [10.4] and equation [10.3] for the deep part of the ocean, $2 \leqslant Z < 5$ km, and a reduced range:

$-2° < T < 8.38°C$

$34.68 < S < 36.10‰$ [10.6]

Friedrich and Levitus' formulae yield very good coincidence with the Knudsen–Ekman data.

The coefficients of polynomials [10.5] and [10.3] are given in Table A10.

11. ON THE COMPARATIVE EXACTNESS OF THE VARIOUS EQUATIONS OF STATE OF SEA WATER

In conclusion to this chapter, let us note anew that the numerical values of the relationships which determine the equation of state of sea water have long since been verified and defined more precisely, and that more than once. In addition to the results of the experiments considered above, independent of the Knudsen–Ekman data (Wilson and Bradley, 1968; Cox et al., 1970), we may point out that Thompson and Wirth (1931) had already carried out an independent determination of σ_0 on the basis of the analysis of 36 samples collected in the Pacific Ocean and Indian Ocean to depths of 1,500 m; the data obtained by them exceed the values of σ_0, as defined by the second formula of [3.2], by approximately 0.02 unit (Fofonoff, 1962). Tilton and Taylor (1937) proposed a formula more exact than [3.5] for the determination of the conventional density of pure water, based on the determinations of density performed by Chappius (1907).

Further, the relationship of the density of sea water to temperature and salinity was investigated again by Bein et al. (Bein et al., 1935), who established that the determination of density following Knudsen's tables yields a systematic error at temperatures above 20° which reaches at 30° the value $\Delta\sigma_T = 0.02$. Table III gives an idea of the differences in the data on density according to Knudsen and Bein.

Newton and Kennedy (1965) carried out experiments to study the compressibility of sea water based on more perfected methods unlike those used earlier; series of experiments were carried out for waters with a salinity of 0.00; 30.52; 34.99 and 41.03‰, as a result of which the values of specific volume were calculated with an accuracy of up to $\pm 7 \cdot 10^{-5}$. The results of the experiments are presented in the form of four tables (for the salinities mentioned) of the specific volumes with accuracy to 10^{-4} for temperatures of 0, 5, 10, 15, 20 and 25°C and pressures 1, 100, 200, ..., 1,300 bar (at intervals of 100 bar), i.e., for 84 values in each of the tables (a total of 336 values of specific volume). A comparison of Newton and Kennedy's data with the Knudsen–Ekman data shows that for oceanic salinities the difference between the two sets of data rises with an increase in pressure to 900–1,000 bar, reaching $3 \div 4 \cdot 10^{-7}$, while Newton and Kennedy's data appear low as compared with the classical data. The authors quoted do not propose an equation of state.

TABLE III

Conventional density σ_T of sea water for different values of temperature and salinity according to Knudsen (upper figures) and according to Bein (lower figures) [1]

$S(^o/oo)$ $T(^\circ C)$	5	10	20	30	40
0	−0.132	8.014	16.065	24.101	32.163
	−0.132	8.021	16.066	24.098	32.177
10	−0.273	7.562	15.321	23.079	30.878
	−0.273	7.561	15.320	23.083	30.888
20	−1.770	5.857	13.416	20.983	28.595
	−1.770	5.860	13.422	20.988	28.597
30	−4.327	3.147	10.568	18.007	25.504
	−4.327	3.171	10.596	18.026	25.498

[1] The figures in this table are taken from the *Oceanological Tables* (Zubov, 1957a, table 10) and from table of σ_T of Bein given in Landolt–Bornstein's handbook (1952, p. 428).

The direct instrument measurements of the coefficients of thermal expansion of sea water, carried out by Bradshaw and Schleicher (1970), and independently by Caldwell and Tucker (1970), give the best agreement with the corresponding quantity calculated according to the Knudsen–Ekman equation of state and also according to Crease's equation (Section 8), indirectly linked with the Knudsen–Ekman equations, and somewhat poorer agreement with Wilson and Bradley's equation (Section 6).

Finally, the results of the most recent measurements of the specific volume of sea water, performed by Kremling (1972) with the help of an electron oscillator, are in good agreement with formula [8.1] of Cox et al. (Cox et al., 1970). In the ranges $0 < T < 25^\circ$ and $9 < S < 39‰$ the mean error amounts to $\Delta\sigma_T = 0.008$.

A comparison of Kremling's results with the Knudsen–Ekman data, in particular, shows that Knudsen's (1901) data for σ_T are somewhat low; the mean value of the error amounts to approximately $\Delta\sigma_T = -0.013$, which agrees with Thompson and Wirth's (1931) data and a similar opinion expressed by Carritt and Carpenter (1959), who considered that the error amounts to approximately $\Delta\sigma_T = -0.02$.

This comparison of the exactness of the various equations of state may be concluded at the present stage with the following quotation: the experience of recent years "exemplifies once again the great quality of standards measurements in the early years of the century" (Crease, 1971).

CHAPTER 3

FUNDAMENTALS OF THE THERMODYNAMICS OF SEA WATER

12. SEA WATER AS A SUBJECT OF THERMODYNAMICS

In Chapter 2 the equation of state of sea water was considered mainly from the oceanographic point of view; we virtually did not touch upon questions of thermodynamics, including the study of the equation of state as one of its particular problems, if we do not count the fact that in Section 1 the fundamental thermodynamic function — specific internal energy ϵ — was briefly considered. However, a knowledge of thermodynamics is of vital importance for the study of the ocean and for the development of methods of thermohaline analysis of its waters. The methods of thermodynamics are directly related to the study of the following problems, at least: the determination of the heat capacity of sea water at constant pressure and constant volume; the study of the adiabatic processes in the ocean, principally in its deep parts; the determination of the speed of propagation of sound in the sea; finally — and this is the most important — the study of such *irreversible thermodynamic processes* as heat conduction, diffusion and thermodiffusion. The study of heat conduction and diffusion, as we shall see below, represents the basis of the analytical theories of T-S curves; so far as thermodiffusion is concerned, this phenomenon practically has not been studied in the ocean. A more particular thermodynamic problem, comprised in T-S analysis, is the study of the contraction on mixing of sea waters (Sections 21, 22).

The complexity in the study of the thermodynamics of sea water consists of the fact that sea water represents a *multicomponent system,* the parameters of which are determined by its saline composition. For such a system an equation of state in general form can be written for the isotropic medium as:

$$v = v(p, T, \sum_i x_i) \tag{12.1}$$

where:

$$x_i = m_i / \sum_i m_i \tag{12.2}$$

is the mass of the i-th component; m_i its mass, $M = \sum_i m_i$ the mass of the entire system.

On the other hand, it is known that the state of a system is determined by

$k + 1$ thermodynamic parameters, where k is the number of different components. Thus, a detailed study of the thermodynamics of such a complex system as sea water constitutes an extremely difficult problem.

However, the problem is substantially simplified if we accept the hypothesis of the constancy of the saline composition of sea water, and consider sea water in such an approximation as a *two-component,* or binary system; a solvent (water) — a single solute, or component ("sea salt") *. Such an approximation was introduced in the works of Eckart (1962) and Fofonoff (1962). Other limitations, usual for thermodynamics in the study of general questions, consist of the fact that the medium is considered as isotropic, without polarization and magnetization, without displacement in the geopotential field (Haase, 1963). In addition, we will not consider in the present brief exposition phase transitions, which have no vital importance in *T-S* analysis, referring the reader in this connection to the work of Fofonoff mentioned.

Many questions of the thermodynamics of the ocean can be solved even if it is considered as a *single-component* system, and therefore we shall begin the exposition of the questions which interest us with the consideration of pricisely such a system.

The fundamental equation of thermodynamics for reversible quasi-static processes in a single-component system with constant composition has the form:

$$d\epsilon = T \, d\eta - p \, dv \qquad\qquad [12.3]$$

where ϵ is specific internal energy, T — absolute temperature, η — specific entropy, p — pressure and v — specific volume **. We will assume this equation to be known from general textbooks, referring the reader, for example, to Guggenheim's course in thermodynamics (1950) and to Chambadal's (1963) monograph, which contains a very detailed discussion and study of such a fundamental concept as entropy. No other information, except a knowledge of equation [12.3], is required to understand what follows. This equation is sometimes simply called the thermodynamic identity for energy, which considerably simplifies the matter. It may be added that equation

---

* "A multicomponent system can be treated approximately as a binary system, one of the components of which is represented by a pure substance, while the second comprises all the other components, combined in a group, the relative composition of which is maintained unchanged" (Sage, 1966, chapter 13).

** Following Eckart (1962) and Fofonoff (1962) and in accordance with formulae [12.1] and [12.2], in this chapter we will use the *specific* values of the fundamental thermodynamic quantities: $\epsilon = U/M$, $\eta = S/M, q = Q/M, v = V/M$, where U, S, Q and V are the generally accepted symbols for internal energy, entropy, amount of heat and volume, M — the mass of the system, as well as specific values of the quantities derived from them. (The generally accepted designation of entropy S should not be confused with the designation of salinity by the same letter!)

[12.3] has a strong analogy with the equation of the balance of turbulent energy: the energy of turbulence ("internal energy") is equal to the total energy of the mean flow minus the work of external forces.

The basic equation of thermodynamics for the particular case of a two-component system — sea water — has the form:

$$d\epsilon = T\,d\eta - p\,dv + \mu\,dS \qquad\qquad [12.4]$$

where μ is "the chemical potential of sea water", and S its salinity. We may note that the introduction here of the term $\mu\,dS$, as well as its introduction below in equations [14.1]—[14.4] precisely for a two-component system — sea water — is not immediately obvious, although it follows formally from a comparison of formulae [1.40] and equation [12.4]. This has been done for reasons of convenience in the exposition, while the corresponding derivation will be given below. What has been said applies also to formula [14.8], which relates to sea water.

The equations [12.3] and [12.4] quoted represent the basis for the further exposition; in the following two sections the formal apparatus of the method of thermodynamic potentials will be developed.

13. FUNDAMENTAL THERMODYNAMIC RELATIONSHIPS FOR A SINGLE-COMPONENT SYSTEM

From the form of the fundamental equation of thermodynamics for a single-component system [12.3] it follows that ϵ represents the *potential* function of the variables η and v — the so-called thermodynamic potential, while $d\epsilon$ is the total differential. This assertion becomes obvious from the comparison of equation [12.3] with the general expression for the total differential of the function $U(x,y)$ of the two variables:

$$dU = P\,dx + Q\,dy \qquad\qquad [13.1]$$

where:

$$P(x,y) = \left(\frac{\partial U}{\partial x}\right)_y$$

$$\qquad\qquad [13.2]$$

$$Q(x,y) = \left(\frac{\partial U}{\partial y}\right)_x$$

Thus, on the basis of [13.2] from equation [12.3] immediately follow the determinations of temperature and pressure through the fundamental thermodynamic function — specific internal energy ϵ (formula [1.40]; Eckart, 1962):

$$T = \left(\frac{\partial \epsilon}{\partial \eta}\right)_v \; ; \qquad p = -\left(\frac{\partial \epsilon}{\partial v}\right)_\eta \qquad\qquad\qquad [13.3]$$

In addition to specific internal energy ϵ, three other fundamental functions (sometimes called characteristic) are introduced in thermodynamics, which are determined by the following expressions (written by means of their specific values) —

enthalpy, or heat content:

$$h = \epsilon + pv \qquad\qquad\qquad\qquad\qquad\qquad\qquad [13.4]$$

free energy:

$$f = \epsilon - T\eta \qquad\qquad\qquad\qquad\qquad\qquad\qquad [13.5]$$

(here $T\eta$ represents the so-called "bound energy") and
the thermodynamic potential (Gibbs' potential):

$$g = \epsilon - T\eta + pv \qquad\qquad\qquad\qquad\qquad\qquad [13.6]$$

Functions h, f and g also represent thermodynamic potentials — the functions of the two corresponding variables from the total number of four (T, η, p and v). The total differentials of these functions are written in the form of the following equations, which are the expression of the first and second principles of thermodynamics (let us write again for convenience also the expression for differential $d\epsilon$):

$$d\epsilon = T \, d\eta - p \, dv \qquad\qquad\qquad\qquad\qquad\qquad [13.7]$$

$$dh = T \, d\eta + v \, dp \qquad\qquad\qquad\qquad\qquad\qquad [13.8]$$

$$df = -\eta \, dT - p \, dv \qquad\qquad\qquad\qquad\qquad\qquad [13.9]$$

$$dg = -\eta \, dT + v \, dp \qquad\qquad\qquad\qquad\qquad\qquad [13.10]$$

Equations [13.8]–[13.10], as well as the expressions themselves of functions [13.4]–[13.6] can be obtained completely formally on the basis of the application of Legendre's transformation to the fundamental equation of thermodynamics [13.7] (Bazarov, 1961); let us show this with the example of function f. Writing the differential of the product $T\eta$:

$$d(T\eta) = T \, d\eta + \eta \, dT$$

and substituting in it the value of the quantity $T \, d\eta$, determined by equation [13.7], we obtain:

$$d(T\eta) - \eta \, dT = d\epsilon + p \, dv$$

or:

$$d(\epsilon - T\eta) = -\eta \, dT - p \, dv$$

Introducing expression [13.5], we obtain equation [13.9]. Equation [13.8] is obtained on the basis of consideration of the differential d(pv) and the application of the transformation also to equation [13.7]; equation [13.10] by the consideration of the differential d(pv) and by the transformation of equation [13.9].

The most convenient characteristic function is the thermodynamic potential g, since it is defined by the parameters of state — temperature and pressure.

From equations [13.7]–[13.10] on the basis of [13.1] and [13.2] there follow the expressions:

$$T = \left(\frac{\partial \epsilon}{\partial \eta}\right)_v = \left(\frac{\partial h}{\partial \eta}\right)_p \qquad [13.11]$$

$$p = -\left(\frac{\partial \epsilon}{\partial v}\right)_\eta = -\left(\frac{\partial f}{\partial v}\right)_T \qquad [13.12]$$

$$v = \left(\frac{\partial h}{\partial p}\right)_\eta = \left(\frac{\partial g}{\partial v}\right)_T \qquad [13.13]$$

$$\eta = -\left(\frac{\partial f}{\partial T}\right)_v = -\left(\frac{\partial g}{\partial T}\right)_p \qquad [13.14]$$

Furthermore, the condition for the existence of the total differential [13.1] is the identity:

$$\left(\frac{\partial P}{\partial y}\right)_x = \left(\frac{\partial Q}{\partial x}\right)_y \qquad [13.15]$$

applying [13.15] to equations [13.7]–[13.10], we obtain the so-called *reciprocal relations*, or *Maxwell's relations*, which play a fundamental role in the method of thermodynamic potentials:

$$\left(\frac{\partial T}{\partial v}\right)_\eta = -\left(\frac{\partial p}{\partial \eta}\right)_v \qquad [13.16]$$

$$\left(\frac{\partial T}{\partial p}\right)_\eta = \left(\frac{\partial v}{\partial \eta}\right)_p \qquad [13.17]$$

$$\left(\frac{\partial \eta}{\partial v}\right)_T = \left(\frac{\partial p}{\partial T}\right)_v \qquad [13.18]$$

$$\left(\frac{\partial \eta}{\partial p}\right)_T = -\left(\frac{\partial v}{\partial T}\right)_p \qquad [13.19]$$

Moreover, on the basis of the relations adduced, the derivatives of the fundamental characteristic functions of the second and third (and also higher) orders can be obtained, as well as the relations among them; the principal second derivatives will be considered in the following section.

The functions ϵ, h, f and g are not the only thermodynamic potentials (Bazarov, 1961, paragraph 22); from equations [13.7]−[13.10] it will be seen that entropy η represents the thermodynamic potential for the independent variables ϵ and v or for the independent variables h and p; specific volume − for the variables ϵ and η, or f and T; pressure − for the variables h and η, or g and T, and, lastly, temperature − for the independent variables f and v, or the variables g and p. Let us take as an example the last of these alternatives. From equation [13.10] it follows that:

$$dT = -\frac{1}{\eta}\,dg + \frac{v}{\eta}\,dp \qquad\qquad\qquad [13.20]$$

here, according to [13.1] and [13.2]:

$$\frac{1}{\eta} = -\left(\frac{\partial T}{\partial g}\right)_p \;; \qquad \frac{v}{\eta} = \left(\frac{\partial T}{\partial p}\right)_g$$

$$v = -\frac{\left(\dfrac{\partial T}{\partial p}\right)_g}{\left(\dfrac{\partial T}{\partial g}\right)_p} = \left(\frac{\partial g}{\partial p}\right)_T$$

which is in conformity with [13.13] and [13.14].

What has been set forth almost fully represents the fundamentals of the formal apparatus of the method of thermodynamic potentials, used for the study of simple (single-component) systems; examples of its application will be considered below. As is seen from the relations adduced, the parameters of state of the system are fully defined by any one of the potentials; accordingly, each of them identically determines its equation of state.

14. FUNDAMENTAL THERMODYNAMIC RELATIONS FOR A TWO-COMPONENT SYSTEM −
    SEA WATER

As has already been said above, we may consider sea water in a general approximation as a two-component system: a solvent (water) and a single solute ("sea salt"). In that case, the fundamental equation of thermodynamics for specific internal energy ϵ, as well as the corresponding equations for the other three fundamental functions − enthalpy h, free energy f and thermodynamic potential g (specific values), will be written in the following form:

$$d\epsilon = T\,d\eta - p\,dv + \mu\,dS \qquad\qquad [14.1]$$

$$dh = T\,d\eta + v\,dp + \mu\,dS \qquad\qquad [14.2]$$

$$df = -\eta\,dT - p\,dv + \mu\,dS \qquad\qquad [14.3]$$

$$dg = -\eta\,dT + v\,dp + \mu\,dS \qquad\qquad [14.4]$$

(equation [12.4] has been repeated for the convenience of the exposition). Here μ is the chemical potential and S is salinity. We see that relations [13.7] – [13.10] represent a particular case of equations [14.1]–[14.4] when S = constant.

Let us now write the general expression of the total differential of the function $U(x, y, z)$ of the three variables:

$$dU = P\,dx + Q\,dy + R\,dz \qquad\qquad [14.5]$$

where:

$$P(x, y, z) = \left(\frac{\partial U}{\partial x}\right)_{y,z}$$

$$Q(x, y, z) = \left(\frac{\partial U}{\partial y}\right)_{x,z} \qquad\qquad [14.6]$$

$$R(x, y, z) = \left(\frac{\partial U}{\partial z}\right)_{x,y}$$

The condition for the existence of the total differential [14.5] are the identities:

$$\left(\frac{\partial P}{\partial y}\right)_{x,z} \equiv \left(\frac{\partial Q}{\partial x}\right)_{y,z}$$

$$\left(\frac{\partial P}{\partial z}\right)_{x,y} \equiv \left(\frac{\partial R}{\partial x}\right)_{y,z} \qquad\qquad [14.7]$$

$$\left(\frac{\partial Q}{\partial z}\right)_{x,y} \equiv \left(\frac{\partial R}{\partial y}\right)_{x,z}$$

On the basis of [14.6], expressions [13.11]–[13.13], valid not only for a single-component, but now also for a two-component system (in case the latter relate to a two-component system, the index S should always be added to them following the parentheses), may be supplemented by the following relations, which determine the chemical potential through the fundamental functions:

$$\mu = \left(\frac{\partial \epsilon}{\partial S}\right)_{\eta,p} = \left(\frac{\partial h}{\partial S}\right)_{\eta,p} = \left(\frac{\partial f}{\partial S}\right)_{T,v} = \left(\frac{\partial g}{\partial S}\right)_{T,p} \qquad [14.8]$$

Further, the four reciprocal relations [13.16]–[13.19], considered in all courses of thermodynamics as applied to a single-component system, may be supplemented on the basis of formulae [14.7] and equations [14.1]–[14.4] by the following eight relations, including salinity and chemical potential (Mamayev, 1968):

$$\left(\frac{\partial \mu}{\partial \eta}\right)_{v,S} = \left(\frac{\partial T}{\partial S}\right)_{\eta,v} \qquad [14.9]$$

$$\left(\frac{\partial \mu}{\partial \eta}\right)_{p,S} = \left(\frac{\partial T}{\partial S}\right)_{\eta,p} \qquad [14.10]$$

$$\left(\frac{\partial \mu}{\partial T}\right)_{v,S} = -\left(\frac{\partial \eta}{\partial S}\right)_{\eta,p} \qquad [14.11]$$

$$\left(\frac{\partial \mu}{\partial T}\right)_{\rho,S} = -\left(\frac{\partial \eta}{\partial S}\right)_{T,S} \qquad [14.12]$$

$$\left(\frac{\partial \mu}{\partial v}\right)_{\eta,S} = -\left(\frac{\partial p}{\partial S}\right)_{\eta,v} \qquad [14.13]$$

$$\left(\frac{\partial \mu}{\partial v}\right)_{T,S} = -\left(\frac{\partial p}{\partial S}\right)_{T,v} \qquad [14.14]$$

$$\left(\frac{\partial \mu}{\partial p}\right)_{\eta,S} = \left(\frac{\partial v}{\partial S}\right)_{\eta,p} \qquad [14.15]$$

$$\left(\frac{\partial \mu}{\partial p}\right)_{T,S} = \left(\frac{\partial v}{\partial S}\right)_{T,p} \qquad [14.16]$$

The reciprocal relations [13.16]–[13.19] continue to remain valid also for a two-component (and generally for a multi-component) system; in this case the index S has to be added to them following the parentheses.

Relation [14.15] was demonstrated already by Gibbs (instead of salinity S he used m – the number of components – see Gibbs, 1950); as applied to sea water Fofonoff (1962) indicated relations [14.11]–[14.12] and [14.15]–[14.16]. Relations [14.9] and [14.10], which include a thermohaline derivative, are of specific interest from the point of view of T-S analysis; they will be considered in somewhat greater detail below (Section 25).

The relations given above in this and in the preceding paragraph make it possible also to obtain a number of derivatives of the second and of higher orders, expressing additional relationships among the thermodynamic parameters which may be of practical interest. Let us consider as an example the second derivatives of the fundamental thermodynamic function–specific internal energy ϵ (Eckart, 1962). Since ϵ is a function of entropy, specific volume and salinity (equation [14.1]), and pressure, temperature and chemical potential in turn depend upon internal energy and the parameters mentioned (equations [1.40]), we can write the expressions of the following differentials:

$$dp = \left(\frac{\partial p}{\partial v}\right)_{\eta,S} dv + \left(\frac{\partial p}{\partial \eta}\right)_{v,S} d\eta + \left(\frac{\partial p}{\partial S}\right)_{\eta,v} dS \qquad [14.17]$$

$$dT = \left(\frac{\partial T}{\partial v}\right)_{\eta,S} dv + \left(\frac{\partial T}{\partial \eta}\right)_{v,S} d\eta + \left(\frac{\partial T}{\partial S}\right)_{\eta,v} dS \qquad [14.18]$$

$$d\mu = \left(\frac{\partial \mu}{\partial v}\right)_{\eta,S} dv + \left(\frac{\partial \mu}{\partial \eta}\right)_{v,S} d\eta + \left(\frac{\partial \mu}{\partial S}\right)_{\eta,v} dS \qquad [14.19]$$

On the basis of equations [1.40] these formulae in turn may be written in the following form, including the second derivatives:

$$dp = \frac{\partial^2 \epsilon}{\partial v^2} dv + \frac{\partial^2 \epsilon}{\partial \eta \, \partial v} d\eta + \frac{\partial^2 \epsilon}{\partial v \, \partial S} dS \qquad [14.20]$$

$$dT = \frac{\partial^2 \epsilon}{\partial \eta \, \partial v} dv + \frac{\partial^2 \epsilon}{\partial \eta^2} d\eta + \frac{\partial^2 \epsilon}{\partial \eta \, \partial S} dS \qquad [14.21]$$

$$d\mu = \frac{\partial^2 \epsilon}{\partial v \, \partial S} dv + \frac{\partial^2 \epsilon}{\partial \eta \, \partial S} d\eta + \frac{\partial^2 \epsilon}{\partial S^2} dS \qquad [14.22]$$

We may point out that the link between the identical mixed second derivatives in these equations is determined by the corresponding reciprocal relations; homogeneous secondary derivatives also have considerable thermodynamic importance.

Substantial interest also attaches to the derivatives of the second and higher orders of another fundamental function – Gibbs' thermodynamic potential g, inasmuch as, according to equation [14.4], it is a function of temperature, salinity and pressure, i.e., of directly measurable quantities (Fofonoff, 1962).

The fundamentals of the formal apparatus of the method of thermodynamic potentials, as applied in the study of a two-component system, are to all intents and purposes also confined to this.

Before proceeding to the application of the apparatus described to the study of some thermodynamic properties of sea water, let us note that many thermodynamic relations between derivatives of the type $(\partial y/\partial x)_z$, where x, y, z are any three quantities from the total number of thermodynamic parameters and fundamental functions, may be obtained fairly rapidly and simply by the use of Jacobians. Without dwelling further on this question, let us note that the method of Jacobians in thermodynamics is set forth briefly by Guggenheim (1950, par. 3.08), who calls this method "elegant", as well as by Batuner and Pozin (1963, chapter XI, par. 10); the latter work contains a key table for the use of Jacobians.

15. CHEMICAL POTENTIAL

The fundamental equation [12.3] is valid for an isotropic medium consisting of only *one* substance, provided mass and composition are constant. However, the number of particles in the system may be variable; both because of mass exchange and chemical reactions within the area considered and mass exchange with the surrounding medium (open system).

In the case of an open system or a system of variable composition, fundamental equation [12.3] can be written in the following generalized form:

$$d\epsilon = T\,d\eta - p\,dv + \sum_i \mu_i\,dx_i \qquad\qquad [15.1]$$

where x_i is the mass fraction of component i, and the quantity μ_i is determined by the expression:

$$\mu_i = \left(\frac{\partial \epsilon}{\partial x_i}\right)_{\eta,v,x_j} \qquad\qquad [15.2]$$

where x_j designates all mass fractions except x_i. The quantity μ_i is called the potential, or chemical potential of the i-th component.

Let us now derive from [15.1] the corresponding equation for a two-component system — sea water, which was written above (formulae [12.4] and [14.1]) without proof. Designating the mass fractions of pure water and salts by x_w and x_s, respectively, we have on the basis of [12.2]:

$$x_w = \frac{m_w}{m_w + m_s}; \qquad x_s = \frac{m_s}{m_w + m_s}; \qquad x_w + x_s = 1 \qquad\qquad [15.3]$$

It is also convenient, expressing salinity in g/g, or in mass fractions, to introduce the following designations:

$$x_s = S; \qquad x_w = w = 1 - S \qquad\qquad [15.4]$$

where the value w may be called the "freshness" of sea water.

Then, for a two-component system, equation [15.1] will be written in the following form:

$$d\epsilon = T \, d\eta - p \, dv + \mu_w \, dx_w + \mu_s \, dx_s \qquad\qquad [15.5]$$

where μ_w and μ_s are the chemical potentials of pure water and of salts respectively; x_w and x_s are their mass fractions. Let us introduce the following designation for the difference of chemical potentials:

$$\mu = \mu_s - \mu_w \qquad\qquad [15.6]$$

Then, substituting in the last term on the right side of [15.5] the value $\mu + \mu_w$ instead of μ_s, we obtain:

$$d\epsilon = T \, d\eta - p \, dv + \mu_w (dx_w + dx_s) + \mu \, dx_s \qquad\qquad [15.7]$$

Since, according to the condition of the conservation of mass [15.3]:

$$dx_w + dx_s = 0 \qquad\qquad [15.8]$$

we finally obtain:

$$d\epsilon = T \, d\eta - p \, dv + \mu \, dS \qquad\qquad [15.9]$$

where $\mu = \partial\epsilon / \partial S$.

The same result can be obtained more briefly, namely:

$$\sum_i^{N=2} \mu_i \, dx_i = \mu_s \, dS + \mu_w \, dw = \mu_s \, dS + \mu_w \, d(1 - S) = \mu_s \, dS - \mu_w \, dS \qquad [15.10]$$

introducing designation [15.6], we obtain for two-component sea water:

$$\sum_i^{N=2} \mu_i \, dx_i = \mu \, dS \qquad\qquad [15.11]$$

Let us derive an important expression for the chemical potential of pure water μ_w in a solution (sea water). For this let us consider the two-phase system water — vapor, which is in conditions of thermodynamic equilibrium; in this case, the chemical potentials of both phases must be identical (this proposition, in agreement with the very definition of potential, is not proved here).

Let us further consider (with a certain approximation) saturated vapor as a perfect gas, obeying Clapeyron's equation of state [2.18]. The reciprocal relation [14.16] for a perfect gas will be written, accordingly, in the form:

$$\left(\frac{\partial \mu}{\partial p}\right)_{T,n} = \left(\frac{\partial v}{\partial n}\right)_{T,p} \qquad [15.12]$$

Substituting in the right side of this expression equation [2.18], we obtain:

$$\frac{\partial \mu}{\partial p} = \frac{RT}{p} \qquad [15.13]$$

and, integrating this expression, we have:

$$\mu = RT \ln p + \text{const.} \qquad [15.14]$$

To determine the constant of integration let us choose some standard state (designated by an asterisk), $p = p^*$, under which $\mu = \mu^*$:

$$\mu^* = RT \ln p^* + \text{const.} \qquad [15.15]$$

Subtracting [15.15] from [15.14] and assuming $p^* = 1$, we obtain the following expression for the chemical potential of a perfect gas:

$$\mu = \mu^* + RT \ln p \qquad [15.16]$$

In the case of a mixture of perfect gases the same expression is valid for each component A:

$$\mu_A = \mu_A^* + RT \ln p_A \qquad [15.17]$$

where:

$$p_A = x_A p \qquad [15.18]$$

is partial pressure (x_A is the mass fraction of substance A).

Substituting [15.18] in [15.17], we obtain:

$$\mu_A = [\mu_A^* + RT \ln p] + RT \ln x_A \qquad [15.19]$$

This substitution brings equation [15.17] into a different standard state; the expression in square brackets represents the chemical potential of pure substance A at pressure p. Designating this expression μ^0 (the index "0" designates "pure substance"), we obtain:

$$\mu_A = \mu_A^0 + RT \ln x_A \qquad [15.20]$$

As a consequence of the equality of the chemical potentials of pure saturated vapor and pure water in a solution (sea water), instead of [15.20] we can finally write:

$$\mu_w = \mu_w^0 + RT \ln x_w \qquad [15.21]$$

Here μ_w is the chemical potential of pure water in a solution (sea water), μ_w^0 is the chemical potential of pure water taken separately. A similar formula may be written also for the chemical potential of the salts μ_s. This elegant derivation is borrowed from Everett's monograph (1963); in this book, incidentally, an analogy is drawn between the chemical potential and mechanical potential (the problem of the loss of water in a vessel with a generatrix in the form of a logarithmic curve).

An important formula (Haase, 1963) follows from expression [15.21]:

$$\left(\frac{\partial \mu_w}{\partial x_w}\right)_{T,p} = \frac{RT}{x_w} \approx RT \quad (x_s \ll 1) \tag{15.22}$$

Solutions for which equation [15.20] is valid for all values of x_A ($0 \leqslant x_A \leqslant 1$), are called *perfect*; solutions for which this equation is valid only if they are strongly diluted are called *ideal diluted* solutions. Since the quantity of salts in sea water is not large, $x_s \ll x_w$, it may be considered, with a certain approximation, as an ideal diluted solution and the methods of the theory of solutions may be applied to its study. In particular, such a property of sea water as contraction on mixing may be considered as a corollary of the deviation of its properties from those of an ideal diluted solution.

16. THE GIBBS–DUHEM EQUATION

The Gibbs–Duhem equation is of great importance for the study of systems with a variable number of particles. Let us derive this equation in general form and then consider it as applied to a two-component system — sea water.

We shall proceed from the fundamental equation of thermodynamics for a multi-component system (equation [15.1]) and the corresponding expression for thermodynamic potential [14.4], which for a multi-component system may be written as:

$$dg = -\eta \, dT + v \, dp + \sum_i \mu_i \, dx_i \tag{16.1}$$

Integrating this expression at constant temperature and constant pressure ($dT = 0$, $dp = 0$), we obtain (Guggenheim, 1950, par. 1.35):

$$g = \sum_i \mu_i x_i \tag{16.2}$$

Accordingly, instead of [13.6] we can write:

$$\epsilon = \sum_i \mu_i x_i + T\eta - pv \tag{16.3}$$

The differential of this expression is equal to:

$$d\epsilon = \sum_i \mu_i \, dx_i + \sum_i x_i \, d\mu_i + T \, d\eta + \eta \, dT - p \, dv - v \, dp \qquad [16.4]$$

Comparing this equation with equation [15.1], we obtain:

$$\eta \, dT - v \, dp + \sum_i x_i \, d\mu_i = 0 \qquad [16.5]$$

This is the generalized Gibbs–Duhem relation.

For two-component sea water the Gibbs–Duhem equation will be written as follows:

$$\eta \, dT - v \, dp + x_w \, d\mu_w + x_s \, d\mu_s = 0 \qquad [16.6]$$

Substituting here instead of μ_s the quantity $\mu + \mu_w$, we obtain:

$$\eta \, dT - v \, dp + (x_w + x_s) \, d\mu_w + x_s \, d\mu = 0 \qquad [16.7]$$

Since $x_w + x_s = 1$ (formula [15.3]), designating $x_s = S$, we obtain the Gibbs–Duhem relation in the form in which it is given by Fofonoff (1962):

$$d\mu_w = -\eta \, dT + v \, dp - S \, d\mu \qquad [16.8]$$

We may obtain also another form of the Gibbs–Duhem relation, namely: substituting in [16.6] instead of μ_w the quantity $\mu_s - \mu$, we obtain:

$$\eta \, dT - v \, dp + (x_w + x_s) \, d\mu_s - x_w \, d\mu = 0 \qquad [16.9]$$

or:

$$d\mu_s = -\eta \, dT + v \, dp + (1 - S) \, d\mu \qquad [16.10]$$

With $dT = 0$ and $dp = 0$ from equation [16.6] we obtain the following particular version of the Gibbs–Duhem equation for a two-component system (sea water):

$$x_w \, d\mu_w + x_s \, d\mu_s = 0 \qquad [16.11]$$

as well as:

$$(1 - S) \frac{\partial \mu_w}{\partial S} + S \frac{\partial \mu_s}{\partial S} = 0 \qquad [16.12]$$

Consequently, if at constant temperature and constant pressure the chemical potential of one of the components increases, the potential of the other component must decrease.

The application of the Gibbs–Duhem equation in form [16.12] will be considered below.

17. PARTIAL QUANTITIES

If we mix n_1 moles of component 1, having a molar mass M_1, and n_2 moles of component 2 with a molar mass M_2, then, according to the principle of the conservation of mass, the mass M of the mixture will be determined as:

$$M = n_1 M_1 + n_2 M_2 \qquad [17.1]$$

But a similar expression written, let us say, for volume V instead of mass M, will prove invalid because of the non-additivity of volume, i.e., its non-conservativeness in the process of mixing (this question will be considered in more detail below; see Sections 21, 22). Therefore, in order to obtain an extremely desirable additive expression of the type [17.1] for volume, the addends of its right side should be defined somehow differently — so that the whole expression may prove valid. Let us carry out, following Guggenheim (1950), the following reasoning:

Let us write an expression of the type [17.1] in the following form:

$$V = n_1 \bar{v}_1 + n_2 \bar{v}_2 \qquad [17.2]$$

and let us prove that it is valid if the quantities $\bar{v}_1$ and $\bar{v}_2$ are defined in the following way:

$$\bar{v}_1 = \left(\frac{\partial V}{\partial n_1}\right)_{T,p,n_2} \qquad [17.3]$$

$$\bar{v}_2 = \left(\frac{\partial V}{\partial n_2}\right)_{T,p,n_1} \qquad [17.4]$$

Let us note immediately that for these quantities we use the small letters v in order to emphasize that they have the dimension of specific volume (cm^3/g); the dash will be explained below. The total change of volume during mixing may then be expressed in the following way:

$$dV = \frac{\partial V}{\partial T}\, dT - \frac{\partial V}{\partial p}\, dp + \bar{v}_1\, dn_1 + \bar{v}_2\, dn_2 \qquad [17.5]$$

while its change at constant temperature and constant pressure due to a change in composition is expressed as:

$$dV = \bar{v}_1\, dn_1 + \bar{v}_2\, dn_2 \qquad [17.6]$$

If the change in volume occurs in the same proportion as change in n_1 and n_2, and at constant relative composition ($n_1/n_2 = $ const.), then we may state:

$$dn_1 = n_1\, d\zeta; \quad dn_2 = n_2\, d\zeta; \quad dV = V\, d\zeta \qquad [17.7]$$

Substituting [17.7] in [17.6] and reducing by $d\zeta$, we obtain expression [17.2].

The quantities $\bar{v}_1$ and $\bar{v}_2$ are called *partial (mole) volumes* of components 1 and 2, respectively, and are symbolized by dashes.

The physical significance of the constancy of composition (concentration) noted above during a change in volume consists of the fact that partial volume is considered when, say, one mole of salts is added to (or removed from) such a large volume of sea water that the concentration is practically unchanged.

Differentiating [17.2]:

$$dV = n_1 \, d\bar{v}_1 + n_2 \, d\bar{v}_2 + \bar{v}_1 \, dn_1 + \bar{v}_2 \, dn_2 \qquad\qquad [17.8]$$

and comparing the expression obtained with [17.5], we obtain:

$$-\frac{\partial V}{\partial T} \, dT + \frac{\partial V}{\partial p} \, dp + n_1 \, d\bar{v}_1 + n_2 \, d\bar{v}_2 = 0 \qquad\qquad [17.9]$$

At constant temperature and constant pressure in particular, we have:

$$n_1 \, d\bar{v}_1 + n_2 \, d\bar{v}_2 = 0 \qquad\qquad [17.10]$$

For a single-mass system from [17.10] we obtain:

$$x_1 \, d\bar{v}_1 + x_2 \, d\bar{v}_2 = 0 \qquad\qquad [17.11]$$

as well as:

$$x_1 \frac{\partial \bar{v}_1}{\partial x} + x_2 \frac{\partial \bar{v}_2'}{\partial x} = 0 \qquad\qquad [17.12]$$

For the *specific* volume of sea water, instead of formulae [17.2], [17.11] and [17.12], we have:

$$v = (1 - S)\bar{v}_w + S\bar{v}_s \qquad\qquad [17.13]$$

$$(1 - S) \, d\bar{v}_w + S \, d\bar{v}_s = 0 \qquad\qquad [17.14]$$

$$(1 - S) \frac{\partial \bar{v}_w}{\partial S} + S \frac{\partial \bar{v}_s}{\partial S} = 0 \qquad\qquad [17.15]$$

where $\bar{v}_w$ and $\bar{v}_s$ are the partial (specific) volumes of water and salts in sea water, respectively.

Differentiating formula [17.13] by salinity and taking into account formula [17.15], we obtain:

$$\frac{\partial v}{\partial S} = \bar{v}_s - \bar{v}_w \qquad\qquad [17.16]$$

Finally, from formulae [17.13] and [17.16] we obtain:

$$\bar{v}_w = v - S \frac{\partial v}{\partial S} \qquad\qquad [17.17]$$

$$\bar{v}_s = v + (1 - S) \frac{\partial v}{\partial S} \qquad\qquad [17.18]$$

Formulae [17.13]–[17.18] are *general in nature* and are applicable to other, not necessarily thermodynamic, but necessarily extensive properties (internal energy, entropy, specific volume, heat capacity, enthalpy, etc.).

In particular, the application of a formula of type [17.15] to the thermodynamic potential g leads, taking account of formula [16.1], to the already well-known version of the Gibbs–Duhem equation [16.11].

Let us consider an example of the application of the Gibbs–Duhem equation [17.15] to the determination of an analytical type of equation of state of sea water, borrowing this example from the work of Duedall and Weyl (1967) *. As these authors point out, in the range of salinities $30 \leqslant S \leqslant 40°/oo$ the quantity $\bar{v}_s$, according to their experiments, can be expressed by the linear formula:

$$\bar{v}_s = b_1 + b_2 S \qquad\qquad [17.19]$$

where b_1 and b_2 are empirical constants. Differentiating this expression by salinity, we obtain:

$$\frac{\partial \bar{v}_s}{\partial S} = b_2$$

Substituting this expression in relation [17.15] and integrating, we obtain:

$$\bar{v}_w = b_0 - b_2 (S^2/2 + S^3/3) \qquad\qquad [17.20]$$

where b_0 is the empirical constant defined in the same range of salinities. Finally, substituting expressions [17.20] and [17.19] in expression [17.13] and simplifying it (simplification consists of neglecting terms of a higher order of S), we obtain the following formula which determines the dependence of the specific volume of sea water on salinity at constant temperature:

$$v = b_0 + (b_1 - b_0) S + b_2 S^2/2 \qquad\qquad [17.21]$$

The values of the constants b_0, b_1 and b_2 were calculated from Knudsen's tables for temperatures of 0, 5, 10, 15, 20 and 25° and are given in the work of Duedall and Weyl mentioned. The values of the quantity of partial volume of salts, $\bar{v}_s$, calculated according to Knudsen's tables, prove to be in good

---

* The description of experiments contained in the work mentioned is interesting by itself as an illustration of the concept of partial volume.

agreement with the values calculated by these authors on the basis of an addition of the partial volumes of the basic salts (components) contained in sea water.

Let us point out in conclusion experiments in studying some other partial quantities as applied to sea water. Thus, in Connors' work (1970) the quantities of partial specific enthalpies of sea water, $\bar{h}_w$ and $\bar{h}_s$, are applied to obtain the dependance of enthalpy on temperature and salinity (Section 18) and for the further analytical calculation of the heat of mixing of waters of different temperature and salinity. The work of Millero (1969) is devoted to the study of partial molal volumes of ions in sea water, and the work of Connors and Weyl (1968) to the partial quantities of electrical conductivity of sea salts in connection with the study of the relationship between conductivity and density. Finally, in Bromley's works (1968, 1970) partial quantities of heat conductivity are studied; in the second of these works there are tables of partial heat conductivities of salts and pure water in sea water.

18. HEAT CAPACITY OF SEA WATER

The *heat capacity** of a system is defined as the quantity of heat necessary to heat the system by one degree (heating in this case has a definite physical similarity with the linear expansion of a bar; Haase, 1963).

For a single-component system we distinguish:

(1) Specific heat capacity at constant volume ($dv = 0$), or the isohoric specific heat capacity, c_v, which according to equation [12.3] is defined as the quantity:

$$c_v = \left(\frac{\partial \epsilon}{\partial T}\right)_v = T\left(\frac{\partial \eta}{\partial T}\right)_v \qquad [18.1]$$

(2) Specific heat capacity at constant pressure ($dp = 0$), or the isobaric specific heat capacity, c_p, which, clearly, is determined from equation [13.8] in the following way:

$$c_p = \left(\frac{\partial h}{\partial T}\right)_p = T\left(\frac{\partial \eta}{\partial T}\right)_p \qquad [18.2]$$

For a two-component system — sea water — we distinguish respectively specific heat capacity at constant volume and constant salinity:

$$c_{v,S} = \left(\frac{\partial \epsilon}{\partial T}\right)_{v,S} = T\left(\frac{\partial \eta}{\partial T}\right)_{v,S} \qquad [18.3]$$

---

* Below we will speak of specific heat capacity, or specific heat (cf. second footnote on p.50).

and specific heat capacity at constant pressure and constant salinity:

$$c_{p,S} = \left(\frac{\partial h}{\partial T}\right)_{p,S} = T\left(\frac{\partial \eta}{\partial T}\right)_{p,S} \tag{18.4}$$

The last two formulae follow from equations [14.1] (with $dv = 0$, $dS = 0$) and [14.2] (with $dp = 0$, $dS = 0$) respectively.

It is clear from what has been set forth that the heat capacity of sea water is a function of all the parameters of its state:

$$c = c(v, S, T, p) \tag{18.5}$$

The relationships between specific heat capacities and pressure and specific volume are determined by the following expressions:

$$\left(\frac{\partial c_{p,S}}{\partial p}\right)_{T,S} = -T\left(\frac{\partial^2 v}{\partial T^2}\right)_{p,S} \tag{18.6}$$

$$\left(\frac{\partial c_{v,S}}{\partial v}\right)_{T,S} = T\left(\frac{\partial^2 p}{\partial T^2}\right)_{v,S} \tag{18.7}$$

These formulae are obtained by the differentiation of the right-hand formulae [18.4] and [18.3], and then by a change in the order of the differentiation and the use of reciprocal relations [10.19] and [10.18] respectively. Formulae [18.6] and [18.7] link heat capacities with the thermal equation of state of the system, determining their right-hand parts; however, the direct calculation of heat capacities according to formulae [18.6] and [18.7] may lead to erroneous results because of the sensitivity of the caloric quantities (heat capacity) to small changes in the parameters of state (Stupochenko, 1956). The empirical formula for the determination of $\partial c_p/\partial p$, according to formula [18.6], is given by Fofonoff (1962) in the form of a polynomial by increasing powers of the parameters of state.

The following derivative is also of interest:

$$\left(\frac{\partial c_{p,S}}{\partial S}\right)_{T,S} = -T\left(\frac{\partial^2 \mu}{\partial T^2}\right)_{p,S} \tag{18.8}$$

which is obtained on the basis of formulae [14.11] and [18.1]. Other derivatives may also be obtained from heat capacities, but they are not of as much interest as those given above.

Let us now consider the connection between isobaric and isohoric specific heat capacities and the relation between them. All the considerations adduced below for a single-component system are also valid for a binary system — sea water, since in the latter case heat capacities are considered at constant salinity

$dS = 0$ (in order to have the considerations formally lead to a binary system, it is necessary to restore the terms containing dS and to provide the derivatives containing indices with still another index S).

Bearing in mind that in the fundamental equation of thermodynamics [12.3] $\epsilon = \epsilon(v, T)$, we may write this equation in the following form (expanding the total differential $d\epsilon$):

$$T \, d\eta = \left(\frac{\partial \epsilon}{\partial T}\right)_v dT + \left[\left(\frac{\partial \epsilon}{\partial v}\right)_T + p\right] dv \qquad [18.9]$$

Differentiating this expression with respect to T, we obtain the equation for specific heat capacity in general:

$$c = T \frac{d\eta}{dT} = \left(\frac{\partial \epsilon}{\partial T}\right)_v + \left[\left(\frac{\partial \epsilon}{\partial v}\right)_T + p\right]\left(\frac{dv}{dT}\right) \qquad [18.10]$$

At constant volume, $dv = 0$, the second term of the right side disappears, and we obtain formula [18.1]:

$$c_v = T\left(\frac{\partial \eta}{\partial T}\right)_v = \left(\frac{\partial \epsilon}{\partial T}\right)_v$$

while at constant pressure, $dp = 0$:

$$c_p = c_v + \left[\left(\frac{\partial \epsilon}{\partial v}\right)_T + p\right]\left(\frac{\partial v}{\partial T}\right)_p \qquad [18.11]$$

This general formula, expressing the link between specific heat capacities c_p and c_v, is unsuitable, since it contains the caloric parameter ϵ. Let us transform it as follows. Considering formula [18.9] as the expression of the total differential $d\eta$ for the variables v and T, i.e., as an expression of the type:

$$d\eta = \left(\frac{\partial \eta}{\partial T}\right) dT + \left(\frac{\partial \eta}{\partial v}\right)_T dv \qquad [18.12]$$

and comparing [18.9] with [18.12], we obtain the following identical relations:

$$\left(\frac{\partial \epsilon}{\partial T}\right)_v = T\left(\frac{\partial \eta}{\partial T}\right)_v \, , \qquad \left(\frac{\partial \epsilon}{\partial v}\right)_T + p = T\left(\frac{\partial \eta}{\partial v}\right)_T \qquad [18.13]$$

Substituting [18.13] in [18.11] and taking the reciprocal relation [13.18] into account, we obtain the following basic formula, expressing the link between the specific heat capacities

$$c_p - c_v = T\left(\frac{\partial p}{\partial T}\right)_v \left(\frac{\partial v}{\partial T}\right)_p \qquad [18.14]$$

This formula in turn can also be transformed. Thus, let us consider the differential of entropy as a function of temperature and pressure (Fofonoff, 1962):

$$d\eta = \left(\frac{\partial \eta}{\partial T}\right)_p dT + \left(\frac{\partial \eta}{\partial p}\right)_T dp \qquad [18.15]$$

This expression can be written, according to formulae [18.2] and [13.19], in the form:

$$d\eta = \frac{c_p}{T} dT - \left(\frac{\partial v}{\partial T}\right)_p dp \qquad [18.16]$$

Let us calculate the isohoric specific heat capacity c_v; for this, having made use of the condition:

$$dv = \left(\frac{\partial v}{\partial T}\right)_p dT + \left(\frac{\partial v}{\partial p}\right)_T dp = 0 \qquad [18.17]$$

having determined from it the differential dp and having substituted it in [18.16], we obtain:

$$d\eta = \frac{c_p}{T} dT + \left[\left(\frac{\partial v}{\partial T}\right)_p^2 \Big/ \left(\frac{\partial v}{\partial p}\right)_T\right] dT$$

whence:

$$c_p - c_v = -T\left(\frac{\partial v}{\partial T}\right)_p^2 \Big/ \left(\frac{\partial v}{\partial p}\right)_T \qquad [18.18]$$

The latter formula is more convenient, particularly from the oceanographic point of view, than formula [18.14], in which it is practically impossible experimentally to determine the derivative $(\partial p/\partial T)_v$. Equation [18.18] can also be obtained by another, shorter way. Multiplying and dividing the right side of [18.14] by $(\partial v/\partial T)_p$, we obtain:

$$c_p - c_v = T \frac{\left(\frac{\partial v}{\partial T}\right)_p^2}{\left(\frac{\partial T}{\partial p}\right)_v \left(\frac{\partial v}{\partial T}\right)_p} \qquad [18.19]$$

whence, on the basis of the identity:

$$\left(\frac{\partial T}{\partial p}\right)_v \left(\frac{\partial v}{\partial T}\right)_p \left(\frac{\partial p}{\partial v}\right)_T = -1 \qquad [18.20]^*$$

we obtain formula [18.18].

---

* Identity [18.20] and the inverse identity [18.21] follow from [2.6] when $dS = 0$.

Finally, still another formula, expressing the difference of heat capacities, can be obtained in the same way from basic formula [18.14]. Multiplying and dividing [18.14] by $(\partial p/\partial T)_v$ and applying the identity:

$$\left(\frac{\partial T}{\partial v}\right)_p \left(\frac{\partial p}{\partial T}\right)_v \left(\frac{\partial v}{\partial p}\right)_T = -1 \qquad\qquad [18.21]$$

we obtain:

$$c_p - c_v = -T\left(\frac{\partial p}{\partial T}\right)_v^2 \bigg/ \left(\frac{\partial p}{\partial v}\right)_T \qquad\qquad [18.22]$$

Let us express the difference $c_p - c_v$ through the coefficients introduced in Section 2. Thus, for example, from formula [18.18] and the determination of the coefficients of thermal (volumetric) expansion [2.7] and compressibility [2.9], there follows the formula:

$$c_p - c_v = \frac{T v_0 a^2}{k} \qquad\qquad [18.23]$$

Finally, *the ratio of specific heat capacities* — a quantity important first of all in the theory of the propagation of sound in the sea, is determined by the following formula, resulting from [18.18]:

$$\gamma = \frac{c_p}{c_v} = \left[1 + \frac{T\left(\frac{\partial v}{\partial T}\right)_p^2}{c_p\left(\frac{\partial v}{\partial p}\right)_T} \right]^{-1} \qquad\qquad [18.24]$$

In solving many problems of oceanography the quantity of specific heat of sea water is considered to be constant (equal to 1 cal. g^{-1} $°C^{-1}$) and in many cases this precision proves adequate. However, in the consideration of a number of important questions connected with the equation of state of sea water, precise knowledge of the specific heat of sea water is absolutely necessary. These problems include, in the first place, the question of the speed of propagation of sound in the sea and the question of the nature of adiabatic processes in the deep parts of the ocean. Precise knowledge of specific heat capacity is of great importance in studying the non-linear properties of sea water, particularly in studying contraction on mixing (Section 22).

The quantity c_p for distilled water is determined according to Jaeger and Steinwehr (1915) by the empirical formula:

$$c_p = 1.00492 - 4.22542 \cdot 10^{-4}\, T + 6.32379 \cdot 10^{-6}\, T^2 \qquad\qquad [18.25]$$

$0 < T < 50°C.$

The data of Osborne et al. (1939) are, apparently, more precise; in particular, they served Cox and Smith as a basis for the compilation of a table of the heat capacity of sea water and, in addition, were verified by the latter two authors *.

The first studies of the heat capacity of sea water were undertaken by Thoulet and Chevallier in 1889; however, their results are unreliable both because of the technical imperfection of their experiments and because of the fact that these authors considered the specific heat of distilled water c_p as not changing with temperature and equal to a constant quantity of 1 cal. g^{-1} °C^{-1}. In spite of this, the data of Thoulet and Chevallier, worked up by Krummel, were used in oceanography for more than fifty years. Comparatively recent, Cox and Smith (1959) published a new table of the values of the specific heat capacity c_p of sea water at various values of temperature, salinity and atmospheric pressure. These authors performed a series of precise laboratory experiments to determine heat capacity. The methodology, on which we will not dwell, is described in detail in the work of Cox and Smith (1959). A total of five series of experiments was performed with water of 10.56; 20.22; 30.055; 34.297 and 39.768‰ salinities at different temperatures, as well as control measurements of the heat capacity of distilled water.

It proved possible to express the result of the experiments by the following formula:

$$c_p = A - 0.005075\,S - 0.000014\,S^2 \qquad\qquad [18.26]$$

where c_p is the specific isobaric heat capacity of sea water of salinity S (g kg^{-1}) in absolute Joules per gram per degree Celsius (J g^{-1} °C^{-1}). A is the specific heat capacity of distilled water at constant pressure, dependent on some conventional (elevated when $S > 0$) temperature T', determined by the formula:

$$T' = (T + 0.7S + 0.0175\,S^2)\ \text{°C}$$

The specific heat of distilled water at constant pressure was taken according to Osborne et al. (1939), and a series of control measurements of the heat capacity of distilled water, referred to above, yielded results coinciding with the data of Osborne et al. Cox and Smith do not give a formula for the determination of the quantity A. However, Fofonoff (1962), relying on the data of Osborne et al., shows that for salinities higher than 30‰ and at a temperature higher than the freezing point the following formula is valid:

$$A = 4.1784 + 8.46 \cdot 10^{-6}\,(T - 33.67)^2 \qquad\qquad [18.27]$$

---

* These authors do not give the formula determining the value of c_p.

From their experimental data Cox and Smith compiled a table of specific heats c_p in absolute Joules per gram per degree Celcius for salinities from 0 to 40‰ and temperatures from −2 to 30°C (Table A12 of the Annex). The possible error in the values of this table amounts to approximately 0.0015 $J g^{-1} {}^{\circ}C^{-1}$ (the experiments themselves were carried out with a precision of up to 0.0001 $J g^{-1} {}^{\circ}C^{-1}$).

Having divided the values of heat capacities by the quantity of the mechanical equivalent of heat, equal to 4.1876 J cal^{-1} (Dorsey, 1968), it is possible to obtain the specific heat values in a more usual dimension, cal. $g^{-1} {}^{\circ}C^{-1}$.

Connors (1970) approximates Cox' and Smith's data by the following simpler formula, drawing upon the results of his experiments in determining the heat of mixing of waters:

$$c_p = 4.2044 - 0.00114\,T - S(6.99 - 0.069\,T) +$$
$$+ S^2 (19.6 - 0.6\,T) \quad J g^{-1} {}^{\circ}C^{-1} \tag{18.28}$$

This formula is valid in the ranges $1.87 < T < 30.84°C$, $10.56 < S < 39.79‰$, and its standard deviation from Cox' and Smith's data amounts in this range to a quantity of 0.0024 $J g^{-1} {}^{\circ}C^{-1}$.

Experiments in determining the specific heat of sea water were also carried out by Bromley et al. (1967), who drew up on the basis of these experiments four alternative formulae of different degrees of accuracy. Let us quote the most accurate of these formulae, namely (salinity in %):

$$c_p = 1.0049 - 0.016210\,S + (3.5261 \cdot 10^{-4}\,S^2) -$$
$$- [(3.2506 - 1.4795\,S + 0.07765\,S^2) \cdot 10^{-4}\,T] +$$
$$+ [(3.8013 - 1.2084\,S + 0.06121\,S^2) \cdot 10^{-6}\,T^2] \tag{18.29}$$

The formula is valid in the ranges $2 < T < 80°C$ and $1 < S < 12\%$, while its confidence limit in these ranges amounts to 0.001 cal. $g^{-1} {}^{\circ}C^{-1}$.

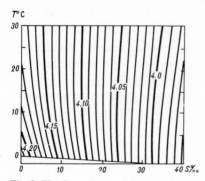

Fig. 8. The dependence of the specific heat of sea water at constant (atmospheric) pressure c_p in absolute Joules per gram per degree Celsius on temperature and salinity (Cox and Smith, 1959; Fofonoff, 1962).

The results of Cox and Smith are shown in Fig. 8, plotted in T-S coordinates (Fofonoff, 1962). The following important fact becomes obvious, in particular, from this figure: the heat capacity of water with salinity of less than approximately 20‰ decreases, whereas beginning with a salinity of about 20‰ it starts to increase. The oceanographic implications of this law have not been studied.

Integrating [18.28] by temperature, Connors (1970) obtains the following empirical equation for determining the specific enthalpy of sea water (cf. formula [18.2]):

$$h(T, S) = 4.2044\,T - 0.00057\,T^2 - S(6.99\,T - 0.0343\,T^2) -$$
$$- S^2(464 - 19.6\,T + 0.3\,T^2) \quad \text{J g}^{-1} \qquad\qquad [18.30]$$

valid in the same ranges as [18.28]: $10 < S < 40$‰ and $0 < T < 30°$C. Here we will not explain the reason for the appearance of the constant of integration in the form of the quantity 464 in the second parenthesis, referring the reader for an exhaustive explanation to the primary source, all the more since equation [18.30] does not yield absolute values for enthalpy, but is applicable only for the calculation of the differences of h. In addition, the equation applies only given the condition of linearity between partial specific enthalpy and salinity (in this connection also, see the primary source).

The nature of the change in heat capacity depending on the change in temperature and salinity reinforces the anomalous nature of the properties of sea water and, in particular, reinforces the effect of contraction on mixing of sea waters. This question will be considered briefly in Section 21.

19. THE SPEED OF SOUND

The theoretical formula for the speed of sound in a compressible liquid may be derived in an elementary way on the basis of the following considerations (Fabrikant, 1949). Let us consider a liquid placed between two pistons in a tube of unit cross-section (Fig. 9). If the liquid is *incompressible*, it behaves like a solid body: any displacement of piston A causes the same instantaneous displacement of piston B. In other words, disturbances in an incompressible liquid are transmitted with infinitely great speed. In the case of a *compressible* liquid these disturbances are transmitted with finite speed, namely: if for the time dt piston A has travelled the distance dl, the longitudinal disturbance (elastic wave) caused by this will be propagated over the distance dx. Thus, dl/dt is the speed of displacement of the liquid particles, $c = dx/dt$ is the speed of displacement of the longitudinal elastic disturbances, which is the *speed of propagation of the sound wave*. Since the displacement

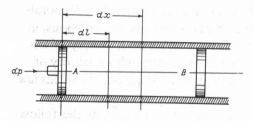

Fig. 9. Derivation of the formula for the speed of propagation of sound (see text).

of the piston dl caused the contraction of the liquid element dx, dl/dx represents the relative linear (or volumetric, the two being the same in this case) contraction of the liquid and Hooke's law is valid, following directly from formula [2.9]:

$$F = dp = -E \frac{dv}{v} = -E \frac{dl}{dx} \qquad\qquad [19.1]$$

where $F = dp$ is the force applied to the piston of unit cross section, and $E = k^{-1}$ is Young's modulus (k is the coefficient of isothermal compressibility).

Let us apply to our argument the theorem of impulses:

$$F \, dt = d(mc) \qquad\qquad [19.2]$$

A liquid mass, set into motion, is equal to ρ dx, while its momentum equals ρ dx (dl/dt); the speed (dl/dt) is considered identical for all particles in view of the smallness of dx. Utilizing the expression of force [19.1], we have, according to [19.2]:

$$E \frac{dl}{dx} \, dt = \rho \frac{dx}{dt} \, dl$$

or:

$$\frac{E}{c} = \rho c$$

whence:

$$c = \sqrt{\frac{E}{\rho}} = \sqrt{\frac{1}{k\rho}} \qquad\qquad [19.3]$$

Young's modulus and the speed of propagation of sound may be expressed by a quantity which is the inverse of the baric gradient of density. For this we make use of the equation of continuity of mass, written in the most general form:

$$\frac{d}{dt}(\rho v) = 0 \tag{19.4}$$

differentiating ρv and dividing by ρv, we obtain:

$$\frac{1}{\rho}\frac{d\rho}{dt} + \frac{1}{v}\frac{dv}{dt} = 0$$

whence:

$$\frac{dv}{v} = -\frac{d\rho}{\rho} \tag{19.5}$$

Substituting [19.5] in [19.1], we obtain the following expressions for Young's modulus:

$$E = \rho\frac{dp}{d\rho} \tag{19.6}$$

Comparing formulae [19.3] and [19.6], we obtain the theoretical formula we are seeking for the speed of sound:

$$c^2 = \left(\frac{dp}{d\rho}\right)_{d\eta=0} \tag{19.7}$$

The process of the propagation of sound is an adiabatic process. Owing to the great frequency of contractions and rarefactions heat does not have time to be dissipated by means of heat conduction and radiation; in practice, each particle behaves as if its store of heat remained constant (Lamb, 1932).

On the other hand, the speed of sound can also be expressed by Laplace's formula (Shuleikin, 1968):

$$c^2 = \frac{\gamma p_0}{\rho_0} \tag{19.8}$$

which, using the relationship:

$$p = \frac{1}{k} \tag{19.9}$$

where k is the coefficient of compressibility, can also be represented in the following form:

$$c^2 = \frac{\gamma}{\rho_{ST_p}k} = \frac{^{\alpha}sT_p\gamma}{k} \tag{19.10}$$

In these formulae γ is the ratio of heat capacities as determined by formula [18.24]. Substituting this ratio, as well as expression [2.9] for coefficient k in formula [19.10], we obtain the following expression for the speed of sound:

$$c^2 = -\frac{\alpha_S^2 T_p}{\dfrac{\partial \alpha}{\partial p} + \Gamma \dfrac{\partial \alpha}{\partial T}} \qquad\qquad [19.11]$$

where:

$$\Gamma = \frac{T}{c_p} \frac{\partial \alpha}{\partial T}$$

is the adiabatic gradient of temperature (see Sections 20 and 27).

Thus, the speed of sound, as well as density, is a function of the parameters of state of sea water:

$$c = c(S, T, p) \qquad\qquad [19.12]$$

and is closely linked with the equation of state of sea water, which, taking account of formula [19.7], may be written in the form:

$$\frac{\mathrm{d}\alpha}{\alpha} = a\,\mathrm{d}T - \beta\,\mathrm{d}S - \frac{\alpha}{c^2}\,\mathrm{d}p \qquad\qquad [19.13]$$

The speed of sound may be considered as still another parameter of state of sea water in situ. The dependence of the speed of sound in sea water, c m/sec, at atmospheric pressure on temperature and salinity is shown in Fig. 10.

Formula [19.11] is applied for the calculation of the speed of sound in the sea; let us call it the *theoretical formula* (sometimes it is referred to not altogether accurately as the Newton—Laplace formula).

The theoretical formula was utilized for the compilation of the well-known Matthews tables of the speed of sound (1927, 1939, 1944), published by the British Admiralty, as well as for the Kuwahara tables (1939), which, as we shall see below, made use precisely of the expanded formula [19.11]. This question is considered below in Section 29.

A comparison of the results of the calculations of the speed of sound according to the theoretical formula, on which Matthews' and Kuwahara's tables are based, with direct measurements, accompanied by simultaneous measurements of temperature and salinity, showed that the theoretical formula produces inaccurate results: at $0 \leqslant T \leqslant 40°C$ and $15 \leqslant S \leqslant 40‰$ the deviation amounts to from 2.5 to 5.0 m/sec. The maximum deviation of the theoretical formula amounts to about 8 m/sec when $S = 0‰$, $T = 0°$. On the average we may consider that the theoretical formula yields a result which is lower by ~3 m/sec (Beyer, 1954; Ganson, 1958). This occurs

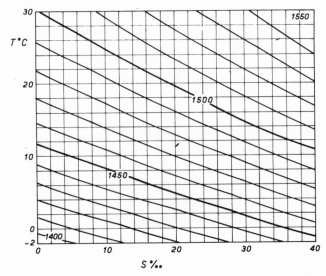

Fig. 10. Speed of sound c m/sec as a function of salinity and temperature at atmospheric pressure. (According to Kuwahara's data, 1939.)

mainly because of the inaccuracy of Ekman's coefficient of isothermal compressibility μ, entering into the calculations of the speed of sound according to the theoretical formula by means of the coefficient k *. It is is perfectly possible, as Beyer (1954) points out, that the lack of constancy of the saline composition of sea waters also exerts an influence on the appearance of deviations in the calculations of the speed of sound according to the theoretical formula; a smaller error is involved in inaccuracy in determining heat capacity c_p at constant pressure **.

Mackenzie pays particular attention to the value $c_{35,0,0}$, equal, according to Kuwahara, to 1445.5 m/sec. As was pointed out by Beyer (1954) and shown on the basis of laboratory experiments by Wilson (1959) and Del Grosso (1959), and also verified by Mackenzie (1960) by measuring the speed of sound during his submersion with Jacques Piccard in the bathyscaph "Trieste" to a depth of 4,000 ft. on May 29, 1959, the quantity $c_{35,0,0}$ according to Kuwahara is less than the true quantity by about 3 m sec^{-1}.

In the last years, methods of accurate laboratoy determination of the speed of sound in pure and sea water have been developed; the results of these experiments have made it possible to undertake the construction of empirical formulae, the accuracy of which could prove greater than the accuracy of the theoretical formula.

* In Del Grosso's work (1959) there is a detailed analysis and criticism of Ekman's methods in calculating the coefficient of compressibility μ.
** Additional light should be shed on this point by the introduction of the new values for heat capacity c_p according to Cox and Smith (par. 18) in the theoretical formula.

The first result of accurate laboratory investigations was Del Grosso's empirical formula (1952):

$$c = 1448.6 + 4.618T - 0.0523T^2 + 0.00023\ T^3 + 1.25(S - 35) -$$

$$- 0.011(S - 35)T + 0.00027 \cdot 10^{-5}(S - 35)T^4 -$$

$$- 2 \cdot 10^{-7}(S - 35)^4(1 + 0.577T - 0.0072T^2) \tag{19.14}$$

Until recently Del Grosso's formula was considered the most accurate: for all temperatures and for salinities above 15‰ the error produced by Del Grosso's formula does not exceed 0.5 m/sec, and amounts basically only to 0.2–0.3 m/sec. Del Grosso's formula is designed for the speed of sound at atmospheric pressure and does not take account of the influence of hydrostatic pressure; detailed tables have been compiled according to it (Bark et al., 1961).

The influence of pressure is taken into account in these tables with the help of the correction Δc_p according to Wilson (see below).

Recent investigations of the speed of sound in sea water taking account of pressure are the works of Wilson (1960a, b, 1962), who proposed two formulae for the speed of sound as a function of temperature, pressure and salinity; we shall dwell below on the second of these which is the more accurate. Wilson describes in detail a first series of experiments in the course of which he undertook instrumental laboratory investigations of the speed and propagation of sound in water samples collected in the Atlantic Ocean in the region of the Bermudas – Key West. Wilson points out that according to the statistical T-S analysis of Montgomery (1958), 99.5% of the waters of the World Ocean lie in the range of temperatures $3 < T < 30°C$, the range of pressures $1.033 < p < 1000$ kg/cm^2 and the range of salinities $33 < S < 37$‰. In accordance with this, the experiments were carried out in these ranges; the water samples, filtered for the purpose of removing organic matter, were diluted to the required salinity by distilled water.

During the first series of experiments 581 measurements of the speed of sound in these ranges were carried out for 15 values of temperature, 8 values of pressure and 5 values of salinity. The methods of measurement were the same as this author had used to measure the speed of sound in distilled water at various pressures; they are described in Wilson's preceding work (1959) *. The results of the measurements were processed on a computer by the method of least squares; Wilson presents these results in the form of an equation, similar to Kuwahara's expansion (Section 29).

---

* The tables for the speed of sound in distilled water at $0 < T < 100°$ and atmospheric pressure, based on the new data, are given in an article by Greenspan and Tshiegg (1959).

Wilson's second formula is as follows:

$$c = 1449.14 + \Delta c_T + \Delta c_p + \Delta c_S + \Delta c_{ST_p} \qquad [19.15]$$

where:

$$\Delta c_T = 4.5721\,T - 4.4532 \cdot 10^{-2} T^2 - 2.6045 \cdot 10^{-4} T^3 +$$
$$+ 7.9851 \cdot 10^{-6} T^4$$
$$\Delta c_p = 1.60272 \cdot 10^{-1} p + 1.0268 \cdot 10^{-5} p^2 +$$
$$+ 3.5216 \cdot 10^{-9} p^3 - 3.3603 \cdot 10^{-12} p^4$$

$$\Delta c_S = 1.39799(S - 35) + 1.69202 \cdot 10^{-3} (S - 35)^2$$

$$\Delta c_{ST_p} = (S - 35)\,(-1.1244 \cdot 10^{-2} T + 7.7711 \cdot 10^{-7} T^2 + \qquad [19.16]$$
$$+ 7.7016 \cdot 10^{-5} p - 1.2943 \cdot 10^{-7} p^2 + 3.1580 \cdot 10^{-8} pT +$$
$$+ 1.5790 \cdot 10^{-9} pT^2) + p(-1.8607 \cdot 10^{-4} T +$$
$$+ 7.4812 \cdot 10^{-6} T^2 + 4.5283 \cdot 10^{-8} T^3) +$$
$$+ p^2(-2.5294 \cdot 10^{-7} T + 1.8563 \cdot 10^{-9} T^2) +$$
$$+ p^3(-1.9646 \cdot 10^{-10} T)$$

This formula is valid within the ranges $-4 < T < 30°C$, $1 < p < 1{,}000$ kg/cm², $0 < S < 37‰$. The quantity 1449.14 m/sec represents the speed of sound at 0°C, 35‰ and pressure 0 kg/cm². At atmospheric pressure, equal to 1.0332 kg/cm², the correction, as may be seen from the second formula (19.16), amounts to approximately 0.16 m/sec. Accordingly, the speed of sound at 0°C, 35‰ and atmospheric pressure amounts to 1449.30 m/sec. The standard deviation of Wilson's second formula with respect to the experimental data amounts to 0.30 m/sec. At the present time it is the most accurate, and tables and nomograms have been prepared from it (Bialek, 1964; *Tables of Sound Speed in Sea Water*, 1962).

Inasmuch as the argument in this (as well as in the first) formula is not depth, but full pressure, the latter may be calculated for a depth of z meters according to the following formula:

$$p = \int_0^z g\rho \, dz + p_a = \sum \bar{g}_\varphi \bar{\rho}_i \Delta z + 10.33 \qquad [19.17]$$

p is expressed in decibars, where $p_a = 10.33$ dbars is the standard surface atmospheric pressure, $\bar{g}_\varphi$ is the mean vertical acceleration of gravity at latitude φ; $\bar{\rho}_i$ is the mean density of the i-th layer of water, Δz is the thickness of the i-th layer.

It is possible to follow another, simpler (but somewhat less accurate) path, namely: to make use of a formula combining pressure and depth. Such a formula, constructed by Polosin (1967) for a standard ocean, was given above [1.8]. The details for calculating the speed of sound in such a way are considered in this author's article.

In conclusion, let us note that in spite of the appearance of accurate empirical formulae for the speed of sound, it is premature to abandon the theoretical formula, since the flaw does not lie in the Newton—Laplace formula itself, which, according to Lamb, enjoys the absolute confidence of the physicists, but in the inaccuracy of the empirical constants contained in the formula, and first of all in the inaccuracy of the Ekman coefficient of isothermic compressibility. The recalculation of these constants, which will be carried out sooner or later, will make it possible substantially to improve the results obtained under the theoretical formula. Speaking, however, of the inaccuracy of Ekman coefficient μ, we must also recognize as inaccurate contemporary methods of calculating α_{ST_p} (according to the tables of Bjerknes, Zubov, Sverdrup et al.), and hence also the calculation of geopotentials and speeds of geostrophic currents, stability, etc., although errors in all these calculations are of considerably less importance than errors in hydroacoustical calculations. All these characteristics are brought together by the equation of state of sea water and, as far as the accurate empirical formulae for the speed of sound are concerned, it is possible with their help to solve the "opposite" problem, i.e. to determine indirectly the coefficient of compressibility (Crease, 1962).

20. KELVIN'S FORMULA

Let us consider the question of the adiabatic change in temperature with a change in depth. Fofonoff (1962) points out that if a layer of water is fully mixed so that two particles within this layer, brought to one and the same pressure, are in no way different from each other, then in this case the layer of water considered must have constant entropy and salinity. Thus, from formula [18.16], expressing the differential of entropy as a function of temperature, salinity and pressure, the following formula, where $d\eta = 0$ and $dS = 0$, is directly derived:

$$\Gamma = \left(\frac{dT}{dp}\right)_\eta = \frac{T}{c_p} \frac{\partial v}{\partial T} \qquad [20.1]$$

This is Kelvin's formula, which expresses the adiabatic vertical temperature gradient in the ocean. Proudman (1953) gives a detailed derivation of Kelvin's formula by the method of thermodynamic cycles (circular processes). In the present exposition Kelvin's formula is obtained on the basis of a consecutive development of the method of thermodynamic potentials.

Kelvin's formula is of great importance in physical oceanography and is used in such questions as the determination of potential temperatures (this question is connected with the problem of the study of deep circulation in the sea), the calculation of stability, the speed of propagation of sound, etc. Kelvin's formula and calculations from it are considered in detail in Section 27.

21. NON-LINEAR PROPERTIES OF SEA WATER

As is well known, sea water is a substance possessing numerous anomalous properties — extremely high heat capacity, heat of melting, heat of vaporization, etc. Without going into the physical and chemical nature of the properties of sea water, the study of which goes beyond the limits of the present work, one may consider that the direct cause of the principal anomalous properties of sea water is the *non-linear dependence on temperature and salinity* of its density (specific volume) and heat capacity, as well as of the gradients of thermal expansion and saline "contraction" $\partial\alpha/\partial T$ and $\partial\alpha/\partial S$; this is immediately obvious from a consideration of the graphs of the functions mentioned (Figs. 1, 8, 10, 14 and 15).

Due to the existence in sea water of the temperature of maximal density (the extreme of the isosteres on the *T-S* diagram), the coefficient of thermal (volumetric) expansion $a = (1/\alpha_0)(\partial\alpha/\partial T)$ in the region of temperatures and salinities, limited on the *T-S* diagram by the lines of the temperature of maximal density and the temperature of freezing (Fig. 1), is negative. In greater detail:

(1) for *pure water* ($S = 0\%_0$):
      $a < 0$ at $0 < T < 4°C$
      $a = 0$ at $T = 4°C$
      $a > 0$ at $T > 4°C$
(2) for *brackish waters* ($0 < S < 24.7\%_0$):
      $a < 0$ at $T < \theta$
      $a = 0$ at $T = \theta$
      $a > 0$ at $T > \theta$
(3) for *sea waters* ($S > 24.7\%_0$):
      $a > 0$ at all temperatures.

Inasmuch as the volumetric expansion of water is connected with adiabatic processes, *at $a < 0$ water on contraction is adiabatically cooled*, and not heated; this fact is proven for pure water on the basis of the second principle of thermodynamics (see, for example, Bazarov, 1961).

This anomaly of sea water — the anomaly of adiabatic processes — has not yet attracted the attention of investigators, apparently for the reason that the waters falling within the area where $a < 0$, are not encountered in the ocean at great pressures (although a comparable situation can be observed in the deep parts of the Black Sea — this question requires further study), and this anomaly is of purely theoretical interest.

The main anomalies of significance in the dynamics of ocean waters are revealed during the study of the process of mixing of two homogeneous water masses having different temperatures and salinities. Let us consider briefly at least three of the most important effects caused by the anomalous properties of sea water:

(1) *The heat of mixing (dilution) effect.* If the temperature and salinity are regarded as additive, i.e., as in the case of a linear dependence between the resulting property and the proportions of mixing, the curve of the binary systems (Anosov and Pogodin, 1947) will be linear (Fig. 11a). However, neither temperature nor salinity are additive properties. Let us consider for example two homogeneous water masses having extreme values for ocean conditions (30°, 35‰ for one mass and 0°, 0‰ for the other). The formula for the mixing of two equal masses, i.e., the formula for a binary solution, has the following form for the determination of temperature:

$$\bar{T} = \frac{c_1 T_1 + c_2 T_2}{c_1 + c_2} \qquad\qquad [21.1]$$

here c is specific heat. Substituting the values of the specific heat capacities of these waters, equal to 3.999 and 4.217 J g^{-1} °C^{-1} respectively, in the formula [21.1], we obtain that the temperature of the resulting mixture will become equal not to $\bar{T} = \frac{1}{2}(T_1 + T_2) = 15°$, as would occur if temperature were additive, but to $T_f = 14.602°$. The difference $T_f - \bar{T} = 0.398°C$ represents the *heat of mixing (dilution)* of the waters; it may be seen from this example that its quantity is not at all negligible. The heat effect of the mixing of sea waters is considered in greater detail in the work of Okubo (1951) and Connors (1970). Thus, temperature is not an additive property (Fig. 11b). The same thing probably also occurs for salinity owing to the degree of dissociation of sea water as an electrolyte; however, here the non-linear effects are probably small. On the other hand, we shall see below (Section 22) that enthalpy, $h \sim c_p T$, is an additive property.

In general T-S analysis, neglecting the quantity of heat of dilution, temperature and salinity are considered for the sake of simplicity as additive properties (see Section 30).

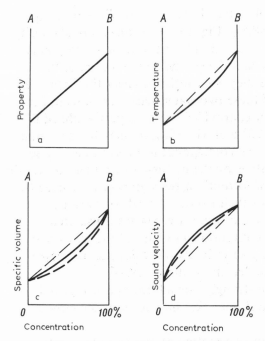

Fig. 11. The curve of an additive property (a) and curves of the most important non-additive proper-ties of sea water (b)–(d). The curves shown in broken lines in (c) and (d) take account of the non-additivity of temperature.

(2) *The effect of contraction on mixing.* The property of contraction may be easily studied in a simple example which we borrow from Zubov (1947, pp. 40–41). Let us consider two homogeneous water masses, A and L, at

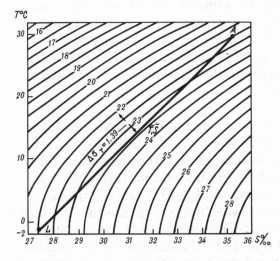

Fig. 12. In explanation of the effect of contraction on mixing of two water masses.

atmospheric pressure and with the following characteristics: $T_A = 30°$, $S_A =$ 35.36‰, and $T_L = -1.5°$, $S_L = 27.38‰$ (Fig. 12). The conventional density of both water masses is identical: $\sigma_T = 22.02$; the first of them corresponds roughly to the waters of the Gulf Stream, the second to the waters of the Labrador current in the region where they are contiguous near the Great New-foundland Bank. If equal volumes of these two water masses are mixed, the temperature and salinity of the mixture, provided these properties are additive, will be equal respectively to 14.25° and 31.18‰. As may be seen from Fig. 12, conventional density $\sigma_T = 23.38$ corresponds to these values; we see that this value for conventional density is greater by $\Delta\sigma_T = 1.36$ than the density of each of the waters mixed. A *"densification"* of the waters has occurred (if we are thinking in terms of density) or their *contraction* (if we consider specific volume). For the two extreme uniform water masses considered above (30°, 35‰ and 0°, 0‰) the conventional density of the mixture of equal volumes of these waters will be greater by $\Delta\sigma_T = 1.78$ than the mean arithmetical value of the conventional density of these waters.

Thus, density (specific volume) is a strongly non-additive property, and the effect of contraction is reinforced if one takes into account the non-additivity of temperature (Fig. 11c).

The physical and chemical nature of contraction on mixing of sea water has virtually not been investigated. It is likely that this effect is to be explained by the reorganization of polymeric clusters of water molecules in specific electrolytic conditions (in this connection see Horne, 1970, 1972). From the oceanographic point of view the phenomenon of contraction on mixing of sea water has been studied in the greatest detail by Zubov (1938, 1947, 1957b) and by Zubov and Sabinin (1958), and, following Zubov, by a number of other investigators. Bulgakov (1960) and Fonofoff (1961) have studied this phenomenon from the theoretical point of view. The important general results of Fofonoff will be considered immediately in the following paragraph. In addition, some development of the theory will be presented in Section 36, while contraction on mixing of the real water masses of the ocean will be considered in Sections 55—57.

(3) *The effect of acceleration of sound on mixing.* For the mixture of two equal volumes of water at extreme conditions (30°, 35‰, and 0°, 0‰) the speed of sound is 9.31 m/sec greater than the mean arithmetical speed. Precisely, the speed of sound (at atmospheric pressure) for the first of the waters amounts to 1545.66 m/sec, for the second to 1402.28 m/sec, and for the mixture of equal volumes of these water to 1483.28 m/sec, whereas the mean arithmetical value amounts to 1473.97 m/sec. Thus, the speed of sound is also a substantially non-additive property (Fig. 11d). The study of the effect of the acceleration of sound during mixing, pointed out for the first time in the

present work (in any event, the author is not aware of any studies devoted to this question) is of great importance for the study of the speed of sound in the World Ocean.

The non-linear effects considered above have a great influence on the processes of mixing, formation and transformation of water masses, as well as on the static conditions in the real ocean, stratified as to density.

22. CONTRACTION ON MIXING OF SEA WATERS

In the preceding section, the effect of contraction on mixing of two water masses was considered in one particular example. The study and evaluation of the influence of this effect, as we shall see in subsequent chapters, play an important part in the T-S analysis of sea waters. For this reason, before proceeding to the study of this effect in the oceanographic context, let us present a general consideration of contraction on mixing from the thermodynamic and mathematical points of view, following the work of Fofonoff (1961) in the exposition of this paragraph.

Let us consider the process of mixing of two homogeneous water masses m_1, m_2, having initial temperatures T_1, T_2 and salinities S_1, S_2 at constant pressure p. Given complete mixing of these water masses, a homogeneous water mass m is formed with temperature T and salinity S. According to the condition of the conservation of masses and salts, we have:

$$\Delta m = m - (m_1 + m_2) = 0 \qquad [22.1]$$

$$m\Delta S = mS - (m_1 S_1 + m_2 S_2) = 0 \qquad [22.2]$$

The change in volume (contraction) will then be defined as:

$$\Delta V = mv - (m_1 v_1 + m_2 v_2) \qquad [22.3]$$

where v_1, v_2 and v are the specific volumes of the original water masses and their mixture, respectively.

Further, according to the condition of the conservation of energy, we have:

$$m\Delta\epsilon = m\epsilon - (m_1 \epsilon_1 + m_2 \epsilon_2) = -p\,dV \qquad [22.4]$$

where ϵ_1, ϵ_2 and ϵ are the respective values of the specific internal energy of the water masses.

Substituting [22.3] in [22.4] and taking account of formula [13.4], we obtain:

$$m\Delta(\epsilon + pv) = m(\epsilon + pv) - [m_1(\epsilon_1 + pv_1) + m_2(\epsilon_2 + pv_2)] = 0 \qquad [22.5]$$

or:

$$m\Delta h = mh - (m_1 h_1 + m_2 h_2) = 0 \tag{22.6}$$

where h_1, h_2 and h are the respective values of the specific enthalpy of the water masses. Formula [22.6] expresses the condition of the conservation of enthalpy.

Let us expand the specific volume, as a function of enthalpy and salinity, $v(h, S)$, into a Taylor series in the neighborhood of a certain point (h_0, S_0), the coordinates of which correspond to equilibrium values, limiting ourselves to terms of the second order:

$$v = v_0 + \left(\frac{\partial v}{\partial h}\right)_0 (h - h_0) + \left(\frac{\partial v}{\partial S}\right)_0 (S - S_0) +$$

$$+ \tfrac{1}{2} \left[\left(\frac{\partial^2 v}{\partial h^2}\right)_0 (h - h_0)^2 + 2\left(\frac{\partial^2 v}{\partial h\, \partial S}\right)_0 (h - h_0)(S - S_0) + \right.$$

$$\left. + \left(\frac{\partial^2 v}{\partial S^2}\right)_0 (S - S_0)^2 \right] \tag{22.7}$$

Substituting [22.7] in [22.3] and taking account, during the transformation of the formula, of [22.1], [22.2] and [22.6], we obtain the following analytical expression for contraction, ΔV:

$$\Delta V = -\frac{m_1 m_2}{2m^2} \left[\frac{\partial^2 v}{\partial h^2} (h_2 - h_1)^2 + 2\frac{\partial^2 v}{\partial h\, \partial S} (h_2 - h_1)(S_2 - S_1) + \frac{\partial^2 v}{\partial S^2} (S_2 - S_1)^2 \right] =$$

$$= -\frac{m_1 m_2}{2m^2} \delta^2_{h,S} v \tag{22.8}$$

where $\delta^2 v$ is the abbreviated designation of the differential form contained in [22.8] in square brackets.

The transition from enthalpy to temperature leads to the following formula:

$$\Delta V = -\frac{m_1 m_2}{2m^2} \left[\delta^2 v - \frac{1}{c_p} \frac{\partial v}{\partial T} \delta^2 h \right] \tag{22.9}$$

where:

$$\delta^2 h = \frac{\partial c_p}{\partial T} (T_2 - T_1)^2 + 2\frac{\partial c_p}{\partial S} (T_2 - T_1)(S_2 - S_1) + \frac{\partial^2 h}{\partial S^2} (S_2 - S_1)^2 \tag{22.10}$$

The additional term $[-(1/c_p)(\partial v/\partial T) \delta^2 h]$ characterizes that small part of contraction on mixing which is determined by the temperature deviation from the linear law of mixing represented in Fig. 11a; this temperature deviation, equal to $[-(1/c_p)\,\delta^2 h]$, is determined by the nature of the dependence of heat capacity on temperature and salinity and represents the heat of mixing of water masses, which was illustrated by a particular example in the preceding section *.

Let us consider further, following Fofonoff, the quantity of contraction per unit of mass provided that the small mass of water δm mixes with the large mass of water m. Calculating the limit of expression [22.8] when $\delta m \to 0$, we obtain:

$$\delta v = \lim_{\delta m \to 0} \left\{ \frac{\Delta V}{\delta m} \right\} = -\tfrac{1}{2}\delta^2_{h,S}v \approx -\tfrac{1}{2}\delta^2_{T,S}v \qquad [22.11]$$

With the help of this formula Fofonoff shows how the quantity of contraction on mixing depends on the angle of slope of the straight line of mixing (shown in Fig. 12) to the axes of the T-S diagram (and, consequently, also to the isosteres).

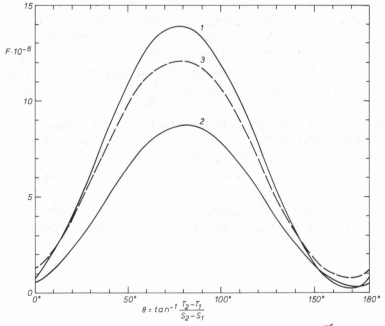

Fig. 13. Dependence of the parameter of contraction on mixing, $F \cdot 10^{-6}$, on the angle of slope θ of the straight line of mixing to the abscissa axis of the T-S diagram (at atmospheric pressure). $1 = 0°C, 35°/_{00}$; $2 = 16°C, 35°/_{00}$; $3 = 8°C, 20°/_{00}$. (From Fofonoff, 1961, with additions.)

* In a subsequent work, Fofonoff (1962) extends the above considerations to the general case of a nonequilibrium unbalanced system, considering, in particular, the question of the speed of production of entropy in the process of mixing and the attainment of equilibrium. This question is not considered here since the work mentioned is well known to oceanographers.

Assuming:

$$T_2 - T_1 = r \sin \theta$$

$$S_2 - S_1 = r \cos \theta$$

and substituting these expressions in [22.11], we obtain:

$$\delta v = -\frac{r^2}{2} \left[\frac{\partial^2 v}{\partial T^2} \sin^2 \theta + 2 \frac{\partial^2 v}{\partial T \partial S} \sin \theta \cos \theta + \frac{\partial^2 v}{\partial S^2} \cos^2 \theta \right] =$$

$$= -\frac{r^2}{2} F(\theta; T, S, p) \qquad\qquad\qquad\qquad [22.12]$$

(instead of these expressions Fofonoff gives a more cumbersome analytical expression which is not considered here).

In Fig. 13 examples are shown of the dependence of the parameter of contraction on mixing, $10^{-6} F(\theta; T, S, p)$, on the angle of slope of the straight line of mixing to the abscissa axis of the T-S diagram for three types of sea water. Curve 1, plotted by Fofonoff, corresponds to standard sea water ($0°$, $35‰$), curve 2 to the "Central" Water of the Pacific Ocean (see Fig. 96), curve 3 to the waters of the Black Sea at a depth of about 100 m. The values of the second derivatives of specific volume, necessary for calculations according to formula [22.12], are taken from the accurate tables of Fofonoff and Froese (1958).

From the figure it is clear that the greatest contraction on mixing is observed at angles θ of the order of $80°$, i.e., for waters which are highly different for temperature and much less so for salinity; the least contraction occurs at angles θ of the order of $170°$, i.e., for waters which are very different in salinity and little different in temperature.

This qualitative conclusion proves highly useful in the practical study of the contraction on mixing of the real water masses of the ocean (see Sections 55–57).

CHAPTER 4

PARTIAL DERIVATIVES OF THE EQUATION OF STATE

23. EXPANSION OF FUNCTION $\alpha = \alpha(S, T, p)$ INTO TAYLOR'S SERIES

Expanding the function $\alpha(S, T, p)$ into Taylor's series (for three variables) in the neighborhood of point $(35, 0, 0)$ with an accuracy to terms of the third order of smallness, we obtain:

$$\alpha(S, T, p) = \alpha(35, 0, 0) + \left(\frac{\partial \alpha}{\partial S} \, dS + \frac{\partial \alpha}{\partial T} \, dT + \frac{\partial \alpha}{\partial p} \, dp \right) +$$

$$+ \frac{1}{2} \left(\frac{\partial^2 \alpha}{\partial S^2} \, dS^2 + \frac{\partial^2 \alpha}{\partial T^2} \, dT^2 + \frac{\partial^2 \alpha}{\partial p^2} \, dp^2 + 2 \frac{\partial^2 \alpha}{\partial S \partial T} \, dS \, dT + 2 \frac{\partial^2 \alpha}{\partial S \partial p} \, dS \, dp + \right.$$

$$\left. + 2 \frac{\partial^2 \alpha}{\partial T \partial p} \, dT \, dp \right) + \frac{1}{6} \left(\frac{\partial^3 \alpha}{\partial S^3} \, dS^3 + \frac{\partial^3 \alpha}{\partial T^3} \, dT^3 + \frac{\partial^3 \alpha}{\partial p^3} \, dp^3 + 3 \frac{\partial^3 \alpha}{\partial S^2 \partial T} \, dS^2 \, dT + \right.$$

$$+ 3 \frac{\partial^3 \alpha}{\partial T^2 \partial S} \, dT^2 \, dS + 3 \frac{\partial^3 \alpha}{\partial S^2 \partial p} \, dS^2 \, dp + 3 \frac{\partial^3 \alpha}{\partial T^2 \partial p} \, dT^2 \, dp +$$

$$\left. + 3 \frac{\partial^3 \alpha}{\partial p^2 \partial S} \, dp^2 \, dS + 3 \frac{\partial^3 \alpha}{\partial p^2 \partial T} \, dp^2 \, dT + 6 \frac{\partial^3 \alpha}{\partial S \partial T \partial p} \, dS \, dT \, dp \right) + ... \qquad [23.1]$$

In order to represent this expansion in the form of finite differences, let us first consider functions $\alpha(S, 0, 0)$, $\alpha(35, T, 0)$ and $\alpha(35, 0, p)$ as functions of *one* variable, S, T and p respectively. Expanding these functions into a Taylor's series for the corresponding variable and designating the differences between them and the quantity of specific volume of standard ocean water $\alpha(35, 0, 0)$ by δ_S, δ_T and δ_p respectively, we obtain:

$$\delta_S = \frac{\partial \alpha}{\partial S} \, dS + \frac{1}{2} \frac{\partial^2 \alpha}{\partial S^2} \, dS^2 + \frac{1}{6} \frac{\partial^3 \alpha}{\partial S^3} \, dS^3 + ... \qquad [23.2]$$

$$\delta_T = \frac{\partial \alpha}{\partial T} \, dT + \frac{1}{2} \frac{\partial^2 \alpha}{\partial T^2} \, dT^2 + \frac{1}{6} \frac{\partial^3 \alpha}{\partial T^3} \, dT^3 + ... \qquad [23.3]$$

$$\delta_p = \frac{\partial \alpha}{\partial p} \, dp + \frac{1}{2} \frac{\partial^2 \alpha}{\partial p^2} \, dp^2 + \frac{1}{6} \frac{\partial^3 \alpha}{\partial p^3} \, dp^3 + ... \qquad [23.4]$$

These three differences thus include nine terms of the expansion [23.1]. Further, considering the expressions [23.2–23.4] as functions of T, p and S respectively, drawing up the second differences δ_{ST}, δ_{Sp} and δ_{Tp} and expanding them into a Taylor's series, we will obtain expressions including nine more terms of the expansion [23.1]. Thus, having drawn up differences of the first, second and third order:

$$\delta_p \quad = \alpha_{35,0,p} - \alpha_{35,0,0}$$

$$\delta_S \quad = \alpha_{S,0,0} - \alpha_{35,0,0}$$

$$\delta_T \quad = \alpha_{35,T,0} - \alpha_{35,0,0}$$

$$\delta_{ST} \ = (\alpha_{S,T,0} - \alpha_{35,T,0}) - (\alpha_{S,0,0} - \alpha_{35,0,0})$$

$$\delta_{Sp} \ = (\alpha_{S,0,p} - \alpha_{35,0,p}) - (\alpha_{S,0,0} - \alpha_{35,0,0})$$

$$\delta_{Tp} \ = (\alpha_{35,T,p} - \alpha_{35,0,p}) - (\alpha_{35,T,0} - \alpha_{35,0,0})$$

$$\delta_{STp} = [(\alpha_{S,T,p} - \alpha_{35,T,p}) - (\alpha_{S,T,0} - \alpha_{35,T,0})] -$$
$$- [(\alpha_{S,0,p} - \alpha_{35,0,p}) - (\alpha_{S,0,0} - \alpha_{35,0,0})]$$

[23.5]

we come to the conclusion, that expansion [23.1] may be represented in the following form (Bjerknes and Sandström, 1910):

$$\alpha_{STp} = \alpha_{35,0,0} + \delta_S + \delta_T + \delta_p + \delta_{ST} + \delta_{Sp} + \delta_{Tp} + \delta_{STp} \qquad [23.6]$$

This dependence can also be written in the form:

$$\alpha_{STp} = \alpha_{35,0,p} + \delta \qquad [23.7]$$

where:

$$\delta = \delta_S + \delta_T + \delta_{ST} + \delta_{Tp} + \delta_{STp} \qquad [23.8]$$

is the expression of the anomaly of specific volume [1.32].

Let us combine quantity $\alpha(35,0,0) = 0.97264$ [or $v(35,0,0) = 72.64$] with corrections δ_S, δ_T and δ_{ST} in quantity α_T (or v_T):

$$\alpha_T = \alpha_{35,0,0} + \delta_S + \delta_T + \delta_{ST}$$

$$v_T = v_{35,0,0} + 10^3 (\delta_S + \delta_T + \delta_{ST})$$

[23.9]

Then we can finally write:

$$\alpha_{STp} = \alpha_T + \delta_p + \delta_{Sp} + \delta_{Tp} + \delta_{STp}$$

$$v_{STp} = v_T + 10^3 (\delta_p + \delta_{Sp} + \delta_{Tp} + \delta_{STp})$$

[23.10]

The latter formula is applied for tabular calculations of specific volumes in situ in practice.

The relationship between anomaly δ [1.32] and conventional specific volume v_{STp} can be obtained from the following considerations. It follows from formulae [23.8] and [23.9] that:

$$\delta = 10^{-3}(v_T - v_{35,0,0}) + \delta_{Sp} + \delta_{Tp} + \delta_{STp}$$

Comparing this expression with [23.10] we obtain:

$$\delta = 10^{-3}(v_{STp} - v_{35,0,0}) - \delta_p \qquad [23.11]$$

and:

$$v_{STp} - v_{35,0,0} = 10^3(\delta + \delta_p) \qquad [23.12]$$

Density in situ can be expressed by a formula similar to [23.7], namely:

$$\rho_{STD} = \rho_{35,0,D} - \epsilon \qquad [23.13]$$

where:

$$\epsilon = \epsilon_S + \epsilon_T + \epsilon_{ST} + \epsilon_{SD} + \epsilon_{TD} + \epsilon_{STD} \qquad [23.14]$$

is the anomaly of density. This formula, also proposed by Bjerknes and Sandström (1912), differs from the formula for the calculation of specific volume in situ by the fact that, instead of depth expressed in decibars, depth appears in it expressed in *dynamic meters*. The replacement in expressions [23.13] and [23.14] of α by ρ, δ by ϵ and p by D, respectively, is convenient, bearing in mind the relationships between density and specific volume (formula [1.9]) and pressure, dynamic depth, specific volume and density (formula [1.35]). Differentiating formula [1.9] we obtain:

$$d\rho = - d\alpha/\alpha^2 , \qquad [23.15]$$

whence, bearing in mind expressions [23.7] and [23.13], we obtain (Bjerknes and Sandström, 1912, p.32) the following relationship between ϵ and δ:

$$\epsilon = - \delta/\alpha^2 \qquad [23.16]$$

In a completely similar way to formulae [23.10] we may write:

$$\rho_{STD} = 1 + 10^{-3}\sigma_T + \epsilon_D + \epsilon_{SD} + \epsilon_{TD} + \epsilon_{STD}$$

$$\sigma_{STD} = \sigma_T + 10^3(\epsilon_D + \epsilon_{SD} + \epsilon_{TD} + \epsilon_{STD}) \qquad [23.17]$$

In such form these formulae were written for the first time by Hessenberg and Sverdrup (1915b). It should be pointed out that the formulae given for the calculation of σ_{STD} are rarely used (Sverdrup et al., 1942). Tables of quantities $\rho_{35,0,D}$, as well as of the corrections to this quantity entering into formula [23.1], may be found in Bjerknes and Sandström (1912, tables 16 H to 23 H). These tables are repeated in part in an article by Hesselberg and Sverdrup (1915a).

The expansion of function $\alpha(S, T, p)$ in a Taylor series is important not only for convenience in calculating specific volumes but also for analytical purposes (Chapter 3). The first derivatives $\partial\alpha/\partial S$, $\partial\alpha/\partial T$ and $\partial\alpha/\partial p$ will be considered immediately below (Sections 24 and 26).

24. THERMAL EXPANSION AND SALINE "CONTRACTION"

In this section we consider the first partial derivatives $\partial\alpha/\partial T$ and $\partial\alpha/\partial S$ entering into the Taylor expansion both at atmospheric pressure and in situ, as well as (briefly) the second partial derivatives $\partial^2\alpha/\partial S^2$, $\partial^2\alpha/\partial T^2$ and $\partial^2\alpha/\partial S\partial T$ at atmospheric pressure. The first partial derivative $\partial\alpha/\partial p$ (the baric gradient of specific volume) is considered separately in Section 26.

Thus, let us turn to the question of calculating specific volume "gradients" for temperature and salinity $\partial\alpha/\partial T$ and $\partial\alpha/\partial S$, as well as the corresponding density gradients $\partial\rho/\partial T$ and $\partial\rho/\partial S$. Let us call the first of these gradients the gradient of thermal expansion and the second the gradient of saline contraction, by analogy with the coefficients of thermal expansion and saline contraction.

Knowledge of the gradients of thermal expansion and saline contraction is absolutely essential for the study of those phenomena in which a change of specific volume takes place, or where account must be taken of such a change for one or two variables: temperature and salinity. In particular, this is necessary for: (a) calculation of the speed of propagation of sound in sea water; (b) calculation of the vertical stability of waters; (c) calculation of steric fluctuations of level; (d) calculation of line integrals on a T-S plane.

The derivatives of thermal expansion and saline contraction *at atmospheric pressure* are calculated by the differentiation of function $\alpha(T, S)$, i.e., by the differentiation of Knudsen's formulae; the result is expressed by the corresponding gradients of conventional density $\partial\sigma_T/\partial T$ and $\partial\sigma_T/\partial S$.

Differentiating the relationship between ρ, α and σ_T:

$$\rho_T = \sigma_T \cdot 10^{-3} + 1 \qquad \text{and} \qquad \alpha_T = \frac{1}{10^{-3}\sigma_T + 1} \qquad [24.1]$$

we obtain:

$$\frac{\partial\rho_T}{\partial T} = 10^{-3}\frac{\partial\sigma_T}{\partial T} \qquad\qquad [24.2]$$

$$\frac{\partial\rho_T}{\partial S} = 10^{-3}\frac{\partial\sigma_T}{\partial S} \qquad\qquad [24.3]$$

and:

$$\frac{\partial \alpha_T}{\partial T} = -10^{-3}\alpha_T^2 \frac{\partial \sigma_T}{\partial T} \tag{24.4}$$

$$\frac{\partial \alpha_T}{\partial S} = -10^{-3}\alpha_T^2 \frac{\partial \sigma_T}{\partial S} \tag{24.5}$$

The derivatives of conventional specific volume $\partial v_T/\partial T$ and $\partial v_T/\partial S$ may also be obtained by differentiation of the formula linking v_T and σ_T:

$$v_T = \frac{10^6}{\sigma_T + 10^3} - 900 \tag{24.6}$$

as a result of which we obtain:

$$\frac{\partial v_T}{\partial T} = -\frac{10^6 \frac{\partial \sigma_T}{\partial T}}{\sigma_T^2 + 2 \cdot 10^3 \sigma_T + 10^6} \tag{24.7}$$

$$\frac{\partial v_T}{\partial S} = -\frac{10^6 \frac{\partial \sigma_T}{\partial S}}{\sigma_T^2 + 2 \cdot 10^3 \sigma_T + 10^6} \tag{24.8}$$

It is easy to see that formulae [24.7] and [24.8] are similar to formulae [24.4] and [24.5]. Differentiating Knudsen's formula [3.4] by T, we obtain:

$$\frac{\partial \sigma_T}{\partial T} = \frac{d\Sigma_T}{dT} + (\sigma_0 + 0.1324)\left[-\frac{dA_T}{dT} + \frac{dB_T}{dT}(\sigma_0 - 0.1324)\right] \tag{24.9}$$

The derivatives entering into this expression are also obtained by differentiation of Knudsen's formulae [3.5], which determine the quantities Σ_T, A_T and B_T, and equal:

$$\frac{d\Sigma_T}{dT} = -\frac{1}{503.570}\left[215.74 - 2(T - 3.98) - \frac{1,094,910}{(T + 67.26)^2}\right] \tag{24.10}$$

$$\frac{dA_T}{dT} = (4.7867 - 0.196370T + 0.0032529T^2) \cdot 10^{-3} \tag{24.11}$$

and

$$\frac{dB_T}{dT} = (18.030 - 1.6328T + 0.05001T^2) \cdot 10^{-6} \tag{24.12}$$

Differentiating Knudsen's formula [3.4] by S, we obtain:

$$\frac{\partial \sigma_T}{\partial S} = \frac{d\sigma_0}{dS}(1 - A_T + 2B_T\sigma_0) \tag{24.13}$$

where:

$$\frac{d\sigma_0}{dS} = (1.4708 - 0.003140\,Cl + 0.0001194\,Cl^2)\frac{dCl}{dS} \qquad [24.14]$$

and

$$\frac{dCl}{dS} = \frac{1}{1.8050} \qquad [24.15]$$

The values of $\partial\sigma_T/\partial T$ and $\partial\sigma_T/\partial S$ were calculated by Hesselberg and Sverdrup (1915b), Kuwahara (1939) and Ivanov-Frantskevich (1953). The latter author established that there were mistakes in the tables of Hesselberg and Sverdrup; Kuwahara does not give the results of his computations. Therefore the tables of derivatives $\partial\sigma_T/\partial T$ and $\partial\sigma_T/\partial S$, calculated by Ivanov-Frantskevich, should be considered the most reliable; they are published both in his work (1953), and in the *Oceanological Tables* of Zubov (1957a, tables 20 and 26). These tables were calculated by Ivanov-Frantskevich with an accuracy of one decimal place greater than necessary for the calculations, i.e. to $10^{-4}(\partial\sigma_T/\partial_T)$ and $\partial\sigma_T/\partial S$. This was done so that it might be possible to plot graphs for corresponding computations from the tables.

The derivatives $\partial v_T/\partial T$ and $\partial v_T/\partial S$ may be computed from formulae [24.7] and [24.8]. Tables of these derivatives were published by the author (Mamayev, 1954, 1963) and by Fofonoff and Froese (1958). Both these tables are given in the Appendix (Tables A4 and A5).

The relationships of the derivatives of thermal expansion and saline contraction to temperature and salinity at constant pressure are shown (for specific volume) in Figs. 14 and 15. The corresponding graphs for density are not given; let us note merely that the graphs of the derivatives $\partial\alpha/\partial T$ and $\partial\rho/\partial T$ are quite similar, whereas graphs of $\partial\alpha/\partial S$ and $\partial\rho/\partial S$ differ — probably due to the fact that specific volume and density are inverse quantities. It is curious that in the area of the greatest non-linearity of graph $\alpha = f(S, T)$ (Fig. 1), i.e. at low values of S and T, graph $\partial\alpha/\partial S = f(S, T)$ is "most linear" and vice versa. The same is true in respect of density. These peculiarities are interesting for the theoretical study of contraction on mixing of sea waters.

The gradients of thermal expansion and saline contraction in situ can be obtained by differentiation of Bjerknes' formula [23.10]. Differentiating this formula for temperature, we obtain:

$$\left(\frac{\partial\alpha}{\partial T}\right)_{STp} = \left(\frac{\partial\alpha}{\partial T}\right)_{ST} + \frac{\partial\delta_{Tp}}{\partial T} + \frac{\partial\delta_{STp}}{\partial T} \qquad [24.16]$$

where:

$$\left(\frac{\partial\alpha}{\partial T}\right)_{ST} = 10^{-3}\frac{\partial v_T}{\partial T} \qquad [24.17]$$

while differentiating it by salinity, we obtain:

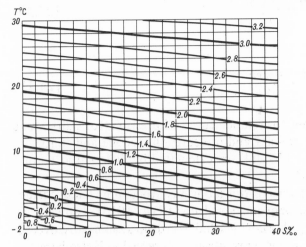

Fig. 14. Graph of function $10^4 \, \partial\alpha/\partial T = f(S, T)$.

$$\left(\frac{\partial \alpha}{\partial S}\right)_{STp} = \left(\frac{\partial \alpha}{\partial S}\right)_{ST} + \frac{\partial \delta_{Sp}}{\partial S} + \frac{\partial \delta_{STp}}{\partial S} \qquad [24.18]$$

where:

$$\left(\frac{\partial \alpha}{\partial S}\right)_{ST} = 10^{-3} \frac{\partial v_T}{\partial S} \qquad [24.19]$$

Similar formulae for the expression of density gradients can be obtained by differentiation of formula [23.18]:

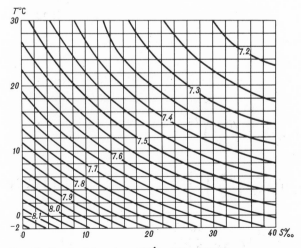

Fig. 15. Graph of function $-10^4 \, \partial\alpha/\partial S = f(S, T)$

$$\left(\frac{\partial \rho}{\partial T}\right)_{STD} = 10^{-3}\frac{\partial \sigma_T}{\partial T} + \frac{\partial \epsilon_{TD}}{\partial T} + \frac{\partial \epsilon_{STD}}{\partial T} \qquad [24.20]$$

and:

$$\left(\frac{\partial \rho}{\partial S}\right)_{STD} = 10^{-3}\frac{\partial \sigma_T}{\partial S} + \frac{\partial \epsilon_{SD}}{\partial S} + \frac{\partial \epsilon_{STD}}{\partial S} \qquad [24.21]$$

The values of the corrections for temperature and pressure, for salinity and pressure and for temperature, salinity and pressure (the second and third terms in formulae [24.20] and [24.21]) are calculated by differentiation of the corrections ϵ (see [23.17]).

We shall not dwell on the details of this laborious procedure. These corrections were calculated by Hesselberg and Sverdrup (1914–1915). Ivanov-Frantskevich (1953), verifying Hesselberg's and Sverdrup's tables, discovered mistakes in them; thus, for example, correction is increased in Hesselberg's and Sverdrup's tables to double the necessary value. Therefore the values of the corrections for pressure were also recalculated by Ivanov-Frantskevich, and his tables included in the *Oceanological Tables* of 1957; thus, the value $[\partial \rho/\partial T]_{STp}$ is calculated from tables 20, 21 and 22, while the value $[\partial \rho/\partial S]_{STp}$ is calculated from tables 26, 27 and 28.

The second derivatives of density and specific volume at constant pressure are determined through the first and second derivatives of conventional density by the following formulae, obtained by secondary differentiation of formulae [24.2]–[24.5]:

$$\frac{\partial^2 \rho}{\partial T^2} = 10^{-3}\frac{\partial^2 \sigma_T}{\partial T^2}$$

$$\frac{\partial^2 \rho}{\partial S^2} = 10^{-3}\frac{\partial^2 \sigma_T}{\partial S^2} \qquad [24.22]$$

$$\frac{\partial^2 \rho}{\partial S \partial T} = 10^{-3}\frac{\partial^2 \sigma_T}{\partial S \partial T}$$

and:

$$\frac{\partial^2 \alpha}{\partial T^2} = -10^{-3}\alpha_T^2\left[\frac{\partial^2 \sigma_T}{\partial T^2} - 2\cdot 10^{-3}\alpha_T\left(\frac{\partial \sigma_T}{\partial T}\right)^2\right]$$

$$\frac{\partial^2 \alpha}{\partial S^2} = -10^{-3}\alpha_T^2\left[\frac{\partial^2 \sigma_T}{\partial S^2} - 2\cdot 10^{-3}\alpha_T\left(\frac{\partial \sigma_T}{\partial S}\right)^2\right] \qquad [24.23]$$

$$\frac{\partial^2 \alpha}{\partial S \partial T} = -10^{-3}\alpha_T^2\left[\frac{\partial^2 \sigma_T}{\partial S \partial T} - 2\cdot 10^{-3}\alpha_T\frac{\partial \sigma_T}{\partial S}\frac{\partial \sigma_T}{\partial T}\right]$$

The second derivatives of conventional density are determined in the following way:

$$\frac{\partial^2 \sigma_T}{\partial T^2} = \frac{d^2 \Sigma_T}{dT^2} + (\sigma_0 + 0.1324) \left[-\frac{d^2 A_T}{dT^2} + (\sigma_0 - 0.1324) \frac{d^2 B_T}{dT^2} \right]$$

$$\frac{\partial^2 \sigma_T}{\partial S^2} = \frac{d^2 \sigma_0}{dS^2} (1 - A_T + 2B_T \sigma_0) + 2B_T \left(\frac{d\sigma_0}{dS} \right)^2 \qquad\qquad \text{[24.24]}$$

$$\frac{\partial^2 \sigma_T}{\partial S \partial T} = \frac{d\sigma_0}{dS} \left(-\frac{dA_T}{dT} + 2\sigma_0 \frac{dB_T}{dT} \right)$$

where the derivatives $d^2 \Sigma_T/dT^2$, $d^2 A_T/dT^2$ and $d^2 B_T/dT^2$ can easily be obtained by secondary differentiation of formulae [24.10], [24.11] and [24.12], and the derivative $d^2 \sigma_0/dS^2$ by secondary differentiation of [24.14].

The secondary derivatives are extremely small in size and necessary only for the most precise oceanographic computations (in particular, for the computation of the speed of propagation of sound). Some idea of them is given by Table IV, the values of the quantities in which are borrowed from the precise tables of specific volume and its derivatives by Fofonoff and Froese (1958).

TABLE IV

Values of second derivatives of the specific volume of sea water at atmospheric pressure

$S(^0/oo)$	0	5	10	15	20	25	30	35	40
$T(^\circ C)$									
$10^4(\partial^2 v_T/\partial T^2) = f_1(S, T, 0)$									
0	183	174	165	157	150	144	138	133	128
10	134	128	122	117	111	106	102	97	93
20	106	103	99	96	92	88	85	81	77
30	90	89	87	86	84	81	78	75	71
$10^4(\partial^2 v_T/\partial S^2) = f_2(S, T, 0)$									
0	23	21	18	16	14	12	10	8	6
10	20	18	16	14	11	9	8	6	4
20	18	16	14	12	10	8	6	5	3
30	17	15	13	11	9	7	5	3	2
$10^4(\partial^2 v_T/\partial S \partial T) = f_3(S, T, 0)$									
0	40	38	36	34	33	31	29	28	27
10	24	23	22	22	21	20	19	19	18
20	15	14	13	13	12	12	11	11	10
30	10	9	8	7	6	5	4	3	2

In conclusion, let us note that approximate values of the first derivatives $\partial\sigma_T/\partial T$ and $\partial\sigma_T/\partial S$ can be obtained from the simplified equation of state [9.6]. Differentiating [9.6] for T and S, we obtain respectively:

$$\frac{\partial\sigma_T}{\partial T} = -0.0735 - 0.00938\,T - 0.002S$$

$$\frac{\partial\sigma_T}{\partial S} = 0.802 - 0.002\,T$$

[24.25]

These formulae can prove useful in various types of approximate calculations.

25. THE THERMOHALINE DERIVATIVE

The coefficient of thermohalinity [2.11] is of specific oceanographic interest, and is an important numerical characteristic of the equation of state of sea water at atmospheric pressure (Mamayev, 1968). It will be seen from [2.11] that the latter at $S_0 = 1$ is numerically equal to the cotangent of the angle φ of inclination of the isostere or isopycnal in the given point (S, T) to the abscissa axis, taken with the opposite sign:

$$\cot\varphi = -\left(\frac{\partial S}{\partial T}\right)_{p,\upsilon}$$

[25.1]

Indeed, having divided the equation of the isopycnal:

$$\frac{\partial\rho}{\partial S}\,dS + \frac{\partial\rho}{\partial T}\,dT = 0$$

[25.2]

by $\partial\rho/\partial S$, we will obtain:

$$dS = -\cot\varphi\,dT$$

whence follows [25.1]. Let us call this function the *thermohaline derivative* and designate it as a partial derivative, unlike the arbitrary value of the cotangent of angle φ in the given T-S point.

Let us consider the thermohaline derivative at constant (atmospheric) pressure. The numerical value of the thermohaline derivative can be obtained by dividing the partial derivative of the equation of state $\partial\alpha/\partial T$ (or $\partial\rho/\partial T$) by the derivative $\partial\alpha/\partial S$ (or $\partial\rho/\partial S$ respectively) *; as a result we will obtain:

---

* In both cases we will obtain one and the same result (this is clear from a comparison of formulae [24.2] and [24.3] with formulae [24.4] and [24.5], in spite of the fact that the values of the derivatives $\partial\alpha/\partial T$ and $\partial\alpha/\partial S$ differ from the values of derivatives $\partial\rho/\partial T$ and $\partial\rho/\partial S$.

$$\frac{\partial S}{\partial T} = \frac{\partial \sigma_T/\partial T}{\partial \sigma_T/\partial S} = \frac{\dfrac{d\Sigma_T}{dT} + (\sigma_0 + 0.1324)\left[-\dfrac{dA_T}{dT} + \dfrac{dB_T}{dT}(\sigma_0 - 0.1324)\right]}{\dfrac{d\sigma_0}{dS}(1 - A_T + 2B_T\sigma_0)} \qquad [25.3]$$

Table A6 gives the values of $\cot\varphi = -(\partial S/\partial T)_\rho$ as a function of salinity and temperature at constant pressure, obtained by dividing the values of the quantities in table 20 by the values of the quantities in table 26 of the *Oceanological Tables* (Zubov, 1957a). In addition, the corresponding relationship is shown in graphs (Figs. 16 and 17), in the first of which is shown the function itself, while the second shows the values of angle φ corresponding to it for a *T-S* diagram with a unit correlation of scales ($1°/1\text{‰} = 1$). Considering the graph (Fig. 16), we come to the conclusion that it reveals an analytical link of a simpler nature between temperature and salinity ("the thermohaline link") in the equation of state of sea water than might have been expected, let us say, from the consideration of the graph $\rho = \rho(S, T)$: the isolines $\cot\varphi$ are almost rectilinear, especially in the region of low temperatures and salinities.

The zero line (the line of the isopycnal extremes in relation to variable T) coincides with the line of temperature θ of maximum density; assuming in formula [25.3] $(\partial S/\partial T)_\rho = 0$, we obtain the equation of line $\theta = \text{const.}$ in *S-T* coordinates, which is more accurate than formulae [5.2] and [5.3].

Furthermore, inasmuch as function $\partial S/\partial T$ characterizes the relation of the quantity of thermal expansion to saline contraction, Fig. 16 enables us to judge the relative influence of these two processes on the formation of the field of density; thus, saline contraction, to a greater extent than thermal expansion, influences density in the region of low salinities (and low temperatures); thermal expansion attains maximum influence in the region of high temperatures and high salinities (this is to be explained by caloric effects).

It is interesting to draw attention to the similarity in the arrangement of the isolines of the gradients of thermal expansion (Fig. 14) and the gradients of thermohalinity (Fig. 16). In both cases the zero isolines coincide; small discrepancies are observed in the region of high salinities and temperatures. This is completely understandable, at least from the formal point of view. Since:

$$\frac{\partial S}{\partial T}\frac{\partial \rho}{\partial S} = \frac{\partial \rho}{\partial T}$$

while the changes of $\partial\rho/\partial S$ are relatively small and do not exceed 10% of the quantity itself throughout the whole oceanic range of salinities and temperatures, within an accuracy of 10% we can write:

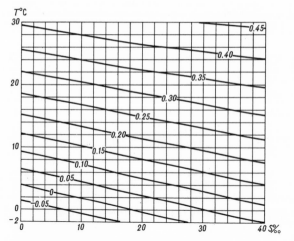

Fig. 16. Graph of function $\cot \varphi = -\partial S/\partial T = f(S, T)$.

$$\frac{\partial \rho}{\partial T} \approx 0.0008 \frac{\partial S}{\partial T}$$

Thus, the gradient of thermohalinity with a good degree of accuracy is re-

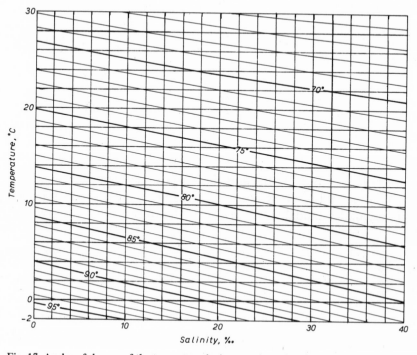

Fig. 17. Angles of slope φ of the tangent to the isopycnals to the abscissa axis of the T-S diagram.

presentative in relation to the quantity of thermal expansion of sea water, and, if necessary, we can judge the latter by the angle of inclination of the iso- pycnal to the abscissa axis.

An interesting example of the application of the thermohaline derivative to prove that mixing may be isopycnal in nature is given by Pingree (1972). Studying small-scale mixing (microstructure) in the deep parts of the ocean in the region to the west of the Iberian Peninsula, Pingree considers the rela- tionship:

$$T' = mS' \qquad\qquad [25.4]$$

where T' and S' are the fluctuating deviations of temperature and salinity from those values of $\overline{T}$ and $\overline{S}$, which form an averaged (flattened) T-S curve, while m is a coefficient which apparently can be calculated from the formula:

$$m^2 = \overline{(T'^2)}/\overline{(S'^2)} \qquad\qquad [25.5]$$

but the theoretical meaning of which still remains to be determined.

By means of an ingenious argument, which we omit here, Pingree shows that if small-scale mixing occurs along isopycnic surfaces, the following con- dition must be fulfilled:

$$m \simeq \left(\frac{\partial T}{\partial S}\right)_\rho \qquad\qquad [25.6]$$

(the sign $\simeq$ means that two quantities are not very different from each other).

Indeed, for two repeated (in one point) stations 6513 and 6521 of the R.V. "Discovery" in the region mentioned above, condition [25.6] is ful- filled, as may be seen from the excerpt from Pingree's table (1972, p. 554)

TABLE V

Values of r, m and $\partial T/\partial S$ (From Pingree, 1972, p. 554)

Range of depths (m)	r	m (°C/°/oo)	$(\partial T/\partial S)_\rho$ (°C/°/oo)
800–1,000	0.96	4.3	4.3
1,000–1,200	0.98	3.9	4.4
1,200–1,400	0.97	4.6	4.7
1,400–1,600	0.98	5.7	5.4
1,600–1,800	0.98	5.3	6.5
1,800–2,000	0.94	6.6	7.4
2,000–2,200	0.90	6.2	8.2

which is given in Table V. The quantity m is calculated from formula [25.5]; in addition, the values of the coefficient of correlation $r = \overline{(T'S')}/(\overline{T'^2 S'^2})^{1/2}$ are given.

Thus, in a region of propagation of Mediterranean waters in the Atlantic not only large-scale, but also small-scale mixing is isopycnic in nature.

26. BARIC COMPRESSIBILITY*

The baric gradient of specific volume $\partial \alpha_{STp}/\partial p$ can be obtained by differentiating formula [4.1] for p; as a result of the differentiation we will obtain:

$$\frac{\partial \alpha_{STp}}{\partial p} = -\alpha_{ST0}\left(\mu + p\,\frac{d\mu}{dp}\right) \qquad [26.1]$$

In this formula:

$$10^8\,\frac{d\mu}{dp} = -\frac{4886 \times 0.000183}{(1+0.000183p)^2} + (105.5 + 9.50T - 0.158T^2)\cdot 10^{-3} - 3pT\cdot 10^{-7}$$

$$+ \frac{\sigma_0 - 28}{10}(32.4 - 0.87T + 0.02T^2)\cdot 10^{-3} - \left(\frac{\sigma_0 - 28}{10}\right)^2 (1.8 - 0.06T)\cdot 10^{-3} \qquad [26.2]$$

(provided that pressure p is expressed in bars). The formula obtained [26.1] makes it possible to establish a correlation between the *true* coefficient of compressibility:

$$k = -\frac{1}{\alpha_{STp}}\,\frac{\partial \alpha_{STp}}{\partial p} \qquad [26.3]$$

and the *mean* coefficient of compressibility μ **. Namely, substituting formula [26.1] in expression [20.3], we will obtain a formula linking up these coefficients:

$$k = \frac{\mu + p\,\dfrac{d\mu}{dp}}{1 - \mu p} \qquad [26.4]$$

From the latter formula, in particular, it will be seen that at atmospheric pressure ($p = 0$) the coefficients are equal: $k = \mu$, while the baric gradient of

---

* Also called isothermic compressibility, inasmuch as it is considered at T = const., i.e., without allowance for the influence of adiabatic processes (see also Section 27).
** Here the letter designation accepted in oceanographic literature has been retained. It should not be confused with chemical potential μ.

specific volume is equal to:

$$\frac{\partial \alpha_{ST0}}{\partial p} = \alpha_{ST0}\mu \qquad [26.5]$$

In the practice of oceanographic analysis the baric gradient of density $\partial \rho / \partial p$ is often applied; let us calculate it:

$$\frac{\partial \rho_{STp}}{\partial p} = \frac{\partial}{\partial p}\left(\frac{1}{\alpha_{STp}}\right) = -\frac{1}{\alpha_{STp}^2}\frac{\partial \alpha_{STp}}{\partial p} = \frac{k}{\alpha_{STp}} \qquad [26.6]$$

or, utilizing formulae [26.2] and [26.3]:

$$\frac{\partial \rho_{STp}}{\partial p} = \frac{\mu + p\dfrac{d\mu}{dp}}{\alpha_{ST0}(1 - \mu p)^2} \qquad [26.7]$$

It may be seen from the latter two formulae that the baric density gradient at atmospheric pressure ($p = 0$) equals:

$$\frac{\partial \rho_{ST0}}{\partial p} = \frac{\mu}{\alpha_{ST0}} \qquad [26.8]$$

It is interesting to elucidate the physical meaning of the baric density gradient in situ. Let us consider the differential of density:

$$d\rho = \frac{\partial \rho}{\partial S}\,dS + \frac{\partial \rho}{\partial T}\,dT + \frac{\partial \rho}{\partial p}\,dp \qquad [26.9]$$

where $T = \theta + \Delta T_A$ is temperature in situ, θ is the potential temperature, ΔT_A is the adiabatic temperature correction, as well as the total derivative:

$$\frac{d\rho}{dp} = \frac{\partial \rho}{\partial S}\frac{dS}{dp} + \frac{\partial \rho}{\partial T}\frac{d}{dp}(\theta + \Delta T_A) + \frac{\partial \rho}{\partial p} \qquad [26.10]$$

Let us consider a homogeneous compressible ideal ocean, in which salinity and *temperature in situ* do not change with depth, i.e., a compressible non-adiabatic ocean. In this case in the latter formula we will have $dT/dp = 0$, $dS/dp = 0$, whence:

$$\left(\frac{d\rho}{dp}\right)_{T,S} = \frac{\partial \rho}{\partial p} \qquad [26.11]$$

Thus, the baric density gradient in situ characterizes the change in density caused only by pressure, i.e., by the compressibility of sea water. Numerically, it characterizes the density gradient in the event that temperature and salinity do not change with depth.

The numerical value of the baric gradient of specific volume (density) by 1 dbar varies within the limits of $350 \div 450 \cdot 10^{-8}$ (Ivanov-Frantskevich, 1956). The values of the true coefficient of compressibility k are given in table 27 of the *Oceanological Tables* (Zubov and Chigirin, 1940).

27. ADIABATIC COMPRESSIBILITY AND POTENTIAL TEMPERATURE

In actual fact, changes in density with depth (pressure) take place adiabatically, which, properly speaking, is expressed by formula [26.10]. In that case, let us consider a homogeneous compressible standard ocean, in which salinity and *potential* temperature do not change with depth: $dS/dp = 0$ and $d\theta/dp = 0$. For such an ocean instead of formula [26.10] we will obtain:

$$\left(\frac{\partial \rho}{\partial p}\right)_A = \left(\frac{\partial \rho}{\partial p}\right)_\theta + \frac{\partial \rho}{\partial T}\left(\frac{dT}{dp}\right)_A \qquad [27.1]$$

(Pollak, 1954). In this formula the derivative $(\partial \rho/\partial p)_\theta$ represents the isothermic (baric) density gradient, $(dT/dp)_A$ the adiabatic temperature gradient determined by Kelvin's formula [20.1].

The temperature gradient in situ (the "total" temperature gradient), according to the formulae given above, equals:

$$\frac{dT}{dp} = \frac{d\theta}{dp} + \left(\frac{dT}{dp}\right)_A \qquad [27.2]$$

The quantity $(\partial \rho/\partial p)_A$ is the baric density gradient in situ; we will call it the *adiabatic gradient of density*. It characterizes only that part of the change in density of sea water which is caused by baric and adiabatic effects. This component does not depend (for all practical purposes) on the thermohaline structure of the ocean; thermohaline compressibility, or stability, will be considered below in Section 28.

Both terms in the right-hand side of formula [27.1] depend on pressure; we have already considered this dependence for the first term and will consider below the dependence of the adiabatic temperature gradient on pressure.

Let us quote the example of a calculation according to formula [27.1] for the simple case $p = 0$ (atmospheric pressure) at $S = 36‰$, $T = 10°$. According to formula [26.8] we can rewrite formula [27.1] in the form:

$$\left(\frac{\partial \rho_{STO}}{\partial p}\right)_A = \frac{\mu}{\alpha_{STO}} + \frac{\partial p}{\partial T}\left(\frac{\partial T}{\partial p}\right)_A \qquad [27.3]$$

The quantities necessary for the calculation are taken from the Oceanological Tables of Zubov (1957a): $\alpha = 0.973$ (table 11), $\mu = 441.5 \cdot 10^8$ (table 14), $\partial \rho/\partial T = -1.74 \cdot 10^{-4}$ (table 20), $(\partial T/\partial p)_A = 1.20 \cdot 10^{-4}$ (table 23).

We obtain:

$$\left(\frac{\partial \rho_{STO}}{\partial p}\right)_A = \left(\frac{441.5}{0.973} - 1.74 \times 1.20\right) \cdot 10^{-8} = 451 \cdot 10^{-8}$$

This computation, fairly simple for $p = 0$ (the sea surface), becomes considerably more complicated in computing for depth. Here we are helped by the formula [19.7] of the theory of sound to which Pollak (1954) drew attention in this aspect, namely:

$$\left(\frac{d\rho_{STp}}{dp}\right)_A = \frac{1}{c^2} \tag{27.4}$$

where c is the speed of propagation of sound in sea water. Thus, for $S = 36‰$ and $T = 10°$, $c = 1487.9$ m/sec; hence:

$$\left(\frac{d\rho}{dp}\right)_A = \frac{1}{(1487.9)^2} = 451 \cdot 10^{-8}$$

From formula [27.4] Pollak (1954) calculated a nomogram for the computation of $(\partial \rho/\partial p)_A$ at atmospheric pressure, as well as a nomogram for calculating the pressure correction. These nomograms, extremely convenient for calculating the adiabatic density gradient in situ, are reproduced in Figs. 18 and 19 (reduced in size); the computation itself, thus, is carried out according to the formula:

$$\left[\left(\frac{\partial \rho}{\partial p}\right)_A\right]_{STp} = \left[\left(\frac{\partial \rho}{\partial p}\right)_A\right]_{ST0} - \Delta \left(\frac{\partial \rho}{\partial p}\right)_A$$

For example, for $S = 35‰$, $T = 2°$ and $p = 5000$ dbar:

$$\left[\left(\frac{\partial \rho}{\partial p}\right)_A\right]_{STp} = (471 - 64) \cdot 10^{-8} = 407 \cdot 10^{-8}$$

The computation of the adiabatic density gradients is used in the simplified computation of the vertical stability of the waters of the sea proposed by Pollak.

Let us now consider the dependence of the adiabatic temperature gradient on pressure. As is already known, the vertical adiabatic temperature gradient in the sea is determined from Kelvin's formula:

$$\Gamma = \left(\frac{\partial T}{\partial p}\right)_A = \frac{T}{c_p} \frac{\partial \alpha}{\partial T} \tag{27.5}$$

where T is absolute temperature (in °K), c_p the specific heat of sea water at constant pressure, in J g^{-1}°C^{-1} (see Section 18). Gradient Γ represents the adiabatic gradient at atmospheric pressure, i.e., without allowing for pressure corrections. In order to obtain the adiabatic gradient in situ it is necessary to introduce quantity $(\partial \alpha/\partial T)_{STp}$ in Kelvin's formula instead of $\partial \alpha/\partial T$, as well

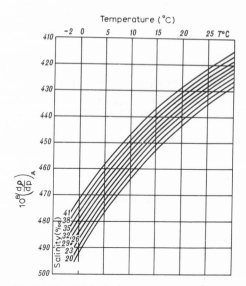

Fig. 18. The adiabatic density gradient $10^8(\partial\rho/\partial p)_A$ as a function of salinity and temperature at atmospheric pressure. (Pollak, 1954.)

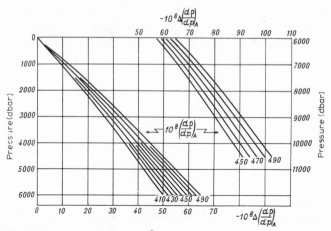

Fig. 19. Pressure correction $-10^8 \Delta(\partial\rho/\partial p)_A$ to the adiabatic density gradient as a function of pressure and of the adiabatic density gradient $10^8(\partial\rho/\partial p)_A$. (Pollak, 1954.)

as to take account of the change in specific heat capacity with a change in pressure, as determined by formula [18.6].

The adiabatic gradient Γ in situ is calculated according to the formula:

$$\Gamma_{STp} = \Gamma_{ST0} + \Delta\Gamma_{Tp} + \Delta\Gamma_{STp} \qquad\qquad [27.6]$$

The gradient Γ_{ST0} at constant pressure is calculated from table 23, corrections $\Delta\Gamma_{Tp}$ and $\Delta\Gamma_{STp}$ are calculated from tables 24 and 25 respectively (Zubov, 1957a).

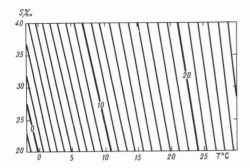

Fig. 20. Adiabatic temperature gradient $\Gamma_{ST0} = (\partial T/\partial p)_A$ in $0.01°/1000$ dbar as a function of salinity and temperature. (Fofonoff, 1959, 1962.)

Instead of tables 23, 24 and 25 it is more convenient to use graphs which can easily be plotted from these tables. These graphs were constructed by Fofonoff (1959, 1962) and are reproduced in Figs. 20 and 21 respectively *. The units Γ in these graphs equal $0.01°C/1,000$ dbar.

Furthermore, Fofonoff (1962) and Bryden (1973) give formulae for the calculation of Γ_{STp}, which represent a polynomial by increasing powers of parameters of state, convenient for computer calculations.

Potential temperature (θ), after Helland-Hansen (1912), is that temperature

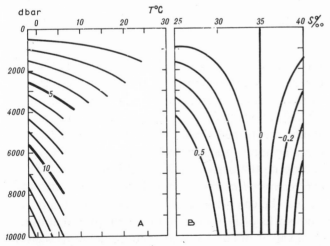

Fig. 21. Correction $\Delta\Gamma_{Tp}$ to the adiabatic temperature gradient for temperature and pressure (A), correction $\Delta\Gamma_{Sp}$ to the adiabatic temperature gradient for salinity and pressure (B) in the same units as in Fig. 20. (Fofonoff, 1959, 1962.)

* Fig. 21 represents only part of table 25 for temperature $0°$.

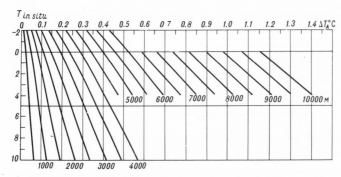

Fig. 22. Adiabatic correction ΔT_A to temperature T_m in rising from a depth of m meters in the ocean (σ_0 = 28.00).

which water will assume adiabatically if pressure ceases; in other words, if a particle is raised from depth (in situ) to the surface of the sea without an exchange of heat (or salts) with the environment (see Section 20). Potential temperature is determined by the following formula (Fofonoff, 1962; Bryden, 1973):

$$\theta = T_{\text{in situ}} - \Delta T_A = T_{\text{in situ}} - \int_{p\,\text{in situ}}^{p_0} \Gamma(\theta', S, p)\, dp \qquad [27.7]$$

where:

$$\theta' = T_{\text{in situ}} - \int_{p\,\text{in situ}}^{p_0} \Gamma(\theta, S, p)\, dp \qquad [27.8]$$

(Γ is the adiabatic temperature gradient). It should be made clear that an adiabatic change in temperature (the integral in [27.7]) depends on temperature, which itself changes with a change in pressure p in the process of integration — hence formula [27.8], in which $\theta_0 = T_0 = T_{\text{in situ}}$ (Bryden, 1973).

On the basis of the formulae mentioned above Ekman (1914) * and Helland-Hansen (1930) compiled tables and graphs for the determination of adiabatic corrections and potential temperatures; Helland-Hansen's tables underly the corresponding *Oceanological Tables* of Zubov (1957a). In addition, potential temperatures can be calculated from the corresponding polynomial formulae given in the works of Fofonoff and Bryden quoted above.

To illustrate the numerical values of adiabatic corrections, a graph is given (Fig. 22) for the determination of potential temperature according to the

---

* Schott's (1914) tables of the adiabatic changes in temperature from a given depth to the surface were published at the same time. As Helland-Hansen points out (1930), Schott's tables are inaccurate since temperature is taken in them as constant, whereas, as has already been pointed out, the latter decreases adiabatically as one gets closer to the surface.

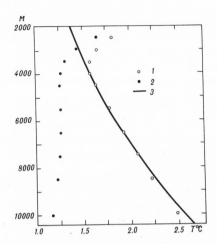

Fig. 23. The distribution with depth of the in situ (1) and potential (2) temperatures at station No.262 of the "Snellius", made in the Mindanao Trench. Curve 3 corresponds to the adiabatic temperature change with depth at a potential temperature $\theta = 1.25°C$. (Fofonoff, 1959.)

formula:

$$\theta = T_{\text{in situ}} - \Delta T_A$$

This graph is constructed from the corresponding table of Helland-Hansen.

As Sverdrup et al. (1942) and Fofonoff (1959) point out, the vertical distribution of temperature in the depressions of the World Ocean is very close to the adiabatic, and the instability of the waters, of which Schott had spoken (1914), apparently practically does not exist. Fig. 23 shows the distribution of deep and potential temperatures at the station "Snellius" No. 262, made in the Mindanao Trench on May 15–16, 1930 ($\varphi = 9°40'N$, $\lambda = 126°51'E$, depth 10,068 m). The curve in the figure corresponds to the adiabatic change in temperature with depth for potential temperature 1.25°C, and this curve coincides with the points of potential temperatures, indicating the adiabatic (equilibrium) stability of the waters in the Mindanao depression.

The concept of potential temperature is essential in T-S analysis; in investigating deep waters on T-S diagrams (and not only on them) instead of temperature, it is preferable (and more correct) to utilize potential temperature, which is an indicator of a certain water mass (Zubov, 1938, et al.).

28. VERTICAL STABILITY

Let us consider in the sea a certain small particle of water with density ρ and unit volume ($v = 1$), brought out of a state of vertical equilibrium, i.e., displaced under the influence of some impetus upwards or downwards over a vertical distance Δz. In a sea which is not uniform for density this particle enters an environment with greater or less density $\rho' = \rho \pm \Delta\rho$, and Archimedes' force will begin to act on it, equal per unit of mass to:

$$F = g(\rho' - \rho) = \pm g\Delta\rho \qquad [28.1]$$

The further behavior of the particle considered under the influence of Archimedes' force will depend on the type of density stratification which may be, if axis z is directed downwards, *positive* (density increases with depth, $\Delta\rho > 0$), *negative* (density decreases with depth, $\Delta\rho < 0$) and *neutral* or indifferent (density does not change with depth, $\Delta\rho = 0$). With positive stratification the particle, displaced up or down, will strive to return to its initial position; with negative stratification, it will continue to move further away from the initial position; with balanced stratification the particle will remain in any position.

The vertical acceleration of the movement of the particle (m is the mass of the particle) equals:

$$\frac{d^2z}{dt^2} = \frac{F}{m} = \pm g\frac{\Delta\rho}{\rho} \qquad [28.2]$$

Related to the unit of vertical distance, this acceleration equals:

$$\pm \frac{g}{\rho}\frac{\Delta\rho}{\Delta z} \qquad [28.3]$$

while the expression:

$$gE_z = \lim_{\Delta z \to 0} \frac{g}{\rho}\frac{\Delta\rho}{\Delta z} = \frac{g}{\rho}\frac{d\rho}{dz} \qquad [28.4]$$

represents the *total vertical stability* of the waters of the ocean.

Thus, stability is the acceleration of a particle displaced from its original state of equilibrium, related to a unit of distance, i.e., a kind of "unit acceleration".

It is natural that at $E > 0$ we observe a stable equilibrium, at $E < 0$ an unstable equilibrium and at $E = 0$ an indifferent equilibrium.

The density gradient in formula [28.4] is written in relation to depth; taking into account the factor of compressibility of sea water from the physical point of view, the following criterion is more acceptable:

$$E_p = \frac{1}{\rho}\frac{d\rho}{dp} \qquad [28.5]$$

where p is pressure. Bearing in mind the correlation between depth expressed in m and pressure expressed in dbars, $dp = (g\rho/10)\,dz$ we can write:

$$E_p = \frac{1.02}{\rho} E_z \qquad\qquad [28.6]$$

With an adequate degree of accuracy both criteria equal:

$$E_p = E_z \qquad\qquad [28.7]$$

It is practically a matter of indifference whether the density gradient is considered according to depth or according to pressure; we will make equal use of expression [28.4] or [28.5] depending on the context. Criterion [28.5] was proposed by Pollak (1954).

The density gradient entering into the expression of stability [28.5] is determined by formula [26.10] and may be viewed as consisting of two parts; *adiabatic* compressibility, or the density gradient in a uniform adiabatically compressible ocean — equation [27.1], and *thermohaline* compressibility, or the density gradient in the real ocean, stratified according to temperature and salinity (*excepting* the "ideal" component from the real ocean):

$$\left(\frac{\partial \rho}{\partial p}\right)_{S,\theta} = \frac{\partial \rho}{\partial S}\frac{dS}{dp} + \frac{\partial \rho}{\partial T}\frac{d\theta}{dp} \qquad\qquad [28.8]$$

As was already said above, the adiabatic part of the density gradient practically does not depend on the thermohaline structure of the ocean and is identical for each isobaric surface. Although the absolute quantity of the adiabatic density gradient is relatively large in comparison with the other components of vertical stability (it varies within the limits of $350 \div 450 \cdot 10^{-8}$ sec^2/m^2), it may be omitted in considering stability (Ivanov-Frantskevich, 1956).

Thus, from the oceanographic point of view, we are interested precisely in that part of stability [28.8] which is determined by the distribution of salinity and potential temperature in the sea. This component, or *thermohaline stability*, is studied in the theory and practice of physical oceanography.

Since $\theta = T - \Delta T_A$, instead of [28.8] we obtain:

$$\left(\frac{\partial \rho}{\partial p}\right)_{S,\theta} = \frac{\partial \rho}{\partial S}\frac{dS}{dp} + \frac{\partial \rho}{\partial T}\left[\frac{dT}{dp} - \left(\frac{dT}{dp}\right)_A\right] \qquad\qquad [28.9]$$

In this formula on the left is written the expression for the thermohaline density gradient; sometimes it is designated as $\delta\rho/dp$ in order to emphasize the difference between this derivative and the total density gradient. The substitution in [28.9] of temperature in situ and the adiabatic correction instead of potential temperature is to be explained by the fact that the latter cannot be determined directly from observations. Substituting [28.9] in [28.5], we obtain the generally accepted expression for vertical (we emphasize anew: thermohaline) stability, written with reference to pressure:

$$E_p = \frac{1}{\rho} \left[\frac{\partial \rho}{\partial S} \frac{dS}{dp} + \frac{\partial \rho}{\partial T} \left\{ \frac{dT}{dp} - \left(\frac{dT}{dp} \right)_A \right\} \right] \qquad [28.10]$$

or the expression equation expressed through depth:

$$E_z = \frac{1}{\rho} \left[\frac{\partial \rho}{\partial S} \frac{dS}{dz} + \frac{\partial \rho}{\partial T} \left\{ \frac{dT}{dz} - \left(\frac{dT}{dz} \right)_A \right\} \right] \qquad [28.11]$$

The constant multiplier is usually neglected and the following quantity is calculated:

$$E' = \frac{\partial \rho}{\partial S} \frac{dS}{dp} + \frac{\partial \rho}{\partial T} \left[\frac{dT}{dp} - \left(\frac{dT}{dp} \right)_A \right] \qquad [28.12]$$

Comparing formulae [28.5], [27.1] and [28.8], we obtain the expression:

$$E' = \frac{d\rho}{dp} - \left[\left(\frac{\partial \rho}{\partial p} \right)_\theta + \frac{\partial \rho}{\partial T} \left(\frac{dT}{dp} \right)_A \right] \qquad [28.13]$$

once again linking the total density gradient (total stability) with thermohaline and adiabatic stability (compressibility). Let us note that the adiabatic gradient enters into both these expressions. In turn, stability [28.10] can be represented in the form of the following notation:

$$E = E_T + E_S + E_A \qquad [28.14]$$

where:

$$E_T = \frac{1}{\rho} \frac{\partial \rho}{\partial T} \frac{dT}{dp} \qquad \text{is the temperature stability ,}$$

$$E_S = \frac{1}{\rho} \frac{\partial \rho}{\partial S} \frac{dS}{dp} \qquad \text{is the saline stability and}$$

$$E_A = -\frac{1}{\rho} \frac{\partial \rho}{\partial T} \left(\frac{dT}{dp} \right)_A \qquad \text{is the adiabatic stability .}$$

Such a division makes it possible to bring out the relative influence of each of the parameters on the formation of total thermohaline stability.

It is essential to point out that the partial derivatives of the equation of state $\partial \rho / \partial S$ and $\partial \rho / \partial T$ themselves depend on temperature, salinity and *pressure*, which has already been referred to in detail in Section 24. In its general form, this dependence can be expressed in the following way:

$$\left(\frac{\partial \rho}{\partial S} \right)_{S,T,p} = \left(\frac{\partial \rho}{\partial S} \right)_{S,T,0} + \left[\Delta \left(\frac{\partial \rho}{\partial S} \right) \right]_{S,p} + \left[\Delta \left(\frac{\partial \rho}{\partial S} \right) \right]_{S,T,p} \qquad [28.15]$$

$$\left(\frac{\partial \rho}{\partial T} \right)_{S,T,p} = \left(\frac{\partial \rho}{\partial T} \right)_{S,T,0} + \left[\Delta \left(\frac{\partial \rho}{\partial T} \right) \right]_{T,p} + \left[\Delta \left(\frac{\partial \rho}{\partial T} \right) \right]_{S,T,p} \qquad [28.16]$$

The same is true of the adiabatic gradient of temperature — formula [27.6]. Thus, the baric effect in the calculation of stability is not finally eliminated; only the basic part of this influence is eliminated where pressure appears in "pure form".

The expression:

$$E'_{S,T,0} = \left(\frac{\partial \rho}{\partial S}\right)_{S,T,0} \frac{dS}{dp} + \left(\frac{\partial \rho}{\partial T}\right)_{S,T,0} \frac{dT}{dp} \qquad [28.17]$$

into which the partial derivatives of the equation of state enter without corrections for compressibility (pressure) we will call the *principal part* of stability. Thus, E'_{ST0} represents stability brought to atmospheric pressure (we disregard the adiabatic gradient in the principal part also).

Taking account of formula [27.4], linking the adiabatic density gradient with the speed of sound, we can write the following important expression for stability, obtained by Pollak (1954):

$$E = \frac{1}{\rho}\left(\frac{d\rho}{dp} - \frac{1}{c^2}\right) \qquad [28.18]$$

Thus, the effect of adiabatic compressibility can be eliminated from total stability by means of the speed of sound. Formula [28.18], identical to formula [28.13], proves to be preferable, however, from the practical point of view, since the equation for the speed of sound in the sea $c = c(S,T,p)$ has been studied at the present time with greater precision than the equation of state $\rho = \rho(S,T,p)$.

There are also other expressions for thermohaline stability; thus, on the basis of formulae [23.18] and [23.10], one can construct identical expressions, which look respectively as follows:

$$E = 10^{-3}\frac{d\sigma_T}{dz} + \frac{\partial \epsilon_{SD}}{\partial S}\frac{dS}{dz} + \frac{\partial \epsilon_{TD}}{\partial T}\frac{dT}{dz} + \frac{\partial \rho}{\partial T}\left(\frac{dT}{dz}\right)_A \qquad [28.19]$$

(Hesselberg and Sverdrup, 1915b; Sverdrup et al., 1942), and:

$$E = \frac{d\alpha_T}{dz} + \frac{\partial \delta_{Sp}}{\partial S}\frac{dS}{dz} + \frac{\partial \delta_{Tp}}{\partial T}\frac{dT}{dz} - \frac{\partial \alpha}{\partial T}\left(\frac{dT}{dz}\right)_A \qquad [28.20]$$

The practical ways of calculating stability are considered in detail in the works of Ivanov-Frantskevich (1953, 1956) and Mamayev (1963).

29. PARTIAL DERIVATIVES OF THE SPEED OF SOUND

To compute the speed of sound from the theoretical formula [19.11] and to compile sound speed tables the expansion of function in a Taylor series is used, exactly as is done with respect to specific volume (Section 23). We shall not consider this question in as great detail as was the case for specific volume, but shall demonstrate, following Kuwahara (1939), only the principle of this computation in the example of the derivative $\partial c/\partial S$, entering into the Taylor expansion.

Differentiating formula [19.11] by S, we obtain:

$$\frac{\partial c}{\partial S} = \frac{1}{2\sqrt{H}} \left(2\frac{\partial \alpha}{\partial S} - \frac{\alpha}{H}\frac{\partial H}{\partial S} \right) \qquad [29.1]$$

where by H in the given context is designated the denominator of formula [19.11]:

$$H = - \left[\frac{\partial \alpha}{\partial p} + \frac{T}{c_p} \left(\frac{\partial \alpha}{\partial T}\right)^2\right]$$

In turn:

$$\frac{\partial H}{\partial S} = - \frac{\partial^2 \alpha}{\partial S \partial p} + \frac{T}{c_p} \left[\frac{1}{c_p}\frac{\partial c_p}{\partial S}\left(\frac{\partial \alpha}{\partial T}\right)^2 - 2\frac{\partial \alpha}{\partial T}\frac{\partial^2 \alpha}{\partial S \partial T}\right] \qquad [29.2]$$

The derivatives entering into the last three formulae are easily determined from Knudsen's and Ekman's formulae: of them, the quantities $\partial \alpha/\partial S$, $\partial \alpha/\partial T$ and $\partial^2 \alpha/\partial S \partial T$ are determined by formulae [24.4], [24.5] and [24.23], while the remaining are:

$$\frac{\partial \alpha}{\partial p} = 10^{-5}\frac{\partial \alpha_{STp}}{\partial p} = -10^{-5}\alpha_{ST0}\mu \,* \qquad [29.3]$$

$$\frac{\partial^2 \alpha}{\partial S \partial p} = 10^{-5}\frac{\partial^2 \alpha_{STp}}{\partial S \partial p} = -10^{-5}\left(\mu\frac{\partial \alpha_{ST0}}{\partial S} + \alpha_{ST0}\frac{\partial \mu}{\partial S}\right) \qquad [29.4]$$

$$10^9\frac{\partial \mu}{\partial S} = - [\{147.3 - 2.72T + 0.04T^2 - p\cdot 10^{-4}(32.4 - 0.87T + 0.02T^2)\}\cdot 10^{-1}$$

$$+ 2\cdot 10^{-2}(\sigma_0 - 28)\{4.5 - 0.1T - p\cdot 10^{-4}(1.8 - 0.06T)\}]\frac{\partial \sigma_0}{\partial S} \qquad [29.5]$$

The formula for $\partial c_p/\partial S$ is not given here.

Thus, the formula for the speed of sound [19.11] can be finally written in the form:

$$c_{STp} = c_{35,0,0} + \Delta c_T + \Delta c_S + \Delta c_p + \Delta c_{STp} \qquad [29.6]$$

where c_{STp} is the speed of sound in situ, $c_{35,0,0}$ is the speed of sound in "standard" water ($T = 0°$, $S = 35\%_0$), Δc_T is the speed of sound temperature

---

* Here the multiplier 10^{-5} serves to convert CGS units (dyn/cm^2) into decibars.

correction, Δc_S is the speed of sound salinity correction, Δc_p is the speed of sound pressure correction, $\Delta c_{STp} = \Delta c_{ST} + \Delta c_{Sp} + \Delta c_{Tp}$ is the total speed of sound correction for temperature, salinity and pressure.

In his tables Kuwahara combines the quantities $c_{35,0,0}$ and Δc_p in one quantity of sound speed in a "standard ocean".

Thus, formula [29.6] can be written:

$$c_{STp} = c_{35,0,p} + \Delta c_T + \Delta c_S + \Delta c_{STp} \qquad [29.7]$$

Kuwahara's formula, as we see, is similar to Bjerknes' for the calculation of specific volume in situ; this circumstance makes Kuwahara's tables compact and, at the same time, the speed of sound obtained from them may be considered more accurate than that derived from Matthews' tables.

Mackenzie (1960) undertook a detailed analysis of Kuwahara's tables and proposed empirical formulae for individual terms of Kuwahara's overall formula for the determination of c_{STp}; these empirical expressions may be introduced into computers and considerably speed up the calculating process. The differences in the determinations of the speeds of sound from Kuwahara's tables and from approximate expansions of terms of Kuwahara's formula in series (by the method of least squares) are basically confined within limits of 0.1 m/sec. Mackenzie's results were simplified by Bellas (1961) and also Horton (1961), who proposed the following simple formula:

$$c = 1399 + 1.31S + 4.592T - 0.0444T^2 + 0.182h \qquad [29.8]$$

where h is depth in meters. We do not dwell on the results of Mackenzie, Bellas and Horton, which were considered in detail in another monograph of the author (Mamayev, 1963), but merely quote the last formula which, like the simplified equation of state of sea water [9.6], can prove useful for analytical purposes.

Let us now consider the differential of the speed of sound as a function of three variables:

$$dc = \frac{\partial c}{\partial S}\, dS + \frac{\partial c}{\partial T}\, dT + \frac{\partial c}{\partial p}\, dp \qquad [29.9]$$

The partial derivatives are defined here from the equation $c = c(S, T, p)$; their approximate values can be obtained from Horton's formula and equal $\partial c/\partial S = 1.31$; $\partial c/\partial T = 4.59 - 0.09T$; $\partial c/\partial p = 0.18$ (in meters per second per 1°/oo, 1°C and 1 m respectively). These values must be considered only as determinations of an order of magnitude, since the partial derivatives of the speed of sound are, generally speaking, functions of the parameters of state of sea water. Expression [29.9] is interesting from the point of view of an accurate determination of the acceleration of sound on mixing.

The vertical speed of sound gradient is determined by the expression:

$$\frac{dc}{dz} = \frac{\partial c}{\partial S}\frac{dS}{dz} + \frac{\partial c}{\partial T}\frac{dT}{dz} + \frac{\partial c}{\partial p}\frac{dp}{dz} \qquad [29.10]$$

This expression, according to formula [29.6], can also be approximately written in the following form:

$$\frac{dc}{dz} = \frac{\partial}{\partial S}(\Delta c_S)\frac{dS}{dz} + \frac{\partial}{\partial T}(\Delta c_T)\frac{dT}{dz} + \frac{\partial}{\partial p}(\Delta c_p)\frac{dp}{dz} \qquad [29.11]$$

The last two formulae are similar to the formulae for density, which express vertical stability, and are of great importance for the study of the curves of vertical distribution of the speed of sound, in particular for the study of the conditions of formation of a deep sound channel (in the axis of a sound channel the derivative dc/dz turns into zero).

Let us consider the interesting example of the relation between the thermal part of vertical (density) stability and the "sound stability" which determines the formation of a sound channel. It is known that temperature and hydrostatic pressure exert the main influence on the formation of the vertical distribution of the speed of sound. The influence of salinity, in spite of the fact that the derivative $\partial c/\partial S$ is after all not so small, is considerably less because of the smaller vertical salinity gradients (if we leave aside regions with strong anomalies of salinity along the vertical such as, for example, the area of propagation of Mediterranean waters in the Atlantic Ocean). Taking account of what has been said, formula [29.11] can be written approximately as:

$$\frac{dc}{dz} = \frac{\partial}{\partial T}(\Delta c_T)\frac{dT}{dz} + \frac{\partial}{\partial p}(\Delta c_p)\frac{dp}{dz} \qquad [29.12]$$

More accurate values of the partial derivatives, if we limit ourselves in the Taylor expansion to terms of the first order, are provided, according to Wilson's formula [19.16], by the following quantities:

$\partial(\Delta c_T)/\partial T = 4.57$ cm sec^{-1} °C^{-1}

$\partial(\Delta c_p)/\partial p = 1.60$ cm^3 sec^{-1} kg^{-1}

Substituting in the second term of the right-hand side instead of the derivative dp/dz its approximate value, equal to 1.03 (this follows from the equation of hydrostatics), and using the given values of the partial derivatives, we can obtain from formula [29.12] the value of the critical vertical temperature gradient:

$$\left.\frac{dT}{dz}\right|_{c_{min}} = -2.89 \cdot 10^{-3} \text{ °C/m}$$

(axis z directed downwards). Everywhere that the values of the temperature gradient become less than this critical quantity the formation of an axis of a sound channel can be expected. The example mentioned is borrowed from the work of Garner (1967), who studied the question of oceanic sound channels in the waters around New Zealand.

CHAPTER 5

THE *T-S* DIAGRAM AND ITS PROPERTIES

The representation of the processes of mixing of the water masses of the ocean on a *T-S* diagram, in the field of which various functions can be plotted — density and its derivatives, the speed of sound, etc., as well as their combinations, possesses a number of advantages over other graphic methods, and most of all because on the *T-S* diagram the largest quantity of characteristics of the waters of the ocean prove to be connected: temperature, salinity, density, speed of sound, and others, and at least one of the *parameters* (time, distance or depth, the concentration of waters in the mixture, etc.). The distinctive "capacity" of the *T-S* diagram determines the need for a quantitative study of the analytical properties of *T-S* relations together with the equation of state, represented on the *T-S* diagram by the isopycnic field.

Before proceeding to study *T-S* relations, representing real or idealized objects — the water masses of the ocean or their models, it is necessary to consider both the properties of the *T-S* diagram and the simplest images which arise on the *T-S* diagram while studying the processes of the mixing of waters. These include, first of all: the straight line of mixing, representing the mixing of two water masses, the triangle of mixing (area), necessary for the analysis of the mixing of three water masses, as well as the arbitrary *T-S* curve (the curve of mixing), the origin of which is due to the processes of incomplete mixing of waters.

Let us consider anew (see Section 21) two homogeneous water masses A and B; let the temperature and salinity of the first water mass equal T_1, S_1, and of the second water mass: T_2, S_2. On the *T-S* diagram, these water masses will be determined by points A and B (Fig. 24). The *T-S* points, determining the position of the water masses in the *T-S* coordinates, are called *thermohaline indexes*; borrowing the terminology of physico-chemical analysis, they may also be called *figurative* points (Anosov and Pogodin, 1947).

Let us consider the question of the intermixing of these two water masses in different proportions. In the case of *additivity* of temperature and salinity (see Section 21), the result of the complete mixing of water masses A and B will be represented by a thermohaline index lying on the straight line AB, which is called the *straight line of mixing*. The temperature and salinity of

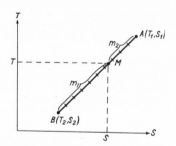

Fig. 24. The straight line of mixing of two water masses.

the mixture will then be determined by the formulae of mixing:

$$T = T_1 m_1 + T_2 m_2 \qquad\qquad [30.1]$$

$$S = S_1 m_1 + S_2 m_2 \qquad\qquad [30.2]$$

where m_1 and m_2 are the proportions (masses) of the first and second water masses taking part in the mixing. In these equations the proportions are expressed in parts of a unit, $m_1 + m_2 = 1$; the percentage expression of the parts of the water masses in the mixture is also used. The resulting T-S point will lie on the straight line of mixing at distances from A and B proportional to m_2 and m_1 respectively (Fig. 24). It should be pointed out that formulae [30.1] and [30.2] are approximate, since no allowance is made in them for

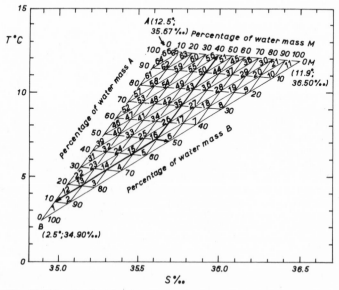

Fig. 25. Triangle of mixing (nomogram for the determination of the percentage content of water masses).

the dependence of specific heat on temperature and salinity (see, for example, Proudman, 1953).

Let us now turn to the case of the mixing of three homogeneous water masses A, B and M, having temperatures and salinities T_1, S_1; T_2, S_2 and T_3, S_3 respectively. On the *T-S* diagram (Fig. 25) the indexes of these three water masses, if they do not lie on one straight line, form a *triangle of mixing*. The product of the complete mixing of these three water masses will have a temperature and salinity also determined by the formulae of mixing:

$$T = T_1 m_1 + T_2 m_2 + T_3 m_3 \qquad\qquad [30.3]$$

$$S = S_1 m_1 + S_2 m_2 + S_3 m_3 \qquad\qquad [30.4]$$

where m_1, m_2 and m_3 are the proportions of the three water masses taking part in the mixing, $m_1 + m_2 + m_3 = 1$. The result of the complete mixing of the three water masses will be represented by a point with the coordinates (T, S), lying within the triangle of mixing. The formulae of mixing constitute the usual formulae for mean-weighted quantities (Zubov and Sabinin, 1958).

The determination from formulae [30.1]−[30.4] of the temperature and salinity of the mixture according to the known values of the ratio of each of the water masses in the mixture represents a *direct* problem; the *inverse* problem consists in the determination of the proportionate (or percentage) composition of each of the waters of the mixture from the known values of the thermohaline indexes of the original water masses and of the water mass of the mixture (Ivanov, 1949).

For the determination of the two unknowns m_1 and m_2 (the mixing of the two water masses), a system of two equations is sufficient:

$$\left.\begin{array}{l} T_1 m_1 + T_2 m_2 = T \\[6pt] m_1 + m_2 \quad = 1 \end{array}\right| \qquad [30.5]$$

the solution of which in the matrix notation (Efimov, 1964) is:

$$\begin{pmatrix} m_1 \\ m_2 \end{pmatrix} = \begin{pmatrix} \dfrac{1}{\Delta} & -\dfrac{T_2}{\Delta} \\[10pt] -\dfrac{1}{\Delta} & \dfrac{T_1}{\Delta} \end{pmatrix} \begin{pmatrix} T \\ 1 \end{pmatrix} \qquad\qquad [30.6]$$

or:

$$m_1 = \frac{T - T_2}{\Delta} \qquad m_2 = \frac{T_1 - T}{\Delta}$$

where:

$$\Delta = \begin{vmatrix} T_1 & T_2 \\ 1 & 1 \end{vmatrix} = T_1 - T_2$$

is the determinant of the system [30.5].

For the determination of the three unknowns m_1, m_2 and m_3 (the mixing of the three water masses), it is necessary to solve the system:

$$T_1 m_1 + T_2 m_2 + T_3 m_3 = T$$

$$S_1 m_1 + S_2 m_2 + S_3 m_3 = S \qquad\qquad\qquad\qquad\qquad\text{[30.7]}$$

$$m_1 + m_2 + m_3 = 1$$

The solution in matrix notation is:

$$\begin{pmatrix} m_1 \\ m_2 \\ m_3 \end{pmatrix} = \begin{pmatrix} \dfrac{S_2 - S_3}{\Delta} & -\dfrac{T_2 - T_3}{\Delta} & \dfrac{T_2 S_3 - T_3 S_2}{\Delta} \\[2ex] -\dfrac{S_1 - S_3}{\Delta} & \dfrac{T_1 - T_3}{\Delta} & -\dfrac{T_1 S_3 - T_3 S_1}{\Delta} \\[2ex] \dfrac{S_1 - S_2}{\Delta} & -\dfrac{T_1 - T_2}{\Delta} & \dfrac{T_1 S_2 - T_2 S_1}{\Delta} \end{pmatrix} \begin{pmatrix} T \\ S \\ 1 \end{pmatrix} \qquad\text{[30.8]}$$

where:

$$\Delta = \begin{vmatrix} T_1 & T_2 & T_3 \\ S_1 & S_2 & S_3 \\ 1 & 1 & 1 \end{vmatrix}$$

is the determinant of the system [30.7]. Thus, for example, for the unkown m_1 we have:

$$m_1 = -\frac{T(S_2 - S_3) - S(T_1 - T_3) + T_1 S_2 - T_2 S_1}{T_1(S_2 - S_3) - S_1(T_2 - T_3) + T_2 S_3 - T_3 S_2} \qquad \text{etc.}$$

We have considered above the question of the analytical determination of the proportions (concentrations) of water masses in the mixture of two and three water masses; this question is also dealt with in the works of Ivanov (1949), Timofeev and Panov (1962) and Baranov (1965). In practice, however, more frequent use is made of the *graphic* method of determining concentrations of water masses from the straight line of mixing and the triangle of mixing, respectively, i.e., from the known coordinates of the indexes of the mixing water masses and of the index of the resulting mixture (T,S). Thus, for example, to the point M with coordinates (T, S) in Fig. 24 there correspond 64% of water mass A and 36% of water mass B.

For convenience in determining the ratio of each of the water masses in the mixing of three water masses, it is necessary to break down each of the sides of the triangle of mixing into ten parts, and to join the points of division by straight lines, parallel to each of the sides of the triangle of mixing. The use of the grid thus obtained for the determination of the percentage content of each of the water masses is clear from Fig. 25: for example, in point *16* we have 10% of water mass A, 50% of water mass B and 40% of water mass M *. Practical methods of determining the percentage composition of the waters in a mixture are considered in more detail in Section 44.

Let us now turn briefly to the question of the mixing of four water masses. The *direct* problem in this case, obviously, is easily solved; the solution of the *inverse* problem is based on the system of equations:

$$T_1 m_1 + T_2 m_2 + T_3 m_3 + T_4 m_4 = T$$
$$S_1 m_1 + S_2 m_2 + S_3 m_3 + S_4 m_4 = S$$
$$P_1 m_1 + P_2 m_2 + S_3 m_3 + P_4 m_4 = P \qquad [30.9]$$
$$m_1 + m_2 + m_3 + m_4 = 1$$

which we shall not consider, referring the reader to Ivanov's article for further details. We see that for a strict solution of the inverse problem two characteristics of the water masses — temperature and salinity — are not enough; the introduction of some third characteristic P is necessary, in which capacity, for example, some stable hydrochemical characteristic may be selected. Ivanov (1949) points out that in this case the nomogram for the determination of the percentage content of the four waters must be spatial. Inasmuch as system [30.9] is linear, this nomogram must have the form of a tetrahedron. However, in the particular case when four water masses mix in pairs and in equal proportions, the construction of a quadrangular nomogram on the T-S plan is possible. This question is considered below, in Section 54.

Thus, in the case of complete (final) mixing of two or three water masses, the T-S index of the mixture, provided temperature and salinity are additive, will lie on the straight line of mixing or within the triangle of mixing. In the

---

* The sides of the triangle of mixing form on the T-S plane a distinctive triangular system of coordinates, the geometric properties of which are considered in Blokh's booklet (1952). The triangles of mixing themselves are widely used not only in oceanography, but also in physical chemistry, geology, metallurgy, etc.

case of partial mixing (the intermediate stage), an aggregate of "intermediate" *T-S* indexes is formed, each of which is related to a definite point of space, in which mixing takes place. These aggregates form the straight *T-S* line (when two waters mix) and the *T-S* curve (when three waters mix); their formation will be qualitatively considered in the following paragraph.

31. *T-S* CURVES OF THE WATERS OF THE OCEAN

Let us take as an example data on the distribution of temperature and salinity with depth at any oceanographical station and let us plot on the *T-S* diagram the *T-S* points for temperature and salinity observed at each of the levels. Noting under the *T-S* points the values of the depths of the corresponding levels and joining these points by a smooth curve, we obtain the *T-S curve of the oceanographical station.* The example of the *T-S* curve for the oceanographical station is represented in Fig. 26. Inasmuch as temperature and salinity are functions of depth:

$$T = \psi(z) \qquad S = \varphi(z) \tag{31.1}$$

and the *T-S* curve itself in the *T-S* coordinates may be considered as the curve

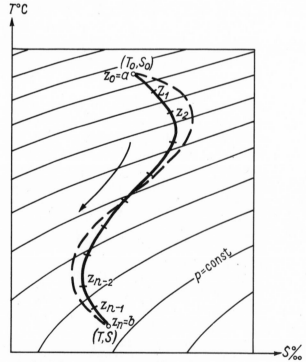

Fig. 26. *T-S* curve of oceanographic station.

of a certain function:

$$T = f(S) \tag{31.2}$$

then expressions [31.1] represent the parametric setting of the *T-S* curve. Function [31.2] is thus an *explicit* function. The *T-S* curve for parameter z can be called the *T-S-z* curve; it is appropriate instead of z to introduce the parameter p (pressure) to obtain the *T-S-p* curve.

The method of interpreting oceanographical observations in the form of *T-S* curves was introduced by Helland-Hansen (1918). The first experiments in constructing *T-S* curves, undertaken by Helland-Hansen and other investigators, showed that the *T-S* curves of many stations lying in one and the same region, often an extremely extensive one, possessed an astonishing similarity (this is dealt with in greater detail in Section 34). It was precisely thanks to this circumstance that the consideration of the *T-S* curve as the curve of a certain function [31.2] attained its physical significance and theoretical justification, forming the fundamental subject of *T-S* analysis.

It has already been stated that, in addition to parameters z or p, *T-S* curves can also be constructed for other parameters: time, distance along the horizontal, etc. The quantity ρ, represented in Fig. 26 by a family of isopycnals, may also be viewed conventionally as a parameter; in this case the parametric equations will prove to be:

$$T = T(\rho) \qquad S = S(\rho)$$

Various methods of parametric setting of *T-S* curves and other *T-S* relations will be considered below in Section 48 when we come to grips with methods. Here, however, we shall briefly consider the question of how the appearance on a *T-S* plane of a curve for parameter z is connected with the process of vertical mixing of waters. The most common case of mixing of water masses in the real conditions of the World Ocean is the vertical mixing of two, three or more water masses superimposed on each other. Therefore, let us consider, following Sverdrup et al. (1942), two and three (Fig. 27) horizontal water masses superimposed on each other. Let the vertical distribution of temperature and salinity in three successive stages of mixing be characterized by curves *1, 2* and *3* in both cases. These curves can reflect both the distribution of temperature and salinity which occurs during the mixing of water masses along one and the same vertical in the ocean and during displacement in space away from the area of the original distribution of the two or three "pure" water masses. In both cases the curves *1* correspond to the initial stage, when two or three unifrom water masses are superimposed on each other.

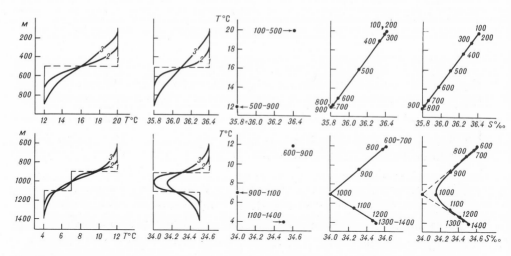

Fig. 27. Schematic representation of the results of mixing of two (above) and three (below) water masses. On the left curves $T(z)$ and $S(z)$ for the initial (*1*) and subsequent (*2* and *3*) stages of mixing are shown; on the right the *T-S* curves corresponding to these three stages are shown (Sverdrup et al., 1942).

Complete mixing may fail to take place between these two or three water masses due to the fact that, apart from the process of mixing, which is striving to equalize temperature and salinity, processes are taking place in the ocean which support the initial temperatures and salinities of the water masses, i.e., processes of the *creation* of water masses (Zubov, 1938). Among the latter, first of all, are included processes of heat exchange with the atmosphere, advection and some others. Therefore, in a stationary situation, the water masses under consideration continue to remain constantly in a state of partial mixing.

Proceeding to the description of the process of mixing on the *T-S* diagram, we obtain a characterization of these three stages of partial mixing first in the form of *T-S* points (stage *1*), then of *T-S* straight lines and *T-S* curves (stages *2* and *3*), represented in the right-hand part of Fig. 27.

Strictly speaking, during partial mixing of real water masses in the ocean the *T-S* points which correspond to different values of parameter z may also not lie on the straight line of mixing or within the triangle — in opposition to complete mixing, when the "final" *T-S* point must satisfy these conditions (in the absence of heat and salt exchange with the environment). This occurs both because of the non-additivity of properties, as has already been stated, and because of the differing intensity of the processes of turbulent heat conduction and diffusion which effect partial mixing of temperatures and salinities (Ivanov, 1949; Stommel, 1962a, b). For simplicity's sake, however, we may disregard these effects — at least in this qualitative exposition.

The most interesting result of the analysis of the mixing of three water masses is the fact that the *T-S* curve fits within the limits of the triangle of mixing in such a way that its branches extend to the *T-S* indexes of the two water masses — surface and deep, while the extreme lies opposite the *T-S* index of the *intermediate* water mass. Simultaneous study of this *T-S* curve (the lower right on the figure) with curves $T(z)$ and $S(z)$ shows that the extreme of the *T-S* curve corresponds to the extreme in the vertical distribution in the given case of salinity.

Inasmuch as we find widespread in the World Ocean the type of intermediate water masses characterized either by an extreme of salinity (such are the Mediterranean Water in the Atlantic and the Red Sea Water in the Indian Ocean), or by an extreme of temperature (such, for example, as the Atlantic water in the Central Arctic Basin), such a type of *T-S* curve appears to be not only interesting but also extremely important from the viewpoint of the analysis of water masses.

Above we have described a purely qualitative and summary picture; from the theoretical point of view, the question of the occurrence of *T-S* curves during the mixing of water masses is considered in Chapter 6.

32. THE LINE INTEGRAL IN THE *T-S* PLANE

The arbitrary curve in the *T-S* plane as a definite characteristic of state of the water masses of the sea possesses a most important analytical property which opens up the possibility of representing the *T-S* plane as the plane potential field and, accordingly, of investigating the equation of state of sea water and the processes of transformation of the waters of the ocean within the framework of the theory of the functions of a complex variable.

The *T-S* curve, on the basis of this property, proves to be bound up with other quantitative criteria, in particular with the vertical stability of waters (Section 28) and the concepts of thermohalinity (Section 25) and *T-S* areas.

Let us consider the line integral along the *T-S* curve [curve $T = f(S)$]:

$$I = \int_{(T_0, S_0)}^{(T, S)} P \, dS + Q \, dT \qquad [32.1]$$

from point (T_0, S_0) to point (T, S), (Fig. 26).

In the case of the equation of state of sea water at atmospheric pressure

$$\rho = \rho(S, T)$$

the total differential of density ρ will be written in the form:

$$d\rho = \frac{\partial \rho}{\partial S}\, dS + \frac{\partial \rho}{\partial T}\, dT \tag{32.2}$$

In this case, the integrand of [32.1] is the total differential [32.3], provided that:

$$P(T, S) = \frac{\partial \rho}{\partial S}$$

$$\tag{32.3}$$

$$Q(T, S) = \frac{\partial \rho}{\partial T}$$

In this case, the following condition is fulfilled identically:

$$\frac{\partial P}{\partial T} \equiv \frac{\partial Q}{\partial S} \tag{32.4}$$

and, consequently, the Leibnitz–Newton formula is valid, extended to the line integral:

$$\rho(S, T) - \rho(S_0, T_0) = \int_{(T_0, S_0)}^{(T, S)} P\, dS + Q\, dT \tag{32.5}$$

The line integral considered can be calculated, "straightening it out" for conversion into an ordinary rectilinear integral by two methods: by parameter and by the length of the arc.

Thus, if we use the parametric representation of the functions *:

$$S = \varphi(z) \qquad T = \psi(z) \tag{32.6}$$

then the line integral:

$$I = \int_{(T_0, S_0)}^{(T, S)} P(S, T)\, dS + Q(S, T)\, dT \tag{32.7}$$

can be calculated by parameter z, using the known formula:

---

* As has already been said, it is also possible to introduce another parameter; the significance of considering precisely parameter z (or p) will become clear later.

$$I = \int_a^b [P\{\varphi(z), \psi(z)\} \; \varphi'(z) + Q\{\varphi(z), \psi(z)\} \; \psi'(z)] \; dz \qquad [32.8]$$

In that case we obtain:

$$I = \int_a^b \left[\frac{\partial \rho}{\partial S} \frac{dS}{dz} + \frac{\partial \rho}{\partial T} \frac{dT}{dz} \right] dz \qquad [32.9]$$

The integrand of [32.9] represents the *principal part* of vertical stability E'_{ST0} (i.e., stability without any corrections for the compressibility of sea water).

Thus, expression [32.9] can be written in the following form:

$$I = \int_a^b E'_{ST0}(z) \; dz \qquad [32.10]$$

We have proven the theorem: *the line integral along the T-S curve from point a* (T_0, S_0) *to point b* (T,S) *by parameter z equals the definite integral of the principal part of stability* (E'_{ST0}) *between z = a and z = b* (Mamayev, 1962).

The quantity:

$$\frac{g}{\rho} I = \frac{g}{\rho} \int_a^b E'_{ST0}(z) \; dz \qquad [32.11]$$

characterizes the store of potential energy of a layer with a thickness from $z = a$ to $z = b$.

Inasmuch as the integrand of [32.1] is a total differential, the value of the line integral I does not depend on the path of integration, but is determined cnly by the position of the final points a and b (broken line in Fig. 26).

For the calculation of the line integral along the *length of the arc* reckoned from the initial point a (T_0, S_0) along the *T-S* curve, let us divide and multiply the integrand of formula [32.7] by the differential of the length of the arc, dl. Then, bearing in mind that:

$$\frac{dS}{dl} = \cos \mu \qquad \frac{dT}{dl} = \sin \mu \qquad [32.12]$$

where μ is the angle of slope of the tangent to the *T-S* curve in the given point n to the abscissa axis, we obtain:

$$I = \int_0^l \left(\frac{\partial \rho}{\partial S} \cos \mu + \frac{\partial \rho}{\partial T} \sin \mu \right) dl \qquad\qquad [32.13]$$

(Mamayev, 1964b, 1965b). Here l is the length of the whole arc ab.

Formula [32.13] is convenient for the numerical computation of the line integral along the straight *T-S* line, when the latter is broken down into intervals Δl, equal in length, corresponding to the parameter m (percentage content) evenly placed along the straight *T-S* line. A computation of this type is considered below in Section 35.

33. THE *T-S* DIAGRAM IN THE LIGHT OF THE THEORY OF A PLANE FIELD

Examining a plane *T-S* diagram, one may express the supposition that there must exist on it a family of lines orthogonal to the family of isopycnals, if only because of the formal consideration of the continuity of the derivatives $\partial \rho / \partial S$ and $\partial \rho / \partial T$. Designating the unknown function by $\gamma(S, T)$ and proceeding to the *dimensionless coordinates* on the *S-T* plane *, we can write the following expression, known from differential geometry, of the orthogonality of two families of curves on a plane, expressing it through the dimensionless *S-T* coordinates:

$$\frac{\partial \rho}{\partial T} \frac{\partial \gamma}{\partial T} + \frac{\partial \rho}{\partial S} \frac{\partial \gamma}{\partial S} = 0 \qquad\qquad [33.1]$$

Writing the differential equation of the isopycnal $\rho(S, T)$:

$$d\rho = \frac{\partial \rho}{\partial S} dS + \frac{\partial \rho}{\partial T} dT = 0 \qquad\qquad [33.2]$$

and the differential equation of the curve $\gamma(S, T)$:

$$d\gamma = \frac{\partial \gamma}{\partial S} dS + \frac{\partial \gamma}{\partial T} dT = 0 \qquad\qquad [33.3]$$

---

* The dimensionless coordinates in this section as well as in Section 37, are defined as $T = T'/T_0$, $S = S'/S_0$, where T' and S' are ordinary (dimensional) temperature and salinity, $T_0 = 1°C$, $S_0 = 1^0/_{00}$ (Proudman, 1953, section 74). This standard designation of T_0 and S_0 is convenient since it retains the values of temperature and salinity commonly used in *T-S* analysis. (See also Mamayev, 1973b.)

we may conclude that the necessary condition for the observance of the condition of orthogonality [33.1] is the validity of the following equations:

$$\frac{\partial \rho}{\partial S} = \frac{\partial \gamma}{\partial T} \qquad \frac{\partial \rho}{\partial T} = -\frac{\partial \gamma}{\partial S} \qquad\qquad [33.4]$$

Indeed, for any curve $\rho_k (S, T)$ the derivative $(dS/dT)_\rho$ will be determined from the equation:

$$\frac{\partial \rho}{\partial S} \frac{dS}{dT} + \frac{\partial \rho}{\partial T} = 0 \qquad\qquad [33.5]$$

and equals:

$$\left(\frac{dS}{dT}\right)_\rho = -\frac{\partial \rho / \partial T}{\partial \rho / \partial S} \qquad\qquad [33.6]$$

Correspondingly, for the curve $\gamma_n (S, T)$ we have:

$$\frac{\partial \gamma}{\partial S} \frac{dS}{dT} + \frac{\partial \gamma}{\partial T} = 0 \qquad\qquad [33.7]$$

whence:

$$\left(\frac{dS}{dT}\right)_\gamma = -\frac{\partial \gamma / \partial T}{\partial \gamma / \partial S} \qquad\qquad [33.8]$$

Having constructed the product of [33.6] and [33.8]:

$$\left(\frac{dS}{dT}\right)_\rho \left(\frac{dS}{dT}\right)_\gamma = \frac{(\partial \rho / \partial T)\,(\partial \gamma / \partial T)}{(\partial \rho / \partial S)\,(\partial \gamma / \partial S)} \qquad\qquad [33.9]$$

and comparing it with equation [33.4], we see that this product is equal to -1 (Milne-Thompson, 1960). For the right-hand parts of [33.9] condition [33.1] directly follows, for the left-hand parts — the condition of perpendicularity of the tangents to curves ρ and γ:

$$\left(\frac{dS}{dT}\right)_\rho \left(\frac{dS}{dT}\right)_\gamma = -1 \qquad\qquad [33.10]$$

Equations [33.4] are the *Cauchy-Riemann equations for the T-S plane*; they are valid if Laplace's equations are satisfied for functions ρ and γ by variables S and T:

$$\frac{\partial^2 \rho}{\partial S^2} + \frac{\partial^2 \rho}{\partial T^2} = 0$$

$$\frac{\partial^2 \gamma}{\partial S^2} + \frac{\partial^2 \gamma}{\partial T^2} = 0$$

[33.11]

Turning to the question of the formal analogy of these expressions, which characterize a *plane thermohaline field,* with the analytical characteristics of a plane field of irrotational motion of an incompressible fluid (a comparison with a plane heat field may look just as valid: the system of orthogonal iso-lines ρ and γ forms an isothermal grid), let us note that function γ, which may be called the *density flux function,* corresponds to the stream function ψ, while the *isopycnic potential* ρ corresponds to the velocity potential φ. Equations [33.4] prove formally similar to the Cauchy-Riemann equations for a plane potential flow:

$$\frac{\partial \varphi}{\partial x} = \frac{\partial \psi}{\partial y} \qquad \frac{\partial \varphi}{\partial y} = -\frac{\partial \psi}{\partial x}$$

[33.12]

which has already been noted earlier, while equations [33.11] prove similar to Laplace's equations for functions φ and ψ in the x, y-plane:

$$\frac{\partial^2 \varphi}{\partial x^2} + \frac{\partial^2 \varphi}{\partial y^2} = 0$$

$$\frac{\partial^2 \psi}{\partial x^2} + \frac{\partial^2 \psi}{\partial y^2} = 0$$

[33.13]

Turning to the inner meaning of the analogy made, we must draw attention to the fact that in hydrodynamics equations [33.13] are valid provided that the condition of continuity is observed; consequently, for a *T-S* plane we must present independent proof of the validity either of equations [33.11] (or of one of them), or of equations [33.28] written below, which are for-mally similar to the equation of continuity for plane flow. We have no such proof: for the *T-S* plane it is difficult to express considerations close to those on the continuity of flow.

The equation of state of sea water according to Knudsen does not, how-ever, satisfy Laplace's equation (the upper equation in [33.11]);the same is true of the simplified equations of state which follow from it. For a model of sea water according to Knudsen (the introduction of this term is justified by the need for making the equation of state more precise and constructing a new model) we have:

$$\frac{\partial^2 \rho}{\partial S^2} + \frac{\partial^2 \rho}{\partial T^2} \approx (-10^{-5}) \div (-2 \cdot 10^{-5})$$

[33.14]

(this sum is determined by the sum of the first two formulae [24.14]).

The error in [33.14] in relation to the upper equation in [33.11] is extremely small and lies fully within the limits of accuracy of the determination of the density of sea water. More than that, it is known from an experiment (Stupochenko, 1956) that the instrumental determination of the equation of state of a system on the basis of the measurement of the parameters of its state leads to different analytical expressions, the divergence among which is most substantially displayed precisely in the second derivatives (let us say, of specific volume or density); therefore, we are entitled to attribute the error in [33.14] to the inaccuracy of Knudsen's model of sea water. What has been said above amounts to affirming that the density of water is a harmonic function of temperature and salinity; this affirmation, however, remains in the nature of a postulate, and, we repeat, requires proof which, probably, can be devised by the methods of thermodynamics.

Continuing this comparison, let us note that to the components of velocity:

$$u = \frac{\partial \varphi}{\partial x} = \frac{\partial \psi}{\partial y} \qquad v = \frac{\partial \varphi}{\partial y} = -\frac{\partial \psi}{\partial x}$$

[33.15]

correspond the components of the "density flux":

$$P = \frac{\partial \rho}{\partial S} = \frac{\partial \gamma}{\partial T} \qquad Q = \frac{\partial \rho}{\partial T} = -\frac{\partial \gamma}{\partial S}$$

[33.16]

while the relationships between the conjugate functions ρ and γ are written in T-S coordinates in the following way:

$$\gamma = \int_{T_0}^{T} \left(\frac{\partial \rho}{\partial S}\right)_{S=S_0} dT - \int_{S_0}^{S} \frac{\partial \rho}{\partial T} dS$$

[33.17]*

The idea of the "density flux", formally recorded, as it were, on the T-S diagram, acquires definite physical meaning, if one considers its components, which may be called the "salt flux" and the "heat flux" (formulae [33.16]); these fluxes are represented in Fig. 28 (the heat flux in the main part of the

---

* This formula can be used for the direct computation of function $\gamma(S, T)$ (with an accuracy to the constant γ_0) from the known expression for function $\rho(S, T)$.

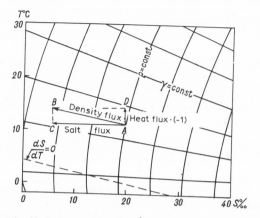

Fig. 28. Orthogonal families of curves of functions ρ and γ on a *T-S* diagram.

T-S diagram is negative). If the analytical considerations set forth above are supplemented by considerations on the distribution of temperature, salinity and density in the ocean (for example, along the vertical) and on the corresponding flows under real conditions, then it is also interesting to consider the distribution of parameter *z* (depth) along the *T-S* curves. Let us imagine that segments *AB*, *AC* and *AD* (Fig. 28) may be viewed as straight *T-S* lines, corresponding to real conditions, with some distribution of parameter *z* along them. In this situation *AD* is the straight *T-S* line in an ocean homogeneous in salinity and stratified in temperature; *AC* is the straight *T-S* line in an ocean homogeneous only for temperature. Then for these straight lines we may consider that $\rho = \rho(T,z)$ and $\rho = \rho(S,z)$ respectively; hence:

$$\frac{\partial \rho}{\partial T} = \frac{\partial \rho}{\partial z} \frac{dz}{dT} \qquad \frac{\partial \rho}{\partial S} = \frac{\partial \rho}{\partial z} \frac{dz}{dS} \qquad\qquad [33.18]$$

(Here derivative $\partial \rho / \partial z$ characterizes the relationship of parameters ρ and z; ρ also appears here as a parameter.)

In this case the heat and salt fluxes in the ocean will prove proportional to the following quantities:

$$\frac{dT}{dz} = \frac{\partial \rho}{\partial z} \left(\frac{\partial \rho}{\partial T}\right)^{-1} \qquad \frac{dS}{dz} = \frac{\partial \rho}{\partial z} \left(\frac{\partial \rho}{\partial S}\right)^{-1} \qquad\qquad [33.19]$$

Formulae [33.19] link the vertical fluxes of heat and salts in the ocean with the real density flux, as well as with the components of the "density flux" on the *T-S* diagram.

The considerations set forth reveal:

(a) An analogy between the *T-S* field and a plane stationary vector field, in particular with a field of plane stationary irrotational motion of an incompressible fluid.

(b) The possibility of investigating the *T-S* diagram and the real and analytical *T-S* correlations represented on it by methods of the theory of functions of a complex variable, in particular by the conformal transformation of the *T-S* diagram.

Let us now consider the plane of a complex variable:

$$z = S + iT \qquad . \qquad \qquad [33.20]$$

and the corresponding function of the complex variable:

$$w = f(z) = \rho + i\gamma \qquad \qquad [33.21]$$

It is obvious that function $w = f(z)$ is regular in the whole *T-S* plane at least to the limit values of salinity and temperature which are encountered in the ocean, since its real and imaginary parts satisfy the Cauchy-Riemann equations [33.4] for the *T-S* plane. Accordingly, function [33.21] performs a conformal transformation of a certain region D in the *T-S* plane into the region D^* in the ρ,γ-plane. If on the plane $z = S + iT$ we take two families of orthogonal lines

$$\rho(S, T) = \rho_0 \qquad \gamma(S, T) = \gamma_0 \qquad \qquad [33.22]$$

where ρ_0 and γ_0 are arbitrary constants, then on the plane $w = \rho + i\gamma$ the families of straight lines:

$$\rho = \rho_0 \qquad \gamma = \gamma_0 \qquad \qquad [33.23]$$

parallel to the coordinate axis, will correspond to them. The conformal transformation of the family of straight lines into lines [33.22] is performed by the function:

$$z = F(w) \qquad \qquad [33.24]$$

which is the inverse in relation to function [33.21]. This transformation is made easier by the introduction of the simplified equation of state of sea water (Section 9); confining ourselves here to stating the problem, we shall not consider it further.

Finally, let us consider another function of the complex variable, namely:

$$w_2 = f(z) = P + iQ \qquad \qquad [33.25]$$

where P and Q are determined by formulae [33.16]. It is obvious that function [33.25] is regular in the *T-S* plane, excluding the line of temperatures of maximum density, or line $dS/dT = 0$ (dashed line in Fig. 28); i.e., the Cauchy-Riemann equations are satisfied for functions P and Q:

$$\frac{\partial P}{\partial S} = \frac{\partial Q}{\partial T} \qquad \frac{\partial P}{\partial T} = -\frac{\partial Q}{\partial S} \qquad\qquad\qquad [33.26]$$

Consequently, we may continue the comparison of the characteristics of the thermohaline field with the corresponding characteristics of a field of irrotational incompressible flow: equations [33.26] prove formally similar to the Cauchy-Riemann equations for the components of velocity u and v:

$$\frac{\partial u}{\partial x} = \frac{\partial v}{\partial y} \qquad \frac{\partial u}{\partial y} = -\frac{\partial v}{\partial x} \qquad\qquad\qquad [33.27]$$

the Laplace equations:

$$\frac{\partial^2 P}{\partial S^2} + \frac{\partial^2 P}{\partial T^2} = 0 \qquad \frac{\partial^2 Q}{\partial S^2} + \frac{\partial^2 Q}{\partial T^2} = 0 \qquad\qquad [33.28]$$

prove similar to the corresponding equations also for u and v, while the identity:

$$\frac{\partial P}{\partial T} \equiv \frac{\partial Q}{\partial S} \qquad\qquad\qquad [33.29]$$

proves similar to the condition of equality to zero of the vertical component of the curl:

$$\omega_z = \frac{\partial v}{\partial x} - \frac{\partial u}{\partial y} = 0 \qquad\qquad\qquad [33.30]$$

The introduction of function [33.26] is equivalent to the consideration of the equation of state in a plane formally analogous to the hodograph plane. The conformal transformation performed by function:

$$z = F(W_2) \qquad\qquad\qquad [33.31]$$

makes it possible to "straighten out" in the plane $P + iQ$ the isolines of the components of the "density flux" $\partial \rho / \partial S$ and $\partial \rho / \partial T$.

The transformations thus considered in most general terms can not only facilitate the introduction of a non-linear equation of state of sea water into theoretical *T-S* analysis, but also reveal new and perhaps very substantial features of the interaction and transformation of the waters of the ocean in studying this question by means of the *T-S* diagram *.

---

* Let us point out once again that all the reasoning in this paragraph (as well as in Section 37) is valid only if the *T-S* diagram is considered as dimensionless.

34. BASIC CHARACTERISTICS OF THE *T-S* CURVE. CONSERVATION OF FORM

The question considered of the integral connection between stability and the *T-S* curve is not enough for a detailed study of the latter, in particular for the study of the basic property of the *T-S* curve — the conservation of its form. Practical experience in constructing *T-S* curves of oceanographical stations indicates their extraordinary similarity for extensive oceanic regions; at the same time, the distribution of salinities and temperatures along the vertical for stations characterized by very similar *T-S* curves can be extremely different. It becomes apparent that we need a separate consideration of *two basic characteristics of the T-S curve* — its form and the distribution of parameter z, for it is only the form of the *T-S* curve which demonstrates the property of conservation. This question is connected with the opinion which has taken root in physical oceanography to the effect that the position in relation to the axes of the coordinates of a *T-S* curve of an oceanographical station, constructed by parameter z, more or less correctly reflects the value of stability E of the water layers of the sea. In confirmation of what has been said one may refer to the monograph of Sverdrup et al. (1942, p. 142), in which the following is stated: "the slope of the observed *T-S* curve relative to the σ_T curves gives immediately an idea of the stability of the stratification".

However, this opinion is not altogether correct; the only thing we can say in looking at a *T-S* curve and its position in relation to the axes of the coordinates and the isolines of σ_T is whether the stability is positive or not. So far as the quantity of stability itself is concerned, it is by no means determined solely by that angle which the *T-S* curve forms in the given point with the isolines of σ_T. It is easy to see this immediately from a specific example. Let the same values of temperature $T = 17°$ and of salinity $S = 36.2‰$ ($\sigma_T = 26.44$) be observed at oceanographical stations A and B at depths of $z = 300$ m and 600 m respectively, while values $T = 9°$ and $S = 35.2‰$ ($\sigma_T = 27.30$) are observed at the same stations and at depths of 400 m and 800 m respectively. The data quoted are close to those observed at stations No. 1638 and 1640 of "Atlantis", carried out in the region of the Gulf Stream (Sverdrup et al., 1942).

The straight *T-S* line connecting on the *T-S* diagram the points (36.2‰; 17°) and (35.2‰; 9°) is the same for both stations; meanwhile, stability $E \approx \Delta\sigma_T/\Delta z$ in the first case amounts to $86 \cdot 10^{-4}$, and in the second to $43 \cdot 10^{-4}$ CGS units. Such a disparity in the values of stability along one and the same *T-S* curve is explained by the irregularity in the change of the parameter z along the latter in each of the two cases considered.

For the purpose of studying this question, let us attempt to separate the effect of parameter z on the results of calculating stability from the effect of the form of the *T-S* curve itself and its position in relation to the axes of the coordinates and the σ_T isolines, and to determine under what conditions the irregular distribution of parameter z along the *T-S* curve does not affect the results of the calculation of the vertical stability of waters (Mamayev, 1964b).

Since the *T-S* curve is given in parametric form, let us write the expression of the differential dl of the arc of the *T-S* curve in the form:

$$dl = \sqrt{\left(\frac{dS}{dz}\right)^2 + \left(\frac{dT}{dz}\right)^2}\ dz \qquad\qquad [34.1]$$

In addition, we shall need the following expressions:

$$\frac{dS}{dl} = \cos\mu \qquad \frac{dT}{dl} = \sin\mu \qquad\qquad [34.2]$$

where μ is the angle of slope of the *T-S* curve in the given point to the abscissa axis.

Replacing in expressions [34.2] differentiation by the length of the arc l by differentiation by parameter z, we can, using [34.1], obtain the following expressions for the derivatives of T and S by parameter z (in other words, for the vertical salinity and temperature gradients):

$$\frac{dS}{dz} = \sqrt{\left(\frac{dS}{dz}\right)^2 + \left(\frac{dT}{dz}\right)^2}\ \cos\mu \qquad\qquad [34.3]$$

$$\frac{dT}{dz} = \sqrt{\left(\frac{dS}{dz}\right)^2 + \left(\frac{dT}{dz}\right)^2}\ \sin\mu \qquad\qquad [34.4]$$

Substituting [34.3] and [34.4] in the expression for the principal part of stability:

$$E' = \frac{\partial\rho}{\partial S}\frac{dS}{dz} + \frac{\partial\rho}{\partial T}\frac{dT}{dz} \qquad\qquad [34.5]$$

(in the expression for E' the effect of pressure and adiabatic processes is disregarded), we obtain:

$$E' = \sqrt{\left(\frac{dS}{dz}\right)^2 + \left(\frac{dT}{dz}\right)^2}\left(\frac{\partial\rho}{\partial S}\cos\mu + \frac{\partial\rho}{\partial T}\sin\mu\right) \qquad\qquad [34.6]$$

The expression [34.6] obtained is interesting by the fact that it isolates the effect of parameter z or, what is the same, of the vertical salinity and temperature gradients, on both the quantity of stability and on the form of the *T-S* curve, and places the sign of inequality between these two criteria of state of the water masses of the ocean, provided that they are considered in differential form. Indeed, the form of the *T-S* curve and its position on the *T-S* plane are determined solely by the binominal appearing in [34.6] in the parentheses; however, where:

$$\frac{\partial \rho}{\partial S} \cos \mu + \frac{\partial \rho}{\partial T} \sin \mu = \text{const.} \qquad [34.7]$$

i.e., given constant form of the *T-S* curve, the quantity of stability E' can assume any values determined by the multipliers appearing under the sign of the square root.

Thus, the opinion expressed, in particular, by Sverdrup and quoted above, requires substantial clarification.

In case of constancy of the *T-S* curves (condition [34.7]) and equal values of stability (E' = const.) for different oceanographical stations, parameter z also can be distributed irregularly on such *T-S* curves. In this case the following condition is imposed on the vertical salinity and temperature gradients:

$$\left(\frac{dS}{dz}\right)^2 + \left(\frac{dT}{dz}\right)^2 = \text{const.} \qquad [34.8]$$

Expression [34.8] represents the normal equation of a circle in Cartesian coordinates, on the axes of which the gradients dS/dz and dT/dz have been laid out.

What has been said above fully justifies the affirmation that *T-S* curves are more stable images of the water masses of the ocean than the curves of vertical distribution of salinity $S(z)$ and temperature $T(z)$. On the other hand, they are more general characteristics, inasmuch as they reflect not only the vertical distribution of salinity and temperature, but also the influence of two other important factors: the equation of state of sea water and some kind of properties of relaxation brought about by the origin of water masses.

With respect to the second factor, let us note that it is confirmed precisely by the surprising similarity of many *T-S* curves in adjacent regions of the ocean; the processes of turbulent heat exchange and diffusion first of all modify the vertical distribution of salinity and temperature, which is manifested in the shift of the values of parameter z along invariable *T-S* curves. The modification of the *T-S* curves themselves is, as it were, a secondary process.

The question of the correlations between stability and the characteristics of the *T-S* curve is considered here in differential form; it is understood that the integral property brought out in Section 32 remains invariable. Formula [32.13] for the calculation of the integral from the length of the arc agrees exactly with this reasoning.

35. THE CALCULATION OF THE LINE INTEGRAL ALONG THE *T-S* CURVE

The concept of the line integral (Section 32) opens up certain possibilities for numerical computations linking up real *T-S* relations with the field of isopycnals which represents the equation of state of sea water on the characteristic diagram. Let us consider first of all the question of the calculation of the line integral:

$$I = \int\limits_{(l)} \frac{\partial \rho}{\partial S}\, dS + \frac{\partial \rho}{\partial T}\, dT = \int\limits_{(l)} d\rho\,(S,T) \qquad\qquad [35.1]$$

along any curve *l*, lying in the *T-S* plane (Mamayev, 1964b). As has already been pointed out, the integral [35.1] is reduced to a definite integral either by the introduction of some parameter or by the representation of [35.1] in the form of an integral for the length of the arc *l*. Let us recall that, in the first case, we come to the definite integral [32.9] and, in the second case, to the definite integral [32.13].

Let us consider the calculation of integral *I* from the example of a particular case — along the straight *T-S* line; in the case of uniform distribution of the parameter along this straight line, formulae [32.9] and [32.13] prove equivalent, as we shall see below.

Generally speaking, the quantity of integral [35.1] is determined by the difference of densities in the final and initial points, and this difference can be determined from the coordinates of the corresponding points with the help of the *Oceanological Tables* (Zubov, 1957a). However, the calculation precisely of expression [35.1] is of definite interest, since it not only considerably increases the accuracy of the determination of densities along the *T-S* line, but also makes it possible more accurately to carry out other calculations, in particular the calculation of contraction on mixing (see below). In addition, in calculating the integral [35.1] we have the possibility both of determining the effect of its temperature and salinity components and of immediately obtaining an idea of the stability of the vertical stratification along the *T-S* curve; the latter follows from the identity of formulae [32.9] and [32.13]. Concerning the technique of calculations, it should be pointed out that, although the partial derivatives $\partial \rho/\partial S$ and $\partial \rho/\partial T$ of the equation of state of sea water $\rho = \rho(S, T)$ at atmospheric pressure are expressed in elementary functions, the method is so cumbersome that taking integral *I* in its final form becomes quite laborious. Therefore, we must proceed along the path of approximate integration, taking the necessary values of derivatives $\partial \rho/\partial S$ and $\partial \rho/\partial T$ from the tables (Zubov, 1957a, tables 20 and 26).

In calculating integral *I* along the straight *T-S* line, it is convenient to consider the latter as a straight line of mixing of water masses $a\ (S_a, T_a)$ and

b (S_b, T_b), and as a parameter, evenly distributed along the straight *T-S* line, to take the percentage content m of one of the water masses in points of the straight *T-S* line (points of the corresponding mixtures). In that case we have:

$$\frac{dS}{dm} = \cos\mu = \frac{S_b - S_a}{100} \qquad [35.2]$$

$$\frac{dT}{dm} = \sin\mu = \frac{T_b - T_a}{100} \qquad [35.3]$$

or

$$\Delta S = \frac{S_b - S_a}{100} \Delta m \qquad [35.4]$$

$$\Delta T = \frac{T_b - T_a}{100} \Delta m \qquad [35.5]$$

Breaking down the straight line of mixing into, let us say, ten equal parts ($\Delta m = 10$), we can write the following expression for the calculation of integral [32.9] or integral [32.13] by the approximate method:

$$I = \sum_{n=1}^{n=10} \left[\left(\overline{\frac{\partial\rho}{\partial S}}\right)_n \Delta S + \left(\overline{\frac{\partial\rho}{\partial T}}\right)_n \Delta T \right] \qquad [35.6]$$

where $(\overline{\partial\rho/\partial S})_n$ and $(\overline{\partial\rho/\partial T})_n$ are the values of the corresponding derivatives in the median points $n_1, n_2, ..., n_{10}$, having coordinates (S_n, T_n), where:

$$S_n = S_a + (S_b - S_a)\frac{m}{100} \qquad [35.7]$$

$$T_n = T_a + (T_b - T_a)\frac{m}{100} \qquad [35.8]$$

and m = 5, 15, 25, ..., 95 are values of parameter m (percentage content) in points $n_1, n_2, ..., n_{10}$; finally,

$$\Delta S = \frac{S_b - S_a}{10} \qquad \Delta T = \frac{T_b - T_a}{10}$$

are the increments of salinity and temperature corresponding to the increment of parameter m by 10%

Table VI gives an example of a calculation according to formula [35.8] for the straight *T-S* line *BM*, the ends of which have coordinates (2.50°; 34.90‰)

TABLE VI

Example of the numerical calculation of an integral I along the straight T-S line

m	S_m	T_m	S_n	T_n	$\dfrac{\partial \sigma T}{\partial S}$	$\dfrac{\partial \sigma T}{\partial T}$
(1)	(2)	(3)	(4)	(5)	(6)	(7)
100	36.50	11.90				
			36.42	11.43	0.778	−0.189
90	36.34	10.96				
			36.26	10.49	0.779	−0.180
80	36.18	10.02				
			36.10	9.55	0.781	−0.169
70	36.02	9.08				
			35.94	8.61	0.783	−0.159
60	35.86	8.14				
			35.78	7.67	0.786	−0.149
50	35.70	7.20				
			35.62	6.73	0.788	−0.138
40	35.54	6.26				
			35.46	5.79	0.790	−0.127
30	35.38	5.32				
			35.30	4.85	0.792	−0.116
20	35.22	4.38				
			35.14	3.91	0.795	−0.105
10	35.06	3.44				
			34.98	2.97	0.797	−0.093
0	34.90	2.50				

m	$\dfrac{\partial \sigma T}{\partial S}\,\Delta S$	$\dfrac{\partial \sigma T}{\partial T}\,\Delta T$	$\Delta \sigma T$	σT	$\bar{\sigma}$	$\sigma T - \bar{\sigma}$
(1)	(8)	(9)	(10)	(11)	(12)	(13)
100	0.124	−0.178	−0.054	$\dfrac{27.788}{27.793}$	27.788	0.005
90	0.125	−0.169	−0.044	27.847	27.796	0.051
80	0.125	−0.159	−0.034	27.891	27.805	0.086
70	0.125	−0.149	−0.024	27.925	27.813	0.112
60	0.126	−0.140	−0.013	27.949	27.822	0.127
50	0.126	−0.130	−0.004	27.962	27.830	0.132
40	0.126	−0.119	+ 0.007	27.966	27.838	0.128
30	0.127	−0.109	+ 0.018	27.959	27.847	0.112
20	0.127	−0.099	+ 0.028	27.941	27.855	0.086
10	0.128	−0.087	+ 0.041	27.913	27.864	0.049
0				27.872	27.872	

and (11.9°; 36.50‰). This straight line on the average corresponds to the straight line of mixing between the Deep and Bottom Water of the Atlantic Ocean B and the Intermediate Mediterranean Water M. In this case, apparently, $\Delta S = +0.16‰$, $\Delta T = +0.94°$, if the direction of the calculation is selected from point B to point M. The explanations to Table VI are as follows:

Column *1* − percentage content m of the water mass in points which are multiples of 10%;

Columns *2* and *3* − salinity and temperature in these points, calculated from formulae [35.7] and [35.8] at $m = 0, 10, 20, ..., 100\%$;

Columns *4* and *5* − salinity and temperature in the points corresponding to the midpoints of the 10% segments. The values of S_n and T_n are also determined by formulae [35.7] and [35.8] where, however, $n = 5, 15, 25, ..., 95\%$;

Columns 6 and 7 — derivatives $\partial\sigma_T/\partial S = 10^3(\partial\rho/\partial S)$ and $\partial\sigma_T/\partial T = 10^3(\partial\rho/\partial T)$ in points $n_1, n_2, ..., n_{10}$, corresponding to the midpoints of the 10% segments. The values of the derivatives are selected, as has already been said, from the *Oceanological Tables* or are taken from graphs constructed from these tables;

Columns 8 and 9 — increments in σ_T due to salinity increment, $(\partial\sigma_T/\partial S)\Delta S$ and due to temperature increment, $(\partial\sigma_T/\partial T)\,dT$, corresponding to the increment in parameter m by 10%. Here $\Delta S = +0.16‰$, $\Delta T = +0.94°$;

Column 10 — total increment in density σ_T:

$$\Delta\sigma_T = \frac{\partial\sigma_T}{\partial S}\,dS + \frac{\partial\sigma_T}{\partial T}\,dT$$

corresponding to the increment in parameter m by 10% and determined as the sum of the numbers in columns 8 and 9;

Column 11 — values of σ_T in points $m_1, m_2, ..., m_{10}$, obtained from the formula:

$$\sigma_T = (\sigma_T)_B + \sum\Delta\sigma_T \qquad [35.9]$$

i.e., by an increasing summation from the bottom up. The value of $(\sigma_T)_B = 27.872$ is determined from the accurate tables of Ivanov-Frantskevich (1956). In addition, value $(\sigma_T)_M = 27.788$ appearing in the first line of column 11 in the numerator is determined from Ekman's (1910) tables as a control. The corresponding value of σ_T, determined from formula [35.9], appears in the same place in the denominator; we see that the error in the calculation from formula [35.9] amounted in the given case to only 0.005. We may point out that the accuracy of this calculation can be increased if the derivatives appearing in columns 6 and 7 are selected with an accuracy greater by one decimal place (the *Oceanological Tables* provide for such a possibility) or if the straight T-S line is broken down into a larger number of segments Δm.

What has been set forth, in particular, constitutes the theoretical foundation of the method of calculating the auxiliary tables for the construction of T-S diagrams (an example of such a table is represented in the Appendix by Table A2). In this case it is necessary to consider function $S(\rho,T)$, the differential of which equals:

$$dS = \frac{\partial S}{\partial\rho}\,d\rho + \frac{\partial S}{\partial T}\,dT \qquad [35.10]$$

At $dT = 0$ (the calculation of the supporting values of salinity, corresponding to whole values of density, is carried out for the given values of temperature, T = const.), the line integral from expression [35.10] turns into an ordinary integral along the line T = const.

$$S(b) = S(a) + \int_{\rho_a}^{\rho_b} \frac{\partial S}{\partial \rho}\, d\rho \qquad\qquad [35.11]$$

In the latter formula, we take $\partial S/\partial \rho = (\partial \rho/\partial S)^{-1}$; the numerical calculation from it is described in works of the author (Mamayev, 1954) and of Burkov et al. (1957).

36. THE REAL AND LINEARIZED ISOPYCNIC FIELDS. CONTRACTION ON MIXING

ı Let us consider two T-S diagrams: one real, the distribution of the isopycnals on which corresponds to the actual equation of state of sea water, and the other, idealized, with a linear field of isopycnals. Without prejudging for the time being the numerical characteristics of the linearized equation of state for the second T-S diagram, let us consider on these T-S diagrams two identical straight T-S lines, having the same coordinates, (i.e., coinciding with each other). Let density $\rho(S_A, T_A)$ in the initial point A (at the end of the straight T-S line) be identical on both T-S diagrams. Calculating the difference of densities between the points having identical coordinates on both straight T-S lines and the initial points, and designating these points N_1 (T_1, S_1) for the real T-S diagram and N_2 (T_2, S_2) for the linear T-S diagram, we obtain:
In the first case:

$$\rho(T_1, S_1) - \rho(T_A, S_A) = \int_{(AN_1)} \frac{\partial \rho}{\partial S}\, dS + \frac{\partial \rho}{\partial T}\, dT \qquad\qquad [36.1]$$

In the second case:

$$\rho(T_2, S_2) - \rho(T_A, S_A) = \int_{(AN_2)} \beta\, dS + \alpha\, dT \qquad\qquad [36.2]$$

In both formulae the line integral [35.1] appears on the right. In formula [36.1], obviously, $\partial \rho/\partial S = f_1(S, T)$; $\partial \rho/\partial T = f_2(S, T)$, while in formula [36.2] $\partial \rho/\partial S = \beta$ = constant, $\partial \rho/\partial T = \alpha$ = constant.

The difference of densities in points N_1 and N_2 equals:

$$\Delta \rho = \rho(T_1, S_1) - \rho(T_2, S_2) =$$

$$= \left[\int_{(AN_1)} \frac{\partial \rho}{\partial S}\, dS + \frac{\partial \rho}{\partial T}\, dT \right] - \left[\int_{(AN_2)} \beta\, dS + \alpha\, dT \right] \qquad\qquad [36.3]$$

or:

$$\Delta\rho = \int\limits_{(AN)} \left(\frac{\partial\rho}{\partial S} - \beta\right) dS - \left(\frac{\partial\rho}{\partial T} - \alpha\right) dT \qquad [36.4]$$

The latter formula expresses the quantity of contraction on mixing of two water masses and shows that this effect is analytically determined by the difference of the derivatives of thermal expansion and saline contraction for non-linear and linear equations of state of sea water. Thus, contraction on mixing represents a basic characteristic of the non-linear (real) equation of state of sea water.

The second term of the right-hand part of expression [36.3] corresponds, generally speaking, to the innumerable quantity of equations of state, since, for the calculation of contraction on mixing, it is enough to know the density only in the end points of the straight T-S line (the straight line of mixing), and, moreover, these densities coincide for the non-linear and linear fields of isopycnals only in the two points mentioned. The inclination of the linear isopycnals may, generally speaking, be arbitrary. However, the values of coefficients α and β cannot be mutually independent; indeed, inasmuch as:

$$\frac{\partial\rho}{\partial T} = \frac{\partial\rho}{\partial S}\frac{dS}{dT} \qquad \text{or} \qquad \alpha = \beta\frac{dS}{dT}$$

where dS/dT is the cotangent of the angle of slope of the isopycnals in the linear field, then, given one of them, we can determine the second only if the slope of the linear isopycnals is known. In other words, a definite linear field of the isopycnals can be constructed only if the values of density are known in three points.

It follows from what has been said that the calculation of contraction on mixing of sea waters places in correspondence to the real field of isopycnals every time a family of linearized fields of isopycnals, determined by the coincidence of densities at the ends of the straight lines of mixing of both fields.

Formula [36.3] or [36.4] makes it possible to carry out the calculation of contraction on mixing with great accuracy; this calculation is illustrated in Table VI considered above. The values of the first integral in the right-hand part of formula [36.3] summed up with the constant of integration $(\sigma_T)_B = 27.872$, are given, as has already been pointed out, in column 11.

The values of the second integral in the right-hand side of formula [36.3] can be obtained from the following considerations. Let us write the integrand

$$d\rho = \left(\frac{\partial\rho}{\partial S}\right)_T dS + \left(\frac{\partial\rho}{\partial T}\right)_S dT$$

whence, having divided it by dS, we obtain:

$$\left(\frac{\partial \rho}{\partial S}\right)_T = \frac{\mathrm{d}\rho}{\mathrm{d}S} - \left(\frac{\partial \rho}{\partial T}\right)_S \frac{\mathrm{d}T}{\mathrm{d}S} \qquad\qquad [36.5]$$

or:

$$\beta = \frac{\mathrm{d}\rho}{\mathrm{d}S} - \alpha \frac{\mathrm{d}T}{\mathrm{d}S} \qquad\qquad [36.6]$$

Hence, in the finite differences:

$$\beta = \frac{\Delta \rho - \alpha \Delta T}{\Delta S} \qquad\qquad [36.7]$$

The final increments of density, salinity and temperature can, in the case of a linear field of isopycnals, be taken on the ends of the straight line of mixing; however, we shall take them 10 times smaller, following the procedure for the calculation of the line integral, determined by the first term of the right-hand part of equation [36.3]. Then:

$$\frac{\Delta \rho}{\Delta l} = \frac{\partial \rho}{\partial S} \frac{\Delta S}{\Delta l} + \frac{\partial \rho}{\partial T} \frac{\Delta T}{\Delta l} \qquad\qquad [36.8]$$

where $\Delta l = 10$ (the straight line of mixing l broken into ten parts). For the numerical example, considered in the preceding paragraph, the integrand of the second integral in the right-hand part of formula [30.3] can be written in the following form:

$$\frac{(\sigma_T)_M - (\sigma_T)_B}{10} = \frac{\partial \sigma_T}{\partial S} \frac{S_M - S_B}{10} + \frac{\partial \sigma_T}{\partial T} \frac{T_M - T_B}{10} \qquad\qquad [36.9]$$

where $(\sigma_T)_M = 27.788$; $(\sigma_T)_B = 27.872$, or, numerically:

$-0.0084 = \beta \times 0.16 + \alpha \times 0.94$

Let us set the value of α arbitrarily, in agreement with what has been said above, namely: let $\alpha = -0.2$. Then:

$$\beta = \frac{-0.0082 + 0.188}{0.16} = 1.1225$$

The increment of $\bar{\sigma}$ in a linear field of isopycnals will be determined by the expression:

$\Delta \bar{\sigma} = \beta \Delta S + \alpha \Delta T$

or, numerically:

$\Delta \bar{\sigma} = 1.1225 \, \Delta S + (-0.2) \Delta T$

Accordingly:

$$\bar{\sigma} = \sigma_B + \sum \Delta \bar{\sigma}$$

The values of $\bar{\sigma}$ are given in column *12* of Table VI; let us note that in practice it is simpler to calculate it from the formula of mixing analogous to formula [30.1] (Zubov and Sabinin, 1958).

Finally, in column *13* are given the values of contraction on mixing $\sigma_T - \bar{\sigma}$
obtained as the difference between the numbers appearing in columns *11* and
12. We see that the values of contraction are obtained with an accuracy to the
third decimal place of σ_T and, if so wished, could be obtained with an accuracy
to the fourth decimal place.

The cotangent of the angle of inclination of the linear isopycnals, corresponding to the example considered, equals $(\Delta S/\Delta T) = 0.178$.

Formula [36.3] is cumbersome in calculating contraction, let us say, in one
specific case; however, for a series of calculations a graph may be constructed
for the calculation of the integrand of the first term of formula [36.3], corresponding to the real field of isopycnals. This facility can be achieved if it is
necessary to carry out calculations of contraction on mixing along different
straight lines of mixing, lying in different coordinates, which are, however,
parallel to each other. Such cases are often encountered in practice, and, as
an example, one may cite the straight T-S lines of what Sverdrup calls the
Central Water Masses of the World Ocean. These straight lines, constructed
from the data of table 89 (Sverdrup et al., 1942), are represented in Fig. 29.
We see that many of these straight T-S lines are parallel to each other; thus,
the cotangents of the angles of inclination to the abscissa axis for the Central
Water Masses of the North Atlantic, South Atlantic, Indian Ocean, the Western
South Pacific and the Eastern North Pacific Ocean amount to approximately
0.125; 0.125, 0.121; 0.121 and 0.127 respectively. Some of these T-S relations,
as may be seen from Fig. 29, are curvilinear.

Therefore, the integrand of the line integral:

$$\Delta \rho = \frac{\partial \rho}{\partial S} \Delta S + \frac{\partial \rho}{\partial T} \Delta T \qquad\qquad [36.10]$$

$\Delta S = 0.1^o/_{oo}$ and $\Delta T = 0.8°$, which corresponds to the North and South
Atlantic, will be suitable also with adequate degree of accuracy for the
three other Central Water Masses enumerated. Table VII gives the
values of function [36.10] with $\Delta S = 0.1‰$ and $\Delta T = 0.8°$ (as above, the
straight line of mixing is broken down into 10 parts in the calculation) for
the different values of salinity and temperature in the range comprising the
Central Water Masses; the same values are represented in Fig. 29 in the form
of a field of isolines. We see from this figure that the quantity of the line

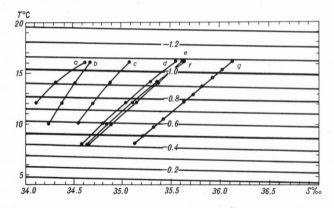

Fig. 29. Graph of function $10^4 [(\partial\rho/\partial S)\, \Delta S + (\partial\rho/\partial T)\, dT]$ at $\Delta S = 0.1\permil$, $\Delta T = 0.8°$. T-S curves for the Central (tropospheric) Water masses are plotted on the graph. a = Eastern North Pacific; b = Western North Pacific; c = Eastern South Pacific; d = Western South Pacific; e = South Atlantic; f = Indian Ocean; g = North Atlantic.

integral at the increments of temperature and salinity taken depends very little on a change in salinity.

Plotting in Fig. 29 any straight T-S line parallel and equal in length to the straight T-S line for the North Atlantic Central Water (the latter is shown in the figure), we can immediately determine the value of the function:

$$10^4\, \Delta\rho = \left(\frac{\partial\rho}{\partial S} \cdot 0.1 + \frac{\partial\rho}{\partial T} \cdot 0.8\right) \cdot 10^4$$

in any point of the straight T-S line. If the length of the straight T-S line is different from the length of the base straight T-S line, which was used for the construction of the graph, then:

TABLE VII

Function $10^4 \Delta\rho(S, T) = 10^4 [(\partial\rho/\partial S)\, \Delta S + (\partial\rho/\partial T)\, \Delta T]$ with $\Delta S = 0.1\permil$, $\Delta T = 0.8°$

$T\,(°C)$ \ $S\,(\permil)$	33	34	35	36	37
5	−0.1036	−0.1247	−0.1450	−0.1652	−0.1854
6	−0.1998	−0.2201	−0.2404	−0.2598	−0.2792
8	−0.3848	−0.4035	−0.4221	−0.4407	−0.4593
10	−0.5598	−0.5769	−0.5939	−0.6117	−0.6287
12	−0.7258	−0.7312	−0.7575	−0.7736	−0.7898
14	−0.8834	−0.8989	−0.9135	−0.9280	−0.9434
16	−1.0352	−1.0490	−1.0628	−1.0765	−1.0902
18	−1.1810	−1.1940	−1.2062	−1.2191	−1.2312
20	−1.3226	−1.3340	−1.3453	−1.3567	−1.3680

$$10^4 \, \Delta\rho = 10^4 \, k\left(\frac{\partial\rho}{\partial S} \, \Delta S + \frac{\partial\rho}{\partial T} \, \Delta T\right)$$

where k is the ratio of the length of the given straight T-S line to the base straight T-S line of the North Atlantic Central Water.

Let us now consider the question of the correlation of the real and linearized fields of isopycnals on the whole T-S diagram. It is known that in the linearization of the equation of state for various analytical purposes we usually take the following mean values of the coefficients of thermal expansion and saline contraction: $\alpha = -2 \cdot 10^{-4}$ and $\beta = 8 \cdot 10^{-4}$ (Hesselberg and Sverdrup, 1915a). Therefore, it is interesting to consider the function:

$$10^4 \, \Delta\rho(S, T) = 10^4 \left[\left(\frac{\partial\rho}{\partial S} \, \Delta S + \frac{\partial\rho}{\partial T} \, \Delta T\right) - (\beta\Delta S + \alpha\Delta T)\right] \qquad [36.11]$$

with the values mentioned of coefficients α and β and where $\Delta S = 1\%_0$ and $\Delta T = 1°$ (the latter values of the increments in salinity and temperature are taken for the sake of uniformity). The values of function [36.11] are given in Table A7 of the Appendix and also represented in the T-S plane in Fig. 30. Considering this figure, we see that it well reflects the relation between the real (non-linear) and the linearized equations of state of sea water at atmospheric pressure. In particular, it constitutes a graph of the errors arising from the replacement of the real equation of state by the linear; these errors are different in the different ranges of temperatures and salinities. If we take as the maximum admissible error, let us say, the value $\delta\rho = 5 \cdot 10^{-5}$, then it may be seen from the figure that the area of admissible replacement of one

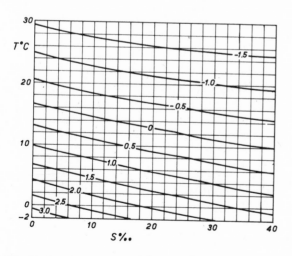

Fig. 30. Graph of function: $10^4 \, \Delta\rho = 10^4 \, [\,\{(\partial\rho/\partial S) \, \Delta S + (\partial\rho/\partial T) \, \Delta T\} - (\beta\Delta S + \alpha\Delta T)]$ with $\beta = 8 \cdot 10^{-4}$; $\alpha = -2 \cdot 10^{-4}$; $\Delta S = 1\%_0$; $\Delta T = 1°C$.

equation by the other extends on the *T-S* diagram in a band lying at 0‰ between 13 and 20° and extending to the limits of 6–15°C at salinity of 40‰.

Fig. 30 is interesting also for the further theoretical study of the phenomenon of contraction on mixing of sea waters.

37. THE TRANSFORMATION OF COORDINATES ON THE *T-S* PLANE

The question of the transformation of *T-S* coordinates arises first of all because the various families of curves in the *T-S* plane, representing the different functions of the equation of state of sea water, form distinctive curvilinear coordinates. The transformation of coordinates can prove highly useful from the theoretical and practical points of view; this question, however, has practically not been investigated. Here we shall confine ourselves, in concluding the consideration of the basic properties of the *T-S* diagram, only to formulating this question in the most preliminary way.

Revolving of the axes of the coordinates

If S' and T' are the new coordinates, while S and T are the old coordinates, the correlation between them is determined by the formulae known from analytical geometry:

$$S' = S \cos \alpha + T \sin \alpha$$
$$T' = -S \sin \alpha + T \cos \alpha$$

[37.1]

where α is the angle of the revolution (counter-clockwise). Formulae [37.1] are valid given an invariable scale.

Revolving of the axes of coordinates is considered by Proudman (1953, section 74) in connection with the question of determining the coefficient of mixing from *T-S* curves (see also Section 47). Proudman introduces the dimensionless coordinates S/S_0 and T/T_0; therefore for the abscissa axis a new dimensionless function arises:

$$F = \frac{S}{S_0} \cos \alpha + \frac{T}{T_0} \sin \alpha$$

[37.2]

At α = constant, F represents a linear combination of S and T (something in the nature of an analog of density), plotted along one of the axes of the new system of coordinates. If in the *S-T* plane a *T-S-z* curve is constructed, for example, then the second axis of the new system of coordinates can, for example, be graduated by parameter z; it is natural that in constructing a regular scale for z in the new system of coordinates a distortion – stretching of the original *T-S* curve is necessary. Revolving the axes of the coordinates enabled Proudman to come up with a derivation of Jacobsen's formula (1927) simpler than the derivation of Jacobsen himself and than Okada's derivation considered in Section 47.

The transformation of curvilinear coordinates

Let us consider a plane related to the dimensionless * coordinates S, T, and another plane, related to certain, also dimensionless, coordinates ζ and η. The transformation of the S-T plane into the ζ-η plane is possible; moreover, if in the plane S-T the system of functions is given as:

$$\zeta = \zeta(S, T)$$
$$\eta = \eta(S, T)$$

[37.3]

then in the plane ζ-η it corresponds to the system of functions:

$$S = S(\zeta, \eta)$$
$$T = T(\zeta, \eta)$$

[37.4]

This transformation is possible if a one-to-one correspondence exists of the areas considered in both planes (see Fikhtengolts, 1956). Moreover, as is known from calculus, the condition must be fulfilled:

$$\frac{D(S, T)}{D(\zeta, \eta)} \frac{D(\zeta, \eta)}{D(S, T)} = 1$$

[37.5]

where the Jacobian:

$$J(\zeta, \eta) = \frac{D(S, T)}{D(\zeta, \eta)} = \begin{vmatrix} \dfrac{\partial S}{\partial \zeta} & \dfrac{\partial S}{\partial \eta} \\ \dfrac{\partial T}{\partial \zeta} & \dfrac{\partial T}{\partial \eta} \end{vmatrix} = \frac{\partial S}{\partial \zeta} \frac{\partial T}{\partial \eta} - \frac{\partial S}{\partial \eta} \frac{\partial T}{\partial \zeta}$$

[37.6]

and its inverse Jacobian $J(S, T)$ should neither of them turn into zero.

In the plane S-T one can consider a whole series of families of functions of type [37.3], for example:

$$\rho = \rho(S, T)$$
$$\gamma = \gamma(S, T)$$

[37.7]

---

* See footnote on p. 130.

or:

$$\frac{\partial \rho}{\partial S} = \zeta(S, T)$$

$$\frac{\partial \rho}{\partial T} = \eta(S, T)$$

[37.8]

(γ is the function of "density flux" — see Section 33). Let us consider the system [37.7] and let us calculate the Jacobian:

$$J(S, T) = \frac{D(\rho, \gamma)}{D(S, T)} = \frac{\partial \rho}{\partial S}\frac{\partial \gamma}{\partial T} - \frac{\partial \rho}{\partial T}\frac{\partial \gamma}{\partial S}$$

[37.9]

On the basis of the Cauchy-Riemann equations [33.4] (on the right) for the plane *S-T* we have:

$$J(S, T) = \left(\frac{\partial \rho}{\partial S}\right)^2 + \left(\frac{\partial \rho}{\partial T}\right)^2 \neq 0$$

[37.10]

Consequently, the system of functions [37.7] transforms the *S-T* plane into the ρ-γ plane, on which are represented two families of curves:

$$S = S(\rho, \gamma)$$

$$T = T(\rho, \gamma)$$

[37.11]

To the system [37.8], represented in the *S-T* plane (these families of curves relating, however, to specific volume, are represented in Figs. 14 and 15), there corresponds a system:

$$S = S\left(\frac{\partial \rho}{\partial S}, \frac{\partial \rho}{\partial T}\right)$$

$$T = T\left(\frac{\partial \rho}{\partial S}, \frac{\partial \rho}{\partial T}\right)$$

[37.12]

On the basis of formulae [33.16] we may say that the system of functions [37.12] transforms the plane *S-T* into a plane analogous to the *hodograph plane*, since the components of the "density flux" $\partial \rho/\partial S$ and $-(\partial \rho/\partial T)$ are analogous to the components of velocity u and v (Section 33).

It is interesting to consider another transformation to the "hodograph plane" carried out by means of the independent variables q and θ — of the "vector of density flux", tangent to the isolines γ, and of the angle of its inclination to the abscissa axis, determined respectively by the formulae:

$$u = q \cos \theta$$

$$v = q \sin \theta$$

[37.13]

In the plane *S-T* to formulae [37.13] there correspond formulae:

$$\frac{\partial \gamma}{\partial T} = q \cos \theta$$

$$\frac{\partial \gamma}{\partial S} = q \sin \theta$$

[37.14]

The modulus of vector *q* is determined by the expression:

$$q = \sqrt{\left(\frac{\partial \rho}{\partial S}\right)^2 + \left(\frac{\partial \rho}{\partial T}\right)^2}$$

[37.15]

and its direction by formula:

$$\tan \theta = \frac{\partial T}{\partial S}$$

[37.16]

moreover:

$$\tan \theta = -\cot \mu = \frac{\partial S}{\partial T}$$

where $\partial S / \partial T$ is the thermohaline derivative (Section 25).

Thus, the thermohaline derivative serves as one of the independent variables in the plane analogous to the hodograph plane; an idea of the latter — the modulus of the "density flux" *q* — is given by Table VIII. This table is calculated from formula [37.15] with the help of tables 20 and 24 of the *Oceanological Tables* of Zubov (1957a).

TABLE VIII

Function $10^7 q = f(S, T)$

$T\,(°C)$ \ $S\,(°/_{00})$	0	10	20	30	40
0	8174	8079	8037	8051	8116
10	7876	7854	7873	7936	8040
20	7890	7892	7929	8002	8113
30	8047	8068	8116	8192	8294

We see that function *q*, unlike function θ (or μ), undergoes very small changes in the ranges $0 < T < 30°$ and $0 < S < 40°/_{00}$.

The relation between coordinates *x* and *y*, the velocity potential and the stream function φ and ψ, as well as variables *q* and θ is established from formulae known in hydro- and aero-mechanics; by analogy formulae can also be constructed for the S-T plane allowing for the differences in scales of salinity and temperature.

As has already been said, the transformation of coordinates can prove useful in the analytical study of T-S relations. Thus, the straight T-S line is represented in the q-θ plane by a smooth curve (it is transformed into a curve); on the contrary, some portions of the T-S curves can be straightened out with appropriate transformation of the coordinates.

Closely connected with this question is the question of the calculation of T-S areas. The calculation of areas on the T-S diagram is necessary, for example, in statistical T-S analysis (Section 58) and in some other cases.

In a dimensionless T-S plane the formula occurs:

$$D = \tfrac{1}{2} \int_{(L)} S\,dT - T\,dS \qquad\qquad [37.17]$$

which expresses area with the help of the line integral along the contour (L) of the T-S area. To use this formula it is necessary to know the function $T = f(S)$ on the smooth (or intermittently smooth) contour of the area.

The corresponding formula for the calculation of the area of a closed region Δ in the ζ-η plane has the form:

$$D = \iint_{(\Delta)} J(\zeta, \eta)\,d\zeta\,d\eta \qquad\qquad [37.18]$$

where $J(\zeta, \eta)$ is the Jacobian determined by formula [37.6].

38. THE FUNCTION OF THE DENSITY FLUX IN OCEANOGRAPHIC T-S ANALYSIS

The use of the function of density flux γ in the practice of oceanographic analysis has also been considered by Veronis (1972); to make such use possible, it is necessary to determine function γ numerically, which was done by the author mentioned. Let us consider briefly some of Veronis' results — the numerical calculation of function γ as well as its possible oceanographic interpretation *.

---

* Concerning Veronis' article (1972) see (Mamayev, 1973b). Veronis designates function γ by the letter τ.

The numerical representation of function γ

Function γ can be calculated with accuracy to γ_0 by direct integration according to formula [33.17], if the analytical expression for density ρ and its derivatives is known. Veronis did not make use of this direct method, but used the finite-difference method of calculation. In spite of the disadvantageous method chosen by Veronis, we shall set it forth here, since it represents the first attempt to calculate function γ.

The differential form of formula [33.17] for a dimensionless *T-S* diagram is as follows:

$$d\gamma = -\frac{\partial \rho}{\partial T}\, dS + \frac{\partial \rho}{\partial S}\, dT \qquad [38.1]$$

Going back to the dimensional coordinates T' and S' (see footnote on p. 130) and dropping the strokes, we have:

$$d\gamma = -\frac{\partial \rho}{\partial T}\frac{T_0}{S_0}\, dS + \frac{\partial \rho}{\partial S}\frac{S_0}{T_0}\, dT \qquad [38.2]$$

This is the formula used by Veronis for the determination of γ (instead of T_0 and S_0 he writes ΔT and ΔS).

Here it should be emphasized, following Veronis, that function γ, the isolines of which on the *T-S* diagram are orthogonal to the isopycnals, is not single-valued and depends on the choice of scales for temperature and salinity, in other words, on the arbitrarily given quantity $\Delta T/\Delta S$. Expressing further formula [33.6] in finite differences, determining from it the value of ΔT (we replace T_0 and S_0 by ΔT and ΔS) and substituting it in [38.2] we obtain:

$$d\gamma = -\left[\frac{\partial \rho}{\partial T}\frac{\Delta T}{\Delta S} + \frac{(\partial \rho/\partial S)^2}{\partial \rho/\partial T}\right]dS \qquad [38.3]$$

Finally, solving [38.3] for dS, Veronis obtains the following formula for calculations:

$$dS = -\left(\frac{\partial \rho}{\partial T}\right)\frac{d\gamma}{\dfrac{\Delta T}{\Delta S}\left(\dfrac{\partial \rho}{\partial T}\right)^2 + \dfrac{\Delta S}{\Delta T}\left(\dfrac{\partial \rho}{\partial S}\right)^2} \qquad [38.4]$$

The calculation of the field of function γ according to this formula is as follows: setting an arbitrary (convenient) value of γ in a certain point of the *T-S* plane and setting increments $\delta\gamma$ along the curve ρ = const., passing through this point, from formula [38.4] we can determine the increments δS, while from formula [33.6], written in finite differences, the corresponding increments δT. By this method the distribution of function γ along one isopycnal is obtained; using the corresponding values of γ, taken at equal intervals, from formula [33.8] and the formula:

$$\frac{\partial \gamma / \partial S}{\partial \gamma / \partial T} \frac{\Delta S}{\Delta T} = -\frac{\partial \rho / \partial T}{\partial \rho / \partial S} \frac{\Delta T}{\Delta S} \qquad\qquad [38.5]$$

we can construct isolines orthogonal to the curves ρ = constant *.

Veronis carried out a calculation of function γ in the ranges $0 \leqslant T \leqslant 30°C$ and $33\%_{oo} \leqslant S \leqslant 37\%_{oo}$, setting as an arbitrary value $\gamma = 5$ in the point: $T = 9.42°C; S = 35\%_{oo}$ ($\sigma_\theta = 27.07$) and as a ratio of scales $\Delta T/\Delta S = 5$ (0.1°C corresponds to 0.02‰). The corresponding *T-S* diagram is represented in Fig. 31 **. The function γ determined by this method changes in the ranges

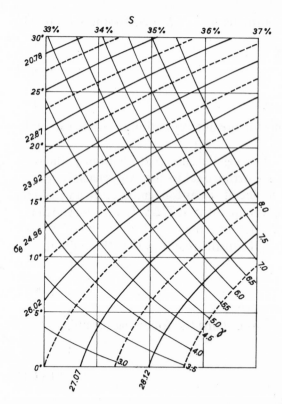

Fig. 31. Numerical representation of a family of curves of function γ, orthogonal to the family of isopycnals, in the oceanic range of temperatures and salinities (Veronis, 1972). Compare with Fig. 28.

---

* Formula [38.5] follows directly from the condition of orthogonality [33.10] and eqs.[33.2] and [33.3].
** In this work Veronis uses potential temperature θ instead of T and potential conventional density σ_θ instead of σ_T, which has substantial oceanographic justification. We shall take up these quantities below.

of temperature and salinity mentioned from 2.5 to 10.0 g l^{-1} (this is seen from the figure), which corresponds to the range of changes in density.

Having calculated by the method described above 100 values of the function for the region of values of temperature and salinity represented in Fig. 31, Veronis approximated this distribution with the help of the least squares method by the following polynomial of the 5th degree:

$$\gamma = \sum_{i=1}^{6} \sum_{j=1}^{6} A_{ij}(\theta - 10°)^{i-1}(S - 35)^{j-1} \qquad [38.6]$$

which is convenient for computer calculations and whose accuracy with respect to the data calculated amounts to 0.002 unit of γ. The coefficients A_{ij} of the polynomial are presented in Table A9 of the Appendix.

Oceanographic application

Let us consider as an example the distribution of function γ, and also of potential conventional density σ_θ on a meridian section in the western part of the Atlantic Ocean (Fig. 32). In the upper kilometer layer of the ocean approximately between latitudes 40°S and 30°N the isolines of γ and σ_θ are roughly parallel; lower down they intersect at a greater or smaller angle. Veronis interprets the parallelism of the curves of γ and σ_θ in the section as

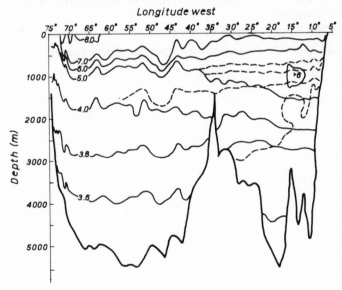

Fig. 32. The distribution of function γ (solid lines) and of potential conventional density σ_θ (broken lines) in a meridional section in the western part of the Atlantic Ocean. In those parts of the section where broken lines are not shown, both families of lines are approximately parallel (Veronis, 1972)

a possible indication of intensive horizontal mixing of water masses, and their intersection as evidence of vertical mixing. Let us add that it is more convenient to speak of the change of one of the functions against the invariable value of the other and vice versa. Indeed, the most general consideration of the *T-S* diagram with the orthogonal families of curves (Fig. 31) leads to the conclusion that a change in function γ along any isopycnal bears witness to the isopycnic process, while a change in σ_θ along any isoline γ testifies to mixing across isopycnic surfaces, i.e., to vertical mixing (or the "density flux" in a stratified ocean, which was mentioned above in Section 33). Turning again to the oceanographic section, we can notice that the greatest change in function γ occurs along the isopycnal $\sigma_\theta = 27.4$. It is known (Section 62) that the transformation of the core of the Antarctic Intermediate Water takes place approximately along just this isopycnic surface due to vertical mixing. This vertical mixing proceeds primarily downward from the core mentioned, which is borne out by the change in density along any of the isolines $\gamma = 3.4$; 3.6; 3.8 and 4.0 — in those parts of them which lie below the isoline $\sigma_\theta = 27.4$ (Fig. 32).

The primarily vertical mixing of the intermediate water mass with the lower water mass (as compared with the upper) was pointed out by the author of the present work earlier on the basis of other considerations (see Sections 55 and 63).

Confining ourselves to this one example, we may agree with Veronis' opinion that the function γ can serve as a yardstick for the mixing of water masses — at least for isopycnic mixing.

With this we conclude the consideration of the basic general properties of the *T-S* diagram and proceed to the systematic exposition of the analytical theories of *T-S* curves.

CHAPTER 6

ANALYTICAL THEORIES OF T-S CURVES

39. FORMULATING THE THEORETICAL PROBLEMS

Equations

The fundamental problem of the analytical theory of T-S curves is the simultaneous solution of the equations which determine the spatial distribution of temperature and salinity, the representation of the results obtained on the T-S diagram and the geometric and analytical investigation of the functional relationships between temperature and salinity so obtained. The foregoing can immediately be made clear in the following way. As a result of the solution of the equations, let the analytical expressions have become known to us which determine the nature of the distribution by depth and in time of temperature and salinity, let us say, in some investigated point of the ocean:

$$T = f(z, t) \qquad S = \varphi(z, t) \qquad\qquad [39.1]$$

In these expressions z (depth) and t (time) represent, as has already been said above, parameters (see Section 31). The simultaneous representation of the results of [39.1] on the T-S plane or, in other words, the transition from the T-z plane and the S-z plane (with t = const.) to the T-S plane presupposes, obviously, obtaining relation:

$$T = \psi(S) \qquad\qquad [39.2]$$

Obtaining the analytical type of expression [39.2] presupposes, obviously, the elimination of parameters z and t. Thus obtaining functional relation [39.2], as well as its investigation and further practical application constitutes the fundamental problem of the analytical theory of T-S curves.

Corresponding equations in the theory of T-S curves prove to be the equations of turbulent heat conduction:

$$\frac{dT}{dt} = k_T \Delta T \qquad\qquad [39.3]$$

and of turbulent salt diffusion:

$$\frac{dS}{dt} = k_S \Delta S \tag{39.4}$$

or rather, their simplified versions; in expressions [39.3] and [39.4]

$$d/dt \equiv \partial/\partial T + u(\partial/\partial x) + v(\partial/\partial y) + w(\partial/\partial z)$$

is the differential Euler operator, $\Delta \equiv \partial^2/\partial x^2 + \partial^2/\partial y^2 + \partial^2/\partial z^2$ is the Laplace operator, k_T and k_S are the coefficients of turbulent heat conduction and turbulent salinity diffusion. It is understood that in notations [39.3] and [39.4] these coefficients are assumed to be identical in the direction of all three axes: $k_{T,x} = k_{T,y} = k_{T,z}$ and $k_{S,x} = k_{S,y} = k_{S,z}$. In addition, $[k_T]$ and $[k_S] = cm^2/sec$.

The following simplified equations of heat conduction and diffusion under-lay the theory of *T-S* curves developed by Shtokman (1938, 1939, 1943a,b, 1944, 1946) and by Ivanov (1943, 1944, 1946, 1949):

$$\frac{\partial T}{\partial t} = k_T \frac{\partial^2 T}{\partial z^2}$$

$$\tag{39.5}$$

$$\frac{\partial S}{\partial t} = k_S \frac{\partial^2 S}{\partial z^2}$$

Each of these equations is identical to the *fundamental equation of heat conduction for a uniform rod* — to the uniform equation (the rod without heat sources) which has been thoroughly studied in the theory of equations of mathematical physics. In accordance with this circumstance, the theory of *T-S* curves is constructed mathematically by analogy with the theory of propagation of heat in rods.

In the theory of *T-S* curves equations [39.5] make it possible to study the non-stationary (i.e., changing in time) distribution of temperature and salinity along the vertical in any point of the sea with subsequent interpretation of the solution in the *T-S* plane, as well as the stationary distribution of temperature and salinity in the two-dimensional case — by depth and along axis x, provided that the entire mass of liquid possesses a constant value of velocity along this axis (Hirano, 1957; Shtokman, 1962). Indeed, the equations which correspond to this case:

$$u \frac{\partial T}{\partial x} = k_T \frac{\partial^2 T}{\partial z^2}$$

$$\tag{39.6}$$

$$u \frac{\partial S}{\partial x} = k_S \frac{\partial^2 S}{\partial z^2}$$

by the substitution:

$$x = \frac{u}{k} \xi$$

are reduced to the equations:

$$\frac{\partial T}{\partial \xi} = \frac{\partial^2 T}{\partial z^2}$$

$$\frac{\partial S}{\partial \xi} = \frac{\partial^2 S}{\partial z^2}$$

[39.7]

identical to equations [39.5].

The simultaneous solution of equations [39.6], given a more general assumption concerning velocity u, namely when the nature of its change with depth is given, represents, as it were, an independent ramification of the theory of T-S curves, possessing a certain value and some advantages (for example, from the point of view of simplification of solutions in studying the transformation of water masses in a sea of finite depth). In this direction almost nothing has been done, aside from the work of the author (Mamayev, 1962a), devoted to the solution of simplified equations:

$$\frac{\partial^2 T}{\partial z^2} = a$$

$$\frac{\partial^2 S}{\partial z^2} = b$$

[39.8]

(the left-hand parts of equations [39.6] are taken as constant) for the purpose of carrying out T-S analysis of moving water masses limited in the vertical direction. The results obtained in the work mentioned will be set forth below in Section 45.

Finally, the problem may be raised of the simultaneous solution of Laplace's equation for temperature and salinity:

$$\Delta T = 0 \qquad \Delta S = 0$$ [39.9]

(here Δ is also a three-dimensional Laplacian); the results of the interpretation of solutions in the T-S plane must in this case form the analytical basis for the method of *volumetric T-S analysis*, which has become current in recent years. The methods of practical volumetric T-S analysis will be set forth below (Section 58); so far as what has been said above with respect to equations [39.9] is concerned, no steps have been undertaken in this direction to our knowledge; this is a third ramification of the theory of T-S curves but nonetheless a most promising one.

It is clear that the development of the analytical theory of T-S curves is not limited to the possibilities of using the equations of heat conduction and diffusion [39.3] and [39.4] or any of their simplified versions. In principle, it is possible to construct and solve any equations describing the spatial and temporal distribution of temperature and salinity with subsequent analysis on the *T-S* plane; there is no doubt that all such results must also be included within analytical theories of *T-S* curves (or, speaking more generally, of *T-S* relationships). Such results already exist, and as an example (perhaps the only one) we may refer to the very promising work of Stommel devoted to the analysis on the *T-S* plane of the processes of thermohaline convection (Stommel, 1961). Stommel's work deserves detailed scrutinity, which it will receive in Section 46; here we will point out that Stommel subjects to *T-S* analysis the result of the solution, for example, of both the following equations (in them T and S are certain constant values of temperatures and salinity, while c and d are the coefficients of heat and salt exchange):

$$\frac{dT}{dt} = c(T - T)$$

$$\frac{dS}{dt} = d(S - S)$$

[39.10]

and of some others, more complex. One cannot fail to conclude that these and many other problems make the analytical theory of *T-S* curves an even more powerful tool of oceanographic analysis than has been thought up to now.

Initial conditions

Our main attention in this chapter will be devoted to the analysis of the solutions of equations [39.5], since it is precisely in this direction that the theory of *T-S* curves has achieved its greatest successes up to the present time. The initial and boundary conditions will be constructed accordingly.

Continuing to use the identical methods for solving problems of the theory of *T-S* curves and problems of mathematical physics, we must bring the problems of the transformation of water masses in oceans of infinite, semi-infinite and finite depth into correspondence with the problems of the propagation of heat in rods of infinite, semi-infinite and finite length. Greatest attention will be paid to *T-S* analysis in infinite and semi-infinite oceans — for reasons which will become clear below. In exactly the same way, the advantage of considering a semi-infinite ocean as compared with an ocean of infinite depth will also be emphasized in the subsequent exposition. Here we will formulate the initial conditions for solving equations [39.5].

The initial (when $t = 0$) distribution by depth of temperature and salinity is established in the form of the existence of parallel plane layers possessing

constant (initial) values of temperature and salinity. Underlying this condition
is the idea, understandable for oceanographic reasons, that the waters of the
World Ocean in any of its points have been formed as a result of the interaction of the original, more or less homogeneous water masses, the appearance
of which is harnessed to the main worldwide sources — the regions of formation of water masses. Here, however, it is necessary to make a qualification:
in spite of the obviousness of this idea, it is by no means possible to affirm
that water masses exist in the ocean in "pure" form, i.e., such water masses
as would be characterized by constant values of temperature and salinity
along the vertical within the limits of a layer more or less substantial in depth.
Such an idea is valid to a certain extent for the bottom water masses: at the
bottom of the ocean vertical homothermy and homohalinity are often encountered within the limits of a more or less considerable thickness. So far as,
say, intermediate waters are concerned, we will hardly find any such homothermy and homohalinity in any significant zone of the ocean (along the
vertical). As an example, Mediterranean water flowing into the Atlantic Ocean
as an undercurrent over the sill of the Strait of Gibraltar has more or less
homogeneous characteristics over this sill. However, vertical gradients of temperature and salinity already exist in the Strait as well, not even to mention
the regions adjacent to the Strait. Nowhere do we observe homogeneousness
along the vertical, and a striking indication of the presence of Mediterranean
Intermediate Water manifests itself correspondingly in a marked extreme of
salinity. However, this extreme in no way upsets the continuity of the curve
of vertical distribution of salinity. A kind of "initial paradox" appears; we
can indicate almost exactly the characteristics, i.e., the constant values of
temperature and salinity of Mediterranean water in situ, at the time of its
origin, as it were; we can also indicate, fairly exactly as well, the thickness of
this homogeneous water mass; however, nowhere can we join these indicators
together by establishing the initial presence of a parallel plane original layer.
The primary water mass slips out of our hands, as it were, and the only thing
that we can do is to surmise the existence of this initial, homogeneous water
mass, somewhere beyond the limits of the region accessible to us: it might
quite well be precisely there, and might quite well lead to that vertical distribution of characteristics which we observe, say, immediately at the outlet
from the Strait of Gibraltar, to return again to this example.

Shtokman (1943b) substantiated even better than has been done above
a theory of T-S curves in which the initial picture took the form of a series of
superimposed parallel plane water masses, and a quotation from his work
therefore seems to us to be in order. He writes: "It should ... be emphasized
that the highly useful concept for oceanography of water masses as some
limited volumes of water possessing individual properties before mixing (this
concept underlies the whole development of the theory of T-S curves) is to a

considerable extent a *conventional* concept. Indeed, an exact reproduction of the concept of water masses could be formed by different layers of a fluid which we had *originally* placed one over the other in some vessel, observing thereafter the gradual equalization of the properties of these layers as a result of mixing and interpreting this process by *T-S* curves. It is not difficult, however, to realize that we could attain a distribution of temperature and salinity of the water identical with the first case also if our vessel originally contained water not divided into layers and possessing identical properties at any point within the vessel. This may be achieved by changing the properties of the water by means of external influences, subjecting the water on the surface to heating, cooling, desalinization or salinization (as a result of deposits or evaporation on the surface of the liquid). As a result of the action of the external factors enumerated, along with the processes of mixing in our vessel, we can achieve such a vertical distribution of temperature as would correspond exactly to the first case (the mixing of several layers of fluid which had been individual in nature at the original moment in time), although in the latter case such layers (water masses) did not in fact exist at the initial moment in time." And further: "... splitting the water of the ocean up into different water masses, we are speaking in point of fact about layers in the sea which do not exist in reality" (Shtokman, 1943).

Although both these arguments start out from different positions — the first from an originally homogeneous sea, and the second from the existence of an original, as it were "instantaneously" created continuous stratification — they bear witness, nevertheless, to the existence of an initial distribution of homogeneous water masses to which reference was made at the beginning of this subsection.

It is relevant to point out that, generally speaking, we could have taken other initial conditions also: say, a linear (by depth) distribution of temperature or salinity at some interval of depth, as is often done in problems of the propagation of heat in rods or as was taken, for example, by Hirano (1957) for salinity in studying the transformation of water masses in the subarctic region of the Northern Pacific Ocean. However, we will not go beyond the framework usually accepted in the theory of *T-S* curves.

Thus, let us return to the initial distribution of water masses in the form of parallel plane layers possessing identical initial values of temperature and salinity. We will consider cases when two, three and four layers exist (and in the case of a sea of semi-infinite depth — also the case of complete absence of layers or, in other words, the existence of one layer), bearing in mind, in the first place, that the vertical structure of the water masses of the ocean is practically limited to the existence of not more than four basic water masses, and in the second place, that, having solutions for this quantity of water masses, we can immediately obtain a solution for a greater (any) number of

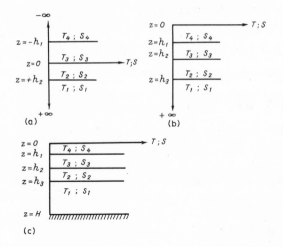

Fig. 33. Four water masses in an ocean of infinite (a), semi-infinite (b) and finite (c) depth (initial conditions).

them, by using general or recurrent formulae which it is easy to construct, when we have solutions for the three versions mentioned (Fig. 33).

The initial conditions are as follows:

(A) *An ocean of infinite depth* (the count starts from one of the interfaces between the water masses; the positive direction z is counted downward):

(a) four water masses (Fig. 33a) —

$$T(z,0) = T_1, \quad S(z,0) = S_1 \quad \text{when} \quad +\infty > z > +h_2$$
$$T(z,0) = T_2, \quad S(z,0) = S_2 \quad \text{when} \quad +h_2 > z > 0,$$
$$T(z,0) = T_3, \quad S(z,0) = S_3 \quad \text{when} \quad 0 > z > -h_1$$
$$T(z,0) = T_4, \quad S(z,0) = S_4 \quad \text{when} \quad -h_1 > z > -\infty$$

[39.11]

(b) three water masses —

$$T(z,0) = T_1, \quad S(z,0) = S_1 \quad \text{when} \quad +\infty > z > h$$
$$T(z,0) = T_2, \quad S(z,0) = S_2 \quad \text{when} \quad h > z > 0$$
$$T(z,0) = T_3, \quad S(z,0) = S_3 \quad \text{when} \quad 0 > z > -\infty$$

[39.12]

(c) two water masses —

$$T(z,0) = T_1, \quad S(z,0) = S_1 \quad \text{when} \quad +\infty > z > 0$$
$$T(z,0) = T_2, \quad S(z,0) = S_2 \quad \text{when} \quad 0 > z > -\infty$$

[39.13]

(B) *An ocean of semi-infinite depth* (the count starts from the surface of the sea, the positive direction z is counted downward):

(a) four water masses (Fig. 33b) —

$$T(z,0) = T_1, \quad S(z,0) = S_1 \quad \text{when} \quad \infty > z > h_3$$
$$T(z,0) = T_2, \quad S(z,0) = S_2 \quad \text{when} \quad h_3 > z > h_2$$
$$T(z,0) = T_3, \quad S(z,0) = S_3 \quad \text{when} \quad h_2 > z > h_1 \qquad [39.14]$$
$$T(z,0) = T_4, \quad S(z,0) = S_4 \quad \text{when} \quad h_1 > z > 0$$

(b) three water masses —

$$T(z,0) = T_1, \quad S(z,0) = S_1 \quad \text{when} \quad \infty > z > h_2$$
$$T(z,0) = T_2, \quad S(z,0) = S_2 \quad \text{when} \quad h_2 > z > h_1 \qquad [39.15]$$
$$T(z,0) = T_3, \quad S(z,0) = S_3 \quad \text{when} \quad h_1 > z > 0$$

(c) two water masses —

$$T(z,0) = T_1, \quad S(z,0) = S_1 \quad \text{when} \quad \infty > z > h$$
$$T(z,0) = T_2, \quad S(z,0) = S_2 \quad \text{when} \quad h > z > 0 \qquad [39.16]$$

(d) one water mass —

$$T(z,0) = T_1, \quad S(z,0) = S_1 \quad \text{when} \quad \infty > z > 0 \qquad [39.17]$$

(C) *An ocean of finite depth* (the count starts from the sea surface, the positive direction z is counted downward, H is the depth of the ocean):

(a) four water masses (Fig. 33c) —

$$T(z,0) = T_1, \quad S(z,0) = S_1 \quad \text{when} \quad H > z > h_3$$
$$T(z,0) = T_2, \quad S(z,0) = S_2 \quad \text{when} \quad h_3 > z > h_2$$
$$T(z,0) = T_3, \quad S(z,0) = S_3 \quad \text{when} \quad h_2 > z > h_1 \qquad [39.18]$$
$$T(z,0) = T_4, \quad S(z,0) = S_4 \quad \text{when} \quad h_1 > z > 0$$

(b) three water masses —

$$T(z,0) = T_1, \quad S(z,0) = S_1 \quad \text{when} \quad H > z > h_2$$
$$T(z,0) = T_2, \quad S(z,0) = S_2 \quad \text{when} \quad h_2 > z > h_1 \qquad [39.19]$$
$$T(z,0) = T_3, \quad S(z,0) = S_3 \quad \text{when} \quad h_1 > z > 0$$

(c) two water masses —

$$T(z,0) = T_1, \quad S(z,0) = S_1 \quad \text{when} \quad H > z > h$$
$$T(z,0) = T_2, \quad S(z,0) = S_2 \quad \text{when} \quad h > z > 0 \qquad [39.20]$$

(d) one water mass —

$$T(z,0) = T_1, \quad S(z,0) = S_1 \quad \text{when} \quad H > z > 0 \qquad [39.21]$$

Boundary conditions

It is obvious that for an ocean of infinite depth as a result of the absence of the effect of any external conditions (the ocean extends upward and downward without limits), boundary conditions are absent. An exception is constituted by the interfaces between water masses for which must be satisfied the condition of equality of flow of heat or salts on both sides of the boundary. For the interface $z = 0$, dividing water masses 1 and 2, these conditions are written in the following way:

$$k_{T,1} \left(\frac{\partial T}{\partial z}\right)_{z=+0} = k_{T,2} \left(\frac{\partial T}{\partial z}\right)_{z=-0}$$

$$k_{S,1} \left(\frac{\partial S}{\partial z}\right)_{z=+0} = k_{S,2} \left(\frac{\partial S}{\partial z}\right)_{z=-0} \qquad [39.22]$$

In addition, it is necessary to observe the conditions:

$$T\Big|_{z=+0} = T\Big|_{z=-0} = T_m$$

$$S\Big|_{z=+0} = S\Big|_{z=-0} = S_m \qquad [39.23]$$

(Shtokman, 1943a; Gröber and Erk, 1955), where T_m and S_m are certain values of temperature and salinity which establish themselves on the dividing line after the beginning of mixing ($t > 0$). The observance of conditions [39.22] is necessary in those cases when the coefficients of heat conduction and diffusion are taken as non-identical in the water masses lying on both sides of the internal boundary considered: $k_{T,1} \neq k_{T,2}$ and $k_{S,1} \neq k_{S,2}$. However, in case the coefficients are taken as identical for all the mixing water masses, conditions [39.22], as well as conditions [39.23], are satisfied automatically, and in the solution we are not obliged to take them into account. The case where the coefficients of heat conduction and diffusion of salts are not identical in the interacting water masses will be considered by us following Gröber and Erk and Shtokman as applied to the mixing of two water masses in an ocean of infinite depth; in this only case we naturally will have to take into account conditions [39.22] and [39.23].

Let us now turn to the cases of the oceans of semi-infinite and finite depth. Remaining within the framework of the mathematical apparatus of the theory of heat conduction applied in the theory of *T-S* curves (in this case, as we shall see below, the solutions of equations [39.5] can be expressed by the integral of probabilities $\Phi(z)$, which makes it much easier in practice to find ready-made solutions), we shall limit ourselves to the consideration of the two simplest cases of boundary conditions for the surface of the sea and for the bottom, namely:

(1) The sea surface is heat-insulated (and "salt-insulated"), i.e., flows of heat and salts do not take place through the surface of the ocean:

$$\frac{\partial T}{\partial z}\bigg|_{z=0} = 0 \qquad \frac{\partial S}{\partial z}\bigg|_{z=0} = 0 \qquad\qquad [39.24]$$

This case corresponds to the "pure" internal mixing of the water masses among themselves, in particular, the effect of heat exchange with the atmosphere is disregarded.

(2) Constant values of temperature and salinity are given on the surface of the sea; any such constant values may be given; however, for simplicity, (and this is quite enough) we will assume that values of temperature and salinity are maintained on the surface equal to the original values of temperature and salinity of the sub-surface water mass:

$$T(0, t) = T_n \qquad S(0, t) = S_n \qquad\qquad [39.25]$$

where $n = 2$, 3 and 4 — the number of the surface water mass. An exception will be the case of the transformation of one water mass in a semi-infinite sea, when arbitrary values of temperature and salinity T_0 and S_0 will be given different from the initial temperature and the initial salinity of this mass. This case, however, is to a greater extent illustrative in nature rather than practical, although it is interesting from the point of view of the methods of *T-S* analysis.

It is clear that the constant values of temperature and salinity on the sea surface are their mean values over many years, developing in each point of the sea under the influence of climatic conditions.

(3) Finally, on the bottom of an ocean of finite depth, just as on the surface, there can be given either conditions:

$$\frac{\partial T}{\partial z}\bigg|_{z=H} = 0 \qquad \frac{\partial S}{\partial z}\bigg|_{z=H} = 0 \qquad\qquad [39.26]$$

or conditions:

$$T\bigg|_{z=H} = T_n \qquad S\bigg|_{z=H} = S_n \qquad\qquad [39.27]$$

The former of these correspond to conditions of absence of flows of heat and salts through the bottom, the latter to the existence of stationary values of temperature and salinity in the bottom layer. It is apparent that in the case of a *deep* finite sea, by virtue of the constancy of the values of the bottom characteristics, conditions [39.27] are preferable.

Diffusion of density

In studying the mixing processes of water masses on the T-S plane, changes in density, as well as in its vertical gradients (stability), which accompany changes in temperature and salinity, have not been considered up to now, aside from the single work of Stommel (1961) quoted above. It is known that in the theory of ocean currents too (it is clear that we are not referring to the theory of geostrophic currents) the influence and distribution of density have also begun to be considered in recent years, after the appearance of the works of Lineikin (1955, 1957), Takano (1955) and Hansen (1956), who independently introduced the equation of diffusion of density into the theory of ocean currents. The difficulties here consist in the non-linearity of the equation of state of sea water; these difficulties have not been overcome up to the present time (in the analytical sense). However, if we approximate the equation of state of sea water by a roughly approximate linear model, it proves possible to obtain the linear equation of diffusion of density, in agreement with the equations of turbulent heat conduction and turbulent diffusion. The works mentioned on the theory of ocean currents, as well as Stommel's work on thermohaline convection, are based on the representation of the equation of state of sea water by a linear formula [9.2]. Differentiating this expression once by t and twice by z, we obtain respectively:

$$\frac{\partial \rho}{\partial t} = -\alpha \frac{\partial T}{\partial t} + \beta \frac{\partial S}{\partial t}$$

$$\frac{\partial^2 \rho}{\partial z^2} = -\alpha \frac{\partial^2 T}{\partial z^2} + \beta \frac{\partial^2 S}{\partial z^2}$$

[39.28]

Let us now multiply the first of equations [39.5] by $-\alpha$ and the second by β, assuming in the first approximation that the coefficients of turbulent heat conduction and turbulent diffusion are equal to each other: $k_T = k_S = k$, and adding the results of these multiplications termwise, we obtain:

$$-\alpha \frac{\partial T}{\partial t} + \beta \frac{\partial S}{\partial t} = k \left(-\alpha \frac{\partial^2 T}{\partial z^2} + \beta \frac{\partial^2 S}{\partial z^2} \right)$$

[39.29]

Comparing the expression obtained with the previous ones, we come to the equation of *diffusion of density* (or rather to one of its simplifications):

$$\frac{\partial \rho}{\partial t} = k_\rho \frac{\partial^2 \rho}{\partial z^2}$$

[39.30]

where the *coefficient of diffusion of density* k_ρ is introduced, which possesses a corresponding physical meaning, obviously only given the condition of equality of coefficients k_T and k_S. The general equation of diffusion of density is written, by analogy with [39.3] and [39.4], in the following way:

$$\frac{d\rho}{dt} = k_\rho \Delta \rho \qquad\qquad\qquad [39.31]$$

Equation [39.30] is identical with equations [39.5], which form the basis of the analytical theory of *T-S* curves; its solution can be obtained every time simultaneously with the solution of these equations. Returning to the question of the lack of correspondence of equation [9.2] to the true equation of state of sea water, we will repeat again that in the necessary conditions at the first stage an intermittently linear approximation of the true equation of state on various zones of the *T-S* diagram is fully possible; the necessary values of coefficients $\alpha(S, T)$ and $\beta(S, T)$ can be selected for this from tables 20 and 26 of the *Oceanological Tables* of Zubov (1957a).

40. THE THEORY OF *T-S* CURVES FOR AN OCEAN OF INFINITE AND SEMI–INFINITE DEPTH

Solutions of the problem

In this section we will consider the solution of the problem of mixing of four, three and two water masses first in an infinite and then in a semi-infinite ocean along the vertical. In this we will consider first the solution for the case of mixing of four water masses, since the other two alternatives can be easily obtained from it as particular cases. This solution, beginning with the general solution of the equation of heat conduction, i.e., in that part of it where function $\Phi(z)$ – the integral of probabilities, is introduced with the corresponding transformations, will be considered in some detail. This has a certain methodological value for the reader who is interested in the further development of the analytical theories of *T-S* curves, all the more since such detailed expositions are rarely found in courses of equations of mathematical physics. An exception is provided by the article of Ivanov (1946), where the solution precisely for the case of four water masses in an infinite sea is considered in detail; thus, the computations quoted in this section are borrowed from this article. Further, for the case of the mixing of four water masses in a semi-infinite ocean we will also cite a detailed solution which is somewhat different from the solution for the infinite ocean. All derivations will be made for temperature, while the identical expressions for salinity in the majority of cases will be omitted for the sake of brevity.

An ocean of infinite depth

The *general* solution of the equation of heat conduction (39.5), given the initial condition $T|_{t=0} = f(z)$, is the following expression (Poisson's integral):

$$T(z,t) = \frac{1}{2\sqrt{\pi kt}} \int_{-\infty}^{+\infty} f(\zeta)\, e^{-\frac{(z-\zeta)^2}{4kt}}\, d\zeta \qquad [40.1]$$

(see, for example, Aramanovich and Levin, 1964, p.162). Bearing in mind that the integrand $f(\zeta)$ is determined for the case of four water masses by initial conditions [39.11], using these conditions we can rewrite expression [40.1] in the following way:

$$T(z,t) = \frac{T_4}{2\sqrt{\pi kt}} \int_{-\infty}^{-h_1} e^{-\frac{(z-\zeta)^2}{4kt}}\, d\zeta + \frac{T_3}{2\sqrt{\pi kt}} \int_{-h_1}^{0} e^{-\frac{(z-\zeta)^2}{4kt}}\, d\zeta +$$
$$+ \frac{T_2}{2\sqrt{\pi kt}} \int_{0}^{+h_2} e^{-\frac{(z-\zeta)^2}{4kt}}\, d\zeta + \frac{T_1}{2\sqrt{\pi kt}} \int_{+h_2}^{+\infty} e^{-\frac{(z-\zeta)^2}{4kt}}\, d\zeta \qquad [40.2]$$

Let us transform this expression by replacing variable ζ; this can be done by means of the following substitutions:

$$\frac{\zeta - z}{2\sqrt{kt}} = \eta; \qquad \zeta = z + 2\eta\sqrt{kt}; \qquad d\zeta = 2\sqrt{kt}\, d\eta \qquad [A]$$

Moreover (Gröber et al., 1955) the limits of integration change in the following way *:

instead of $\zeta = +\infty$ we will have $\eta = +\infty$

instead of $\zeta = -\infty$ we will have $\eta = -\infty$

instead of $\zeta = \pm 0$ we will have $\eta = -z/2\sqrt{kt}$

---

* Beside substitution [A], we can also use one of the following substitutions:

$$\frac{z - \zeta}{2\sqrt{kt}} = \eta; \qquad \zeta = z - 2\eta\sqrt{kt}; \qquad d\zeta = -2\sqrt{kt}\, d\eta \qquad [B]$$

$$\frac{z + \zeta}{2\sqrt{kt}} = \eta; \qquad \zeta = 2\eta\sqrt{kt} - z; \qquad d\zeta = 2\sqrt{kt}\, d\eta \qquad [C]$$

Transformations [A] and [C] are equivalent in the sense that they do not change the sign of limit of integration when it is equal to $+\infty$ or $-\infty$; thus, when $\zeta = +\infty$ we have $\eta = +\infty$, when $\zeta = -\infty$ we have $\eta = -\infty$ (this was indicated above), whereas it follows from [B] that when $\zeta = +\infty$ we have $\eta = -\infty$, when $\zeta = -\infty$ we have $\eta = +\infty$. All three variants lead to the same final result, if we consider the properties of the integral of probabilities, which will now be discussed.

Thus, using substitution [A], we reduce expression [40.2] to the form:

$$T(z,t) = \frac{T_4}{\sqrt{\pi}} \int\limits_{\eta=-\infty}^{\eta=\frac{z+h_1}{2\sqrt{kt}}} e^{-\eta^2} d\eta + \frac{T_3}{\sqrt{\pi}} \int\limits_{-\frac{z+h_1}{2\sqrt{kt}}}^{-\frac{z}{2\sqrt{kt}}} e^{-\eta^2} d\eta +$$

$$+ \frac{T_2}{\sqrt{\pi}} \int\limits_{-\frac{z}{2\sqrt{kt}}}^{-\frac{z-h_2}{2\sqrt{kt}}} e^{-\eta^2} d\eta + \frac{T_1}{\sqrt{\pi}} \int\limits_{-\frac{z-h_2}{2\sqrt{kt}}}^{+\infty} e^{-\eta^2} d\eta \qquad [40.3]$$

Each of the four integrals of the right-hand side of the expression obtained can be transformed by using the special function called the *integral of probabilities* *:

$$\Phi(z) = \frac{2}{\sqrt{\pi}} \int\limits_0^z e^{-\eta^2} d\eta \qquad [40.4]$$

Function $\Phi(z)$ possesses the following properties:

(1) function $\Phi(z)$ is *odd*: $\Phi(-z) = -\Phi(z)$, whereas the integrand function $e^{-\eta^2}$ is *even* (this is completely obvious). The oddness of $\Phi(z)$ follows from the fact that:

$$\Phi(-z) = \frac{2}{\sqrt{\pi}} \int\limits_0^{-z} e^{-\eta^2} d\eta = -\frac{2}{\sqrt{\pi}} \int\limits_{-z}^0 e^{-\eta^2} d\eta =$$

$$= -\frac{2}{\sqrt{\pi}} \int\limits_0^z e^{-\eta^2} d\eta = -\Phi(z) \qquad [40.5]$$

(2) $\Phi(\pm\infty) = \pm 1$ \qquad\qquad [40.6]

(3) $\Phi(0) = 0$ \qquad\qquad [40.7]

---

* It is also called the integral of Gauss's errors, or Laplace's function or Kramp's function. Sometimes this function is designated erf (z) instead of $\Phi(z)$; in such cases by $\Phi(z)$ is meant another function (the function of Laplace quoted), with a different standard designation used in the theory of probabilities (see, for example, Bronshtein and Semendiaev, 1953).

The integral of probabilities is convenient because for it detailed tables exist by means of which it is possible easily and quickly to obtain numerical values of the solutions. Thus, by using the properties of the integral of probabilities expressed by formulae [40.5] and [40.6], we can transform expression [40.3] in the following way: here, consequently, each of the addends of the right-hand side of [40.3] will be broken down into two:

$$T(z, t) = \frac{T_4}{\sqrt{\pi}} \int_{-\infty}^{0} e^{-\eta^2} d\eta + \frac{T_4}{\sqrt{\pi}} \int_{0}^{-\frac{z+h_1}{2\sqrt{kt}}} e^{-\eta^2} d\eta + \frac{T_3}{\sqrt{\pi}} \int_{-\frac{z+h_1}{2\sqrt{kt}}}^{0} e^{-\eta^2} d\eta +$$

$$+ \frac{T_3}{\sqrt{\pi}} \int_{0}^{-\frac{z}{2\sqrt{kt}}} e^{-\eta^2} d\eta + \frac{T_2}{\sqrt{\pi}} \int_{-\frac{z}{2\sqrt{kt}}}^{0} e^{-\eta^2} d\eta + \frac{T_2}{\sqrt{\pi}} \int_{0}^{-\frac{z-h_2}{2\sqrt{kt}}} e^{-\eta^2} d\eta +$$

$$+ \frac{T_1}{\sqrt{\pi}} \int_{-\frac{z-h_2}{2\sqrt{kt}}}^{0} e^{-\eta^2} d\eta + \frac{T_1}{\sqrt{\pi}} \int_{0}^{+\infty} e^{-\eta^2} d\eta =$$

$$= \frac{T_1 + T_4}{\sqrt{\pi}} \int_{0}^{+\infty} e^{-\eta} d\eta + \frac{T_1 - T_2}{\sqrt{\pi}} \int_{0}^{\frac{z-h_2}{2\sqrt{kt}}} e^{-\eta^2} d\eta +$$

$$+ \frac{T_2 - T_3}{\sqrt{\pi}} \int_{0}^{\frac{z}{2\sqrt{kt}}} e^{-\eta^2} d\eta + \frac{T_3 - T_4}{\sqrt{\pi}} \int_{0}^{\frac{z+h_1}{2\sqrt{kt}}} e^{-\eta^2} d\eta \qquad [40.8]$$

Introducing the symbol $\Phi(z)$ in [40.8], we come to the final expression for the vertical distribution of temperature in the case of mixing of four water masses in an ocean of infinite depth:

$$T(z, t) = \frac{1}{2} \left[T_1 + T_4 + (T_1 - T_2) \, \Phi \left(\frac{z - h_2}{2\sqrt{kt}} \right) + \right.$$

$$\left. + (T_2 - T_3) \, \Phi \left(\frac{z}{2\sqrt{kt}} \right) + (T_3 - T_4) \, \Phi \left(\frac{z + h_1}{2\sqrt{kt}} \right) \right] \qquad [40.9]$$

Now it is easy to see that combining any two water masses in one, i.e., proceeding from the case of mixing of four water masses to three water masses, we can obtain the corresponding formula for the case of vertical mixing of three water masses. Thus, eliminating in Fig. 33a the boundary $-h_1$ and assuming $T_3 = T_4$, and also writing h instead of h_2, we will obtain from formula [40.9], in which, obviously, the last term will disappear, the formula for the mixing of three water masses. Acting further in the same way, we can obtain the formula for the distribution of temperature in the case of mixing of two water masses. This may be seen by comparing all three formulae which are written out below as a summary for the mixing of four, three and two water masses in a sea of infinite depth.

Recapitulation of the formulae for an ocean of infinite depth

Four water masses:

$$T(z,t) = \frac{1}{2} \left[T_1 + T_4 + (T_1 - T_2) \, \Phi \left(\frac{z - h_2}{2\sqrt{kt}} \right) + \right.$$

$$\left. + (T_2 - T_3) \, \Phi \left(\frac{z}{2\sqrt{kt}} \right) + (T_3 - T_4) \, \Phi \left(\frac{z + h_1}{2\sqrt{kt}} \right) \right] \qquad [40.9]$$

Three water masses:

$$T(z,t) = \frac{1}{2} \left[T_1 + T_3 + (T_1 - T_2) \, \Phi \left(\frac{z - h}{2\sqrt{kt}} \right) + (T_2 - T_3) \, \Phi \left(\frac{z}{2\sqrt{kt}} \right) \right] \qquad [40.10]$$

Two water masses:

$$T(z,t) = \frac{1}{2} \left[T_1 + T_2 + (T_1 - T_2) \, \Phi \left(\frac{z}{2\sqrt{kt}} \right) \right] \qquad [40.11]$$

Considering the formula for three water masses separately, let us note that for further analysis it will be more convenient for us to use the formula which is obtained if we combine two *intermediate* water masses of a four-layer-sea, placing the starting point of the vertical coordinate in the middle of the new intermediate water mass thus obtained. Then, instead of formula [40.10] we will obtain formula:

$$T(z,t) = \frac{1}{2} \left[T_1 + T_3 + (T_1 - T_2) \, \Phi \left(\frac{z - h}{2\sqrt{kt}} \right) + (T_2 - T_3) \, \Phi \left(\frac{z + h}{2\sqrt{kt}} \right) \right] \qquad [40.12]$$

(Shtokman, 1943a)

The results obtained make it possible for us, on the basis of a comparison of formulae [40.9], [40.10] and [40.11], to write a general, or recurrent, formula for the case of mixing of any number of water masses with any thicknesses. Ivanov (1943) constructed a corresponding general formula; it is also given in the book of Timofeev and Panov (1962). However, we will neither quote this formula nor construct a recurrent formula, recalling what has already been said to the effect that the vertical structure of the waters of the ocean does not exceed, as a rule, the case when four basic water masses exist; in any case, the fundamental analysis of the results obtained will be considered as it applies to the mixing of three water masses.

The same formula for salinity must be placed in correspondence to each of the formulae obtained above for temperature. Below, when we proceed directly to T-S analysis, this is what we will do. Above, however, the formulae for salinity were not written solely in order to save space.

An ocean of semi-infinite depth

The original problem of the theory of T-S curves is the construction of analytical T-S curves which would reproduce as accurately as possible the curves constructed from the results of observations at hydrological stations in the sea. In striving to attain similarity between both curves, the purpose of theory consists not only of constructing an analytical T-S curve the form of which would be similar to the form of the real curve, but also of having the distribution of parameter z (depth) along it reflect the real picture. Only after the observance of both these conditions — similarity in the form of T-S curves and likeness in the distribution along them of parameter z — is it possible to make further theoretical conclusions on the distribution of water masses (first of all by depth), the speed of their transformation and displacement, the percentage ratio of water masses in different points, the quantities of the coefficients of exchange, etc.

In connection with what has been said, attention should be drawn to one general and highly important property of T-S curves of deep water oceanographical stations: the reference marks of parameter z are distributed along them irregularly; the points with values of parameter z taken at even or close intervals are thinned out in the upper part of the T-S curve and fill out along its lower part, asymptotically approaching when $z \rightarrow H$, where H is the depth of the ocean, the T-S index of the deep (bottom) water mass. For illustration in Fig. 34 are given three T-S curves of stations of the research vessel "Carnegie" No. 136, 137 and 147, made in September–October, 1929, in the Northeastern Pacific Ocean (the upper layer of the ocean of a thickness of about 100 m is not considered); the parametric points of depth become more frequent along the T-S curves as they approach the thermohaline index of the Bottom Water of the North Pacific Ocean B (1.30° ; 34.70‰), determined by Sverdrup et al. (1942).

The property noted of T-S curves is perfectly obvious, and is explained by

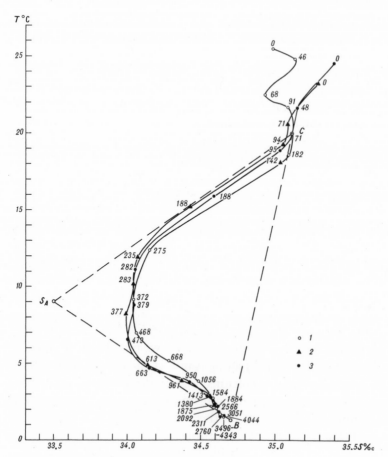

Fig. 34. *T-S* curves of "Carnegie" stations No. 136, 137 and 147, made in September–October 1929 in the Northeastern Pacific Ocean (Mamayev, 1966b)

the great homogeneousness of the deep water mass for temperature and salinity; having imagined the case of an idealized ocean, unlimited in depth, filled to infinity with water homogeneous in temperature and salinity, we will come to the conclusion that the parametric points will, as before, asymptotically approach the *T-S* index of this homogeneous mass, however, no longer when $z \to H$, but when $z \to \infty$.

Comparing mentally the *T-S* curves constructed for a very deep but finite $(z = H)$ and an infinite $(z \to \infty)$ ocean with an identical index (T_0, S_0) in the first case at the bottom and in the second at infinity, we can come to the conclusion that the distribution of parameter z in the neighborhood of point (T_0, S_0) of radius ϵ, provided that ϵ is small enough, will have little effect on the distribution of points z_n in the remaining part of the *T-S* curves. What has been said above is enough for the affirmation that the development of the analytical theory of *T-S* curves for an ocean of infinite depth with subsequent application to a real deep ocean is fully justified.

The solution of the problem

Let us consider the solutions of equations [39.5] for the case of mixing in a semi-infinite ocean of four, three and two water masses at initial conditions [39.14], illustrated in Fig. 33b, and boundary conditions [39.24] and [39.25]*.

As earlier, we shall consider solutions for temperature; the expressions determining the distribution of salinity are identical.

(A) *Constant value of temperature is maintained on the sea surface.* The general solution of the equation of heat conduction [39.5] given initial condition [39.14] and boundary condition $T(0, t) = T_0$ is determined by the expression:

$$T(z, t) = T_0 \left[1 - \Phi \left(\frac{z}{2\sqrt{kt}} \right) \right] + \frac{1}{2\sqrt{\pi kt}} \int_0^\infty f(\zeta) \left\{ e^{-\frac{(z-\zeta)^2}{4kt}} - e^{-\frac{(z+\zeta)^2}{4kt}} \right\} d\zeta$$

$$[40.13]$$

(see, for example, Aramanovich and Levin, 1964, p. 199). Transforming the second addend of the right-hand side in accordance with initial conditions [39.14], we obtain:

$$T(z, t) = T_0 \left[1 - \Phi \left(\frac{z}{2\sqrt{kt}} \right) \right] + \frac{T_4}{2\sqrt{\pi kt}} \int_0^{h_1} \left\{ e^{-\frac{(z-\zeta)^2}{4kt}} - e^{-\frac{(z+\zeta)^2}{4kt}} \right\} d\zeta +$$

$$+ \frac{T_3}{2\sqrt{\pi kt}} \int_{h_1}^{h_2} \left\{ e^{-\frac{(z-\zeta)^2}{4kt}} - e^{-\frac{(z+\zeta)^2}{4kt}} \right\} d\zeta +$$

$$+ \frac{T_2}{2\sqrt{\pi kt}} \int_{h_2}^{h_3} \left\{ e^{-\frac{(z-\zeta)^2}{4kt}} - e^{-\frac{(z+\zeta)^2}{4kt}} \right\} d\zeta +$$

$$+ \frac{T_1}{2\sqrt{\pi kt}} \int_{h_3}^\infty \left\{ e^{-\frac{(z-\zeta)^2}{4kt}} - e^{-\frac{(z+\zeta)^2}{4kt}} \right\} d\zeta \qquad [40.14]$$

---

* In his last work on the theory of *T-S* curves, Ivanov (1949) gives solutions of equations [39.5] for the transformation of two and three water masses in a sea of semi-infinite depth with mixed boundary conditions for temperature and salinity. Ivanov, however, did not focus attention on the reasons for the need to develop a theory for an ocean of semi-infinite depth, and his results were in fact not utilized.

We transform successively eight addends of the type $\exp\left[(z\pm\zeta)^2/4kt\right]$ appearing in this expression in curly brackets, introducing in the cases 1, 3, 5 and 7 the variable:

$$\eta = \frac{\zeta - z}{2\sqrt{kt}}\;;\qquad d\zeta = 2\sqrt{kt}\,d\eta$$

and in cases 2, 4, 6 and 8 the variable:

$$\eta = \frac{z + \zeta}{2\sqrt{kt}}\;;\qquad d\zeta = 2\sqrt{kt}\,d\eta$$

and proceeding to the integral of probabilities [40.4]. Then we will obtain:

$$\frac{T_4}{2\sqrt{\pi kt}}\int_0^{h_1} e^{-\frac{(z-\zeta)^2}{4kt}}\,d\zeta = \frac{T_4}{\sqrt{\pi}}\int_{-\frac{z}{2\sqrt{kt}}}^{-\frac{z-h_1}{2\sqrt{kt}}} e^{-\eta^2}\,d\eta =$$

$$= \frac{T_4}{\sqrt{\pi}}\int_{-\frac{z}{2\sqrt{kt}}}^{0} e^{-\eta^2}\,d\eta + \frac{T_4}{\sqrt{\pi}}\int_0^{-\frac{z-h_1}{2\sqrt{kt}}} e^{-\eta^2}\,d\eta =$$

$$= \frac{T_4}{\sqrt{\pi}}\int_0^{\frac{z}{2\sqrt{kt}}} e^{-\eta^2}\,d\eta - \frac{T_4}{\sqrt{\pi}}\int_0^{\frac{z-h_1}{2\sqrt{kt}}} e^{-\eta^2}\,d\eta =$$

$$= \frac{T_4}{2}\left[\Phi\left(\frac{z}{2\sqrt{kt}}\right) - \Phi\left(\frac{z-h_1}{2\sqrt{kt}}\right)\right] \qquad\qquad [40.15]$$

for the first addend. For the remaining seven addends we can obtain in exactly the same way:

$$\frac{T_4}{2\sqrt{\pi kt}}\int_0^{h_1} e^{-\frac{(z+\zeta)^2}{4kt}}\,d\zeta = \frac{T_4}{2}\left[\Phi\left(\frac{z+h_1}{2\sqrt{kt}}\right) - \Phi\left(\frac{z}{2\sqrt{kt}}\right)\right] \qquad\qquad [40.16]$$

$$\frac{T_3}{2\sqrt{\pi kt}}\int_{h_1}^{h_2} e^{-\frac{(z-\zeta)^2}{4kt}}\,d\zeta = \frac{T_3}{2}\left[\Phi\left(\frac{z-h_1}{2\sqrt{kt}}\right) - \Phi\left(\frac{z-h_2}{2\sqrt{kt}}\right)\right] \qquad\qquad [40.17]$$

$$\frac{T_3}{2\sqrt{\pi kt}}\int_{h_1}^{h_2} e^{-\frac{(z+\zeta)^2}{4kt}}\,d\zeta = \frac{T_3}{2}\left[\Phi\left(\frac{z+h_2}{2\sqrt{kt}}\right) - \Phi\left(\frac{z+h_1}{2\sqrt{kt}}\right)\right] \qquad\qquad [40.18]$$

$$\frac{T_2}{2\sqrt{\pi kt}} \int_{h_2}^{h_3} e^{-\frac{(z-\zeta)^2}{4kt}} \, d\zeta = \frac{T_2}{2} \left[\Phi\left(\frac{z-h_2}{2\sqrt{kt}}\right) - \Phi\left(\frac{z-h_3}{2\sqrt{kt}}\right) \right] \qquad [40.19]$$

$$\frac{T_2}{2\sqrt{\pi kt}} \int_{h_2}^{h_3} e^{-\frac{(z+\zeta)^2}{4kt}} \, d\zeta = \frac{T_2}{2} \left[\Phi\left(\frac{z+h_3}{2\sqrt{kt}}\right) - \Phi\left(\frac{z+h_2}{2\sqrt{kt}}\right) \right] \qquad [40.20]$$

$$\frac{T_1}{2\sqrt{\pi kt}} \int_{h_3}^{\infty} e^{-\frac{(z-\zeta)^2}{4kt}} \, d\zeta = \frac{T_1}{2} \left[1 + \Phi\left(\frac{z-h_3}{2\sqrt{kt}}\right) \right] \qquad [40.21]$$

$$\frac{T_1}{2\sqrt{\pi kt}} \int_{h_3}^{\infty} e^{-\frac{(z+\zeta)^2}{4kt}} \, d\zeta = \frac{T_1}{2} \left[1 - \Phi\left(\frac{z+h_3}{2\sqrt{kt}}\right) \right] \qquad [40.22]$$

Adding according to [40.13], and [40.14], $T_0 [1 - \Phi(z/2\sqrt{kt})]$ with the results [40.15]–[40.22] and introducing for brevity of notation the symbol:

$$\Psi_+(h) = \Phi\left(\frac{z-h}{2\sqrt{kt}}\right) + \Phi\left(\frac{z+h}{2\sqrt{kt}}\right) \qquad [40.23]$$

we obtain finally:

$$T(z,t) = T_0 + (T_4 - T_0)\, \Phi\left(\frac{z}{2\sqrt{kt}}\right) +$$

$$+ \tfrac{1}{2}[(T_3 - T_4)\Psi_+(h_1) + (T_2 - T_3)\Psi_+(h_2) + (T_1 - T_2)\Psi_+(h_3)] \qquad [40.24]$$

Since we assumed above that $T_0 = T_4$, the second term of formula [40.24] disappears and we have:

$$T(z,t) = T_4 + \tfrac{1}{2}[(T_3 - T_4)\Psi_+(h_1) + (T_2 - T_3)\Psi_+(h_2) + (T_1 - T_2)\Psi_+(h_3)] \qquad [40.25]$$

For the case of mixing of three and two water masses in a sea of semi-infinite depth given boundary condition [39.25] we have respectively:

$$T(z,t) = T_3 + \tfrac{1}{2}[(T_2 - T_3)\Psi_+(h_1) + (T_1 - T_2)\Psi_+(h_2)] \qquad [40.26]$$

and:

$$T(z,t) = T_2 + \tfrac{1}{2}(T_1 - T_2)\Psi_+(h_1) \qquad [40.27]$$

Each of the successive formulae [40.26] and [40.27] can be obtained from the preceding one by combining two water masses in one (from the top downward), by lowering the number n etc.; therefore we will not repeat the transformations.

(B) *No heat flow through the sea surface (heat insulation).* The general solution of the equation of heat conduction in this case is written in the form:

$$T(z,t) = \frac{1}{2\sqrt{\pi kt}} \int_0^\infty f(\zeta) \left(e^{-\frac{(z-\zeta)^2}{4kt}} + e^{-\frac{(z+\zeta)^2}{4kt}} \right) d\zeta \qquad [40.28]$$

Transforming it in accordance with initial conditions [39.14] we obtain:

$$T(z,t) = \frac{T_4}{2\sqrt{\pi kt}} \int_0^{h_1} \left(e^{-\frac{(z-\zeta)^2}{4kt}} + e^{-\frac{(z+\zeta)^2}{4kt}} \right) d\zeta +$$

$$+ \frac{T_3}{2\sqrt{\pi kt}} \int_{h_1}^{h_2} \left(e^{-\frac{(z-\zeta)^2}{4kt}} + e^{-\frac{(z+\zeta)^2}{4kt}} \right) d\zeta +$$

$$+ \frac{T_2}{2\sqrt{\pi kt}} \int_{h_2}^{h_3} \left(e^{-\frac{(z-\zeta)^2}{4kt}} + e^{-\frac{(z+\zeta)^2}{4kt}} \right) d\zeta +$$

$$+ \frac{T_1}{2\sqrt{\pi kt}} \int_{h_3}^\infty \left(e^{-\frac{(z-\zeta)^2}{4kt}} + e^{-\frac{(z+\zeta)^2}{4kt}} \right) d\zeta \qquad [40.29]$$

Using the ready-made values of the eight addends of expression [40.29], determined by formulae [40.15]–[40.22] of the preceding exposition, and introducing for brevity the symbol:

$$\Psi_-(h) = \Phi\left(\frac{z-h}{2\sqrt{kt}}\right) - \Phi\left(\frac{z+h}{2\sqrt{kt}}\right) \qquad [40.30]$$

we obtain for the case of mixing of four water masses:

$$T(z,t) = T_1 + \tfrac{1}{2}[(T_3-T_4)\Psi_-(h_1) + (T_2-T_3)\Psi_-(h_2) + (T_1-T_2)\Psi_-(h_3)] \quad [40.31]$$

For the case of mixing of three and two water masses, given boundary condition [39.24], we have respectively:

$$T(z,t) = T_1 + \tfrac{1}{2}[(T_2-T_3)\Psi_-(h_1) + (T_1-T_2)\Psi_-(h_2)] \qquad [40.32]$$

and:

$$T(z,t) = T_1 + \tfrac{1}{2}(T_1-T_2)\Psi_-(h_1) \qquad [40.33]$$

Each of the two successive formulae [40.32] and [40.33] necessarily follows from the preceding one with the reduction in the number of water masses.

Let us now proceed to the practical application of the theory. We must immediately point out the fundamental difficulty of this problem which consists of the fact that the construction of the "geometry" of the T-S curves in this case is considerably more complex than for the case of mixing of water masses in a sea of infinite depth. In the latter case we had the symmetry of three water masses in relation to the core of the intermediate water mass, which enabled us fairly simply to eliminate the parameter (function) $\Phi(z)$ in certain basic cases for the purpose of obtaining the equations of certain T-S lines in explicit form: $T = f(S)$. For the theory of T-S curves in a semi-infinite ocean this is far from being the case; some geometrical constructions can be made only when the thicknesses of the water masses are the same. We will not dwell on these cases, rarely encountered in practice, but will briefly consider the fundamental results of the theory which illustrate its difference from the theory for the infinite ocean. We shall pay particular attention to two cases: (a) the mixing of two water masses given boundary condition [39.24] (the sea surface is "heat insulated" and "salt insulated"); (b) the mixing of three water masses given boundary condition [39.25] (constant values of temperature and salinity are maintained on the sea surface).

The first problem is connected with the problem of the zonal transformation of the water masses of the ocean; the second with the question of the correct determination of the percentage ratio of the water masses from the T-S-z curves.

It should probably be pointed out that the theory of T-S curves has been developed up to the present time on the basis of the hypothesis that the coefficients of turbulent heat conduction and turbulent diffusion are equal. A more correct concept to the effect that turbulent heat conduction is considerably more intensive than the turbulent exchange of salts introduces substantial corrections into the theory when in further analysis we assume that $k_T > k_S$. Remaining on the whole within the first of the hypotheses mentioned, we will, however, in some cases investigate the effect of inequality of the coefficients; this question is too complex for us to be able to give decisive preference to one alternative of the analysis or the other.

41. MIXING OF TWO WATER MASSES IN AN OCEAN OF INFINITE DEPTH

The solution of equations [39.5], given initial conditions [39.13], is expressed, as was already indicated, by formula [40.11] and the analogous formula for salinity. However, this solution represents a particular case of a more general solution which corresponds to the assumption that the coefficients of exchange, taken as identical for heat and salts, are considered non-identical in the upper and lower water masses. This case corresponds to the problems of heat exchange between two heterogeneous bodies, which has been thor-

oughly studied and is classical in the theory of heat conduction (see, for
example, Gröber and Erk, 1955); it was applied to the theory of *T-S* curves
by Shtokman (1943a).

The considerations leading to the solution of this more general problem
must be completed by the physical hypothesis to the effect that, after the
beginning of mixing or, as Gröber points out, "after the degeneration of
separation", on the interface between two water masses some constant
value of temperature (and salinity) is instantaneously established, which
remains unchanged throughout the whole process of heat and salt exchange.
Mathematically, this assumption corresponds to the introduction of boundary condition [39.23] for each of the water masses which now, obviously,
must be considered separately — as two cases of heat and salt transfer in one
semi-infinite water mass. The combination of two solutions for two semi-
infinite water masses in one solution for two water masses in an infinite sea
must be carried out by means of boundary condition [39.22].

The solution for one semi-infinite water mass for temperature and salinity
when $0 < z < +\infty$ has the form:

$$T(z,t) = T_m + (T_1 - T_m) \, \Phi \left(\frac{z}{2\sqrt{kt}} \right)$$

$$S(z,t) = S_m + (S_1 - S_m) \, \Phi \left(\frac{z}{2\sqrt{kt}} \right)$$

[41.1]

where T_1 and S_1 are the initial values of temperature and salinity of a semi-
infinite water mass, T_m and S_m are the given constant values of temperature
and salinity on its boundary.

Considering Fig. 35 (Shtokman, 1943a, fig. 5; Gröber and Erk, 1955,
fig. 66), we must come to the obvious conclusion that the solution of equations [39.5], given initial conditions [39.13] and boundary conditions
[39.22] and [39.23], must be written (separately for each of the water
masses) in the following form:

$$T(z,t)\big|_{z>0} = T_m - (T_m - T_1) \, \Phi \left(\frac{z}{2\sqrt{kt}} \right)$$

[41.2]

$$T(z,t)\big|_{z<0} = T_m - (T_m - T_2) \, \Phi \left(\frac{z}{2\sqrt{kt}} \right)$$

[41.3]

(for the sake of brevity we do not write the analogous expressions for salinity). T_m and S_m represent the values of temperature and salinity established
on the interface after the beginning of mixing. Equations [41.2] and [41.3]
satisfy conditions [39.23], since $\Phi(0) = 0$.

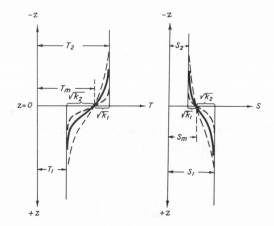

Fig. 35. Diagram of the equalization of temperature and salinity on mixing of two heterogenous ($\sqrt{k_2}/\sqrt{k_1} = 2$) water masses in an ocean of infinite depth, according to Gröber and Erk (1955) and Shtokman (1943a).

For the determination of T_m and S_m, let us use the conditions of continuity of curves $T(z)$ and $S(z)$ on the interface ($z = 0$) — formula [39.22]. The first derivative of the integral of probability is determined by the expression:

$$\frac{d}{dz} \Phi \left(\frac{\pm z}{2\sqrt{kt}}\right) = \pm \frac{1}{\sqrt{\pi kt}} e^{-\frac{z^2}{4kt}} \qquad [41.4]$$

Consequently:

$$k_1 \frac{\partial T}{\partial z}\bigg|_{z=+0} = k_1(T_1 - T_m)\left[\frac{\partial}{\partial z} \Phi \left(\frac{+z}{2\sqrt{k_1 t}}\right)\right]_{z=+0} = \frac{\sqrt{k_1}}{\sqrt{\pi}} (T_1 - T_m)$$

and:

$$k_2 \frac{\partial T}{\partial z}\bigg|_{z=-0} = \frac{\sqrt{k_2}}{\sqrt{\pi}} (T_2 - T_m)$$

Equating these two expressions on the basis of boundary condition [39.22], we obtain:

$$\sqrt{k_1}(T_1 - T_m) = \sqrt{k_2}(T_m - T_2) \qquad [41.5]$$

From this expression and the analogous expression for salinity are determined the values of temperature and salinity T_m and S_m on the interface:

$$T_m = \frac{\sqrt{k_1}\,T_1 + \sqrt{k_2}\,T_2}{\sqrt{k_1} + \sqrt{k_2}} \qquad\qquad [41.6]$$

$$S_m = \frac{\sqrt{k_1}\,S_1 + \sqrt{k_2}\,S_2}{\sqrt{k_1} + \sqrt{k_2}} \qquad\qquad [41.7]$$

Substituting these expressions in formulae [41.2] and [41.3], we obtain finally solutions of equations [39.5] for the case under consideration:

for *water mass I*

$$T(z,t)\big|_{z>0} = \frac{\sqrt{k_1}\,T_1 + \sqrt{k_2}\,T_2}{\sqrt{k_1} + \sqrt{k_2}} + \frac{\sqrt{k_2}}{\sqrt{k_1} + \sqrt{k_2}}\,(T_1 - T_2)\,\Phi\left(\frac{z}{2\sqrt{k_1\,t}}\right)$$

$$S(z,t)\big|_{z>0} = \frac{\sqrt{k_1}\,S_1 + \sqrt{k_2}\,S_2}{\sqrt{k_1} + \sqrt{k_2}} + \frac{\sqrt{k_2}}{\sqrt{k_1} + \sqrt{k_2}}\,(S_1 - S_2)\,\Phi\left(\frac{z}{2\sqrt{k_1\,t}}\right)$$

$$[41.8]$$

for *water mass II*

$$T(z,t)\big|_{z<0} = \frac{\sqrt{k_1}\,T_1 + \sqrt{k_2}\,T_2}{\sqrt{k_1} + \sqrt{k_2}} + \frac{\sqrt{k_1}}{\sqrt{k_1} + \sqrt{k_2}}\,(T_1 - T_2)\,\Phi\left(\frac{z}{2\sqrt{k_2\,t}}\right)$$

$$S(z,t)\big|_{z<0} = \frac{\sqrt{k_1}\,S_1 + \sqrt{k_2}\,S_2}{\sqrt{k_1} + \sqrt{k_2}} + \frac{\sqrt{k_1}}{\sqrt{k_1} + \sqrt{k_2}}\,(S_1 - S_2)\,\Phi\left(\frac{z}{2\sqrt{k_2\,t}}\right)$$

$$[41.9]$$

(Shtokman, 1943a). We see that these formulae differ from formula [40.11], derived for the case when the coefficients of exchange are taken as identical in both water masses. The latter represent a particular case of formulae [41.8] and [41.9], provided that $k_1 = k_2$.

The nature of the equalization of temperature and salinity in the process of mixing of two water masses follows from the same Fig. 35: according to formula [41.5], the points T_m and S_m divide the segments $T_2 - T_1$ and $S_1 - S_2$ into parts inversely proportional to the corresponding values of coefficients $\sqrt{k}$, which Gröber calls the *coefficient of penetration* and Shtokman the *coefficients of accumulation* (of heat and salts) *. When $k_1 = k_2$ the quantities T_m and S_m prove to be arithmetical means: $T_m = (T_1 + T_2)/2$; $S_m = (S_1 + S_2)/2$. The case of equalization of temperatures, given the condi-

---

* One cannot help drawing attention to the problem of the physiological sensation "hot" or "cold", explained by this problem; Gröber draws attention to this problem, and we do not quote his argument only because it is not most directly related to the problem under consideration.

tion of equality of the coefficients of heat conduction in both water masses, was considered for the first time by Defant; assuming $k = 5$ CGS units, he plotted curves for the vertical distribution of temperature at intervals of 1.16; 11.6; 57.9 and 220 days after the beginning of mixing. The initial temperatures were taken by him as equal to 5 and $10°$, and the vertical distance − 400 m upward and downward from the interface. We do not reproduce Defant's figure, analogous to Fig. 35, referring the reader either to the primary source (Defant, 1929) or to the monograph of Shuleikin, where Defant's conclusions are reproduced (Shuleikin, 1953, figs. 182 and 183).

Let us proceed to the interpretation of the results obtained in the T-S plane. Eliminating function $\Phi(z/2\sqrt{kt})$ from formulae [41.8] and [41.9], we obtain:

for *water mass I*

$$T - T_m = \frac{T_1 - T_m}{S_1 - S_m} (S - S_m) \qquad (z > 0) \qquad\qquad [41.10]$$

for *water mass II*

$$T - T_m = \frac{T_2 - T_m}{S_2 - S_m} (S - S_m) \qquad (z < 0) \qquad\qquad [41.11]$$

both the formulae obtained represent equations of straight lines passing through two given points, in the first case − through points (T_1, S_1) and (T_m, S_m), in the second case − through points (T_2, S_2) and (T_m, S_m). Since, according to [41.6] and [41.7], point (T_m, S_m) itself lies on the straight line connecting points $(T_1; S_1)$ and (T_2, S_2), all three points mentioned *lie on one straight T-S line*.

In particular, in the case of homogeneous water masses, when the coefficients of exchange are taken as identical in both masses $(k_1 = k_2)$, function $\Phi(z/2\sqrt{kt})$ must be eliminated from equation [40.11] for temperature and from the analogous equation for salinity. In this case we will obtain the equation of the straight line passing both through points $I(T_1, S_1)$ and $II(T_2, S_2)$ and through the middle of segment $I-II$ − point $[\frac{1}{2}(T_1 + T_2), \frac{1}{2}(S_1 + S_2)]$:

$$T - \frac{T_1 + T_2}{2} = \frac{T_1 - T_2}{2} \left(S - \frac{S_1 + S_2}{2}\right) \qquad\qquad [41.12]$$

In other words, we come to the same result, namely: the process of mixing of two infinite water masses, provided the coefficients of exchange of heat and salts are equal, is represented in the T-S plane in the form of a straight line.

   This affirmation contains the fundamental result of the theory of T-S curves for the case of mixing of two water masses in an infinite sea — provided that the coefficients of heat and salt exchange are equal.

   It is interesting to trace the development of the process of mixing which is expressed in the T-S plane by a different arrangement of points with discrete values of parameter z along the straight T-S line at different moments in time. It is obvious that at the initial moment, when $t = 0$, all these points for the region $+\infty > z > 0$ are grouped in point I, while points z for region $-\infty < z < 0$ are grouped in point II. Indeed, by virtue of the properties of the function $\Phi(z)$ — formula [40.6] — when $t = 0$ we have: when $z > 0$, $\Phi(+z/2\sqrt{kt}) =$ $= \Phi(+\infty) = +1$, and when $z < 0$, respectively $\Phi(-z/2\sqrt{kt}) = \Phi(-\infty) = -1$. Substituting these values in formula [40.11] and its counterpart for salinity, in both cases we will obtain:

when $z > 0$ $T|_{t=0} = T_1$; $S|_{t=0} = S_1$

when $z < 0$ $T|_{t=0} = T_2$; $S|_{t=0} = S_2$

   On the expiry of the initial moment, which takes place instantaneously, the points with the values of parameter z will begin from points I and II to move "inside" the straight T-S line; meanwhile, in the end points an infinite number of discrete points z will continue to remain throughout. This property of representing the mixing of infinite water masses in the T-S plane has been well delineated by Ivanov, who writes: "The ends of the segment — points I and II — are limiting points for the set of points z arranged on it at any time. Indeed, at any t in however small a vicinity of point I or II are arranged infinitely many points $+z$ or $-z$, and only when $t = \infty$ do all the points z instantaneously ... pass to the position of point $z = 0$, concerning which it is by no means possible to say that it is limiting for any fixed t, although it is precisely to it that the points are rushing ... On the contrary, at small distances from the interface, i.e., in the vicinity of point $z = 0$, at fixed t only a finite set of points z will necessarily arrange themselves. The picture will be the opposite, if we consider the set formed by the positions of fixed point z on segment I–II during a change in time t" (Ivanov, 1943, p.38).

   Thus, when $t = \infty$, $\Phi(z/2\sqrt{kt}) = 0$ and we obtain from formulae [41.8] and [41.9] formulae [41.6] and [41.7], representing the coordinates of limiting point M in the T-S plane. Point M divides the straight T-S line I–II into segments also inversely proportional to quantities $\sqrt{k_1}$ and $\sqrt{k_2}$; in the particular case when $k_1 = k_2$, point M divides the straight line I–II into equal halves. Shtokman points out concerning formulae [41.6] and [41.7] that they are analogous to the formulae of mixing of two waters [30.1] and [30.2] with the difference, however, that instead of volumes or vertical heights of water masses (when they are finite) here appear quantities of the coefficients of penetration (accumulation).

To illustrate the process of mixing, in Fig. 36 are represented five straight T-S lines, corresponding to the five successive stages of mixing of the same initial water masses, possessing thermohaline indexes I (2.5°; 34.90‰) and II (24°; 36.3‰); this figure is similar to fig. 1 of Ivanov's work (1943). Water masses I and II represent respectively Deep and Bottom Water of the North Atlantic and the water mass of the Gulf Stream itself and Antilles Current (Jacobsen, 1929); the second of these water masses spreads in the North Atlantic, gradually being transformed, over the first (see Section 43 for greater detail). Let us note that in the graphic representation of the different T-S relationships (Fig. 36 represents one such an example) in the majority of cases, instead of time t, we will use as an argument the parameter $2\sqrt{kt}$ so as to leave ourselves freedom to select the quantity of coefficient k. The relationship between the different moments in time and the different values of coefficient k can easily be obtained by means of a special graph (nomogram), represented in Fig. 37. The straight T-S lines, represented in Fig. 36 have been plotted for the following successive values of this parameter:

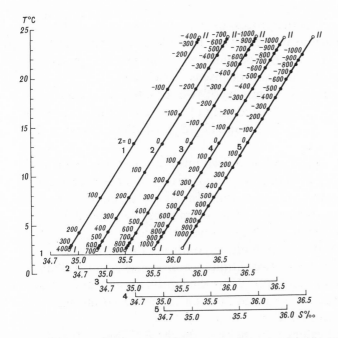

Fig. 36. Straight T-S lines corresponding to the five successive stages of transformation of water masses I (2.5°; 34.9°/oo) and II (24.0°; 36.3°/oo). (1) $2\sqrt{kt} = 2 \cdot 10^4$; (2) $2\sqrt{kt} = 4 \cdot 10^4$; (3) $2\sqrt{kt} = 6 \cdot 10^4$; (4) $2\sqrt{kt} = 8 \cdot 10^4$; (5) $2\sqrt{kt} = 10 \cdot 10^4$.

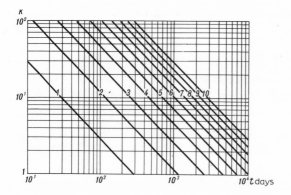

Fig. 37. Relation between coefficient of turbulence k and time t (in days). Isolines of parameter $2\sqrt{kt}\cdot10^{-4}$ are plotted on the graph.

$2\sqrt{kt}\cdot10^{-4}$ = 2, 4, 6, 8 and 10 respectively. Fig. 37 well illustrates the process described above of the development of mixing in time, which is characterized by the gradual movement of the parametric points z from the initial points I and II in the direction of the middle point (13.25°, 35.60‰) with the designation $z = 0$.

The results of the analysis of the process of mixing of two water masses of infinite depth and homogeneous properties are formulated by Ivanov in the form of the following theorems, proof of which was demonstrated by us above (except for the fourth, which will be discussed below):

(1) The T-S diagram in the case of mixing of two infinite water masses I (T_1, S_1) and II (T_2, S_2) always, independently of time, constitutes a segment of the straight line joining points (T_1, S_1) and (T_2, S_2), which represent in the T-S plane water masses I and II.

(2) The points of the straight T-S line of two infinite water masses I and II, corresponding to the values of parameter $+z$ and $-z$, are symmetrical in relation to the middle of segment I–II at any time. In particular, point $z = 0$ coincides with the middle of segment I–II at any moment of time (this theorem relates to the case when $k_1 = k_2$; its proof is obvious).

(3) Points I and II at any time t are limiting points for the set of points z, arranged on the straight T-S line of two infinite water masses I and II. Point M (in Ivanov – the middle of segment I–II) is the limiting point for a set of positions of any definite point z, assumed by it on segment I–II during a successive change of time from 0 to ∞.

(4) Points with identical values z on the two straight T-S lines I–II and I–III, which have a common end in point I, belong (at the same moment in time) to the straight lines parallel to segment II–III.

The latter affirmation is proven in the following way. The coordinates of any point z' on segment $I-II$ are expressed by formulae:

$$T' = \frac{1}{2}\left[T_1 + T_2 + (T_1 - T_2)\,\Phi\left(\frac{z'}{2\sqrt{kt}}\right)\right]$$

$$S' = \frac{1}{2}\left[S_1 + S_2 + (S_1 - S_2)\,\Phi\left(\frac{z'}{2\sqrt{kt}}\right)\right]$$

while the coordinates of any point z'' on the segment $I-III$ are expressed by formulae:

$$T'' = \frac{1}{2}\left[T_1 + T_3 + (T_1 - T_3)\,\Phi\left(\frac{z''}{2\sqrt{kt}}\right)\right]$$

$$S'' = \frac{1}{2}\left[S_1 + S_3 - (S_1 - S_3)\,\Phi\left(\frac{z''}{2\sqrt{kt}}\right)\right]$$

Establishing the same direction for the reckoning of coordinates on both T-S segments, let us draw straight lines through points z' of the first and through the same points z'' of the second segment, and let us calculate the tangent of the angle of slope of these straight lines to the axis S. Obviously, we will obtain for any such straight line:

$$\tan\alpha = \frac{T'' - T'}{S'' - S'} = \frac{T_3 - T_2}{S_3 - S_2} \qquad\qquad [41.13]$$

which proves theorem 4.

The last theorem is important to generalize the case of mixing of two infinite water masses to a larger quantity of infinite water masses by means of the simple graphic methods proposed by Ivanov.

The effect of inequality of the coefficients of turbulence

Up to now we have considered the case where the coefficients of turbulent heat conduction and turbulent diffusion of salinity were deemed to be identical: $k_T = k_S$. Let us now investigate the question of how the inequality of these coefficients affects the results of the T-S analysis of two mixing water masses of infinite depth. In this connection, in accordance with contemporary ideas about the nature of marine turbulence, we will assume that heat exchange is carried out with more intensity than the exchange of salts, i.e., that $k_T > k_S$. From the physical point of view such an assumption is obvious, since turbulent heat exchange is stimulated also by the direct radiation of heat, the direction of which has the same sign; it is clear that this cannot be said about

the process of exchange of solid substances. In so doing, we will not prejudge
the relations between the coefficients of exchange, since this question has
practically not been investigated enough; more than that, the introduction into
T-S analysis of the fact of inequality of the coefficients may, with further
comparison of the results of theory with empirical data, by itself shed light
on this question.

Thus, expression [40.11] is considered again, in which it is now necessary to insert indexes in designating the coefficient k:

$$T(z,t) = \frac{1}{2} \left[T_1 + T_2 + (T_1 - T_2) \, \Phi \left(\frac{z}{2\sqrt{k_T t}} \right) \right]$$

$$S(z,t) = \frac{1}{2} \left[S_1 + S_2 + (S_1 - S_2) \, \Phi \left(\frac{z}{2\sqrt{k_S t}} \right) \right] \qquad [41.14]$$

The question of the effect of inequality of the coefficients of turbulence
on the mixing of water masses was investigated for the first time by Ivanov
(1949), who demonstrated that the result of this mixing was represented in
the *T-S* plane no longer by a straight line but by some curve. Indeed, we can
no longer immediately eliminate function $\Phi(z/2\sqrt{kt})$ from expressions [41.14],
as was done by us above, since by virtue of the inequality of $k_T \neq k_S$ we have
$\Phi(z/2\sqrt{k_T t}) \neq \Phi(z/2\sqrt{k_S t})$, which does not enable us to come to the equation of the straight line passing through the given points. On the other hand,
if, recalling the expression for the derivative by z of the integral of probability [41.4], we calculate the tangent of the angle of slope to the *T-S* curve
mentioned, then, as Ivanov demonstrated, we will obtain:

$$\tan \alpha = \frac{dT}{dS} = \frac{(T_1 - T_2)\sqrt{k_S}}{(S_1 - S_2)\sqrt{k_T}} \, \exp \frac{z^2}{4t} \left(\frac{k_T - k_S}{k_T \, k_S} \right) \qquad [41.15]$$

We see from this expression that the slope of the tangent to the *T-S* curve
depends on z, i.e., changes along the *T-S* curve. It is essential that the form
of this line does not depend on t, but is wholly determined by the relationship of coefficients k_T and k_S; the circumstance that any two successive
stages of mixing are represented in the *T-S* plane by the same *T-S* curve, but
with a different (as also in the case of the straight *T-S* line) arrangement of
parameter z along it is proved in the following way: if $t_1 > t$, then $t_1 = \alpha_1 t$.
Then in expressions [41.14] we have:

$$\Phi \left(\frac{z}{2\sqrt{k_T \alpha_1 t}} \right) = \Phi \left(\frac{\frac{z}{\sqrt{\alpha_1}}}{2\sqrt{k_T t}} \right)$$

i.e., the effect of time on the form of the T-S curve does not really take place (Ivanov, 1949).

In the particular case when $k_T = k_S$, from formula [41.15] we obtain the expression for the tangent of the angle of slope of the tangent:

$$\tan \alpha = \frac{dT}{dS} = \frac{T_1 - T_2}{S_1 - S_2}$$

which in this case coincides with the T-S curve itself.

To study the effect of inequality of the coefficients it is useful to construct T-S curves at various ratios of these coefficients. In Fig. 38A are represented three such T-S curves constructed on the assumption that the value of parameter $2\sqrt{k_S t} \cdot 10^{-4}$ is equal to 2, while the values of parameter $2\sqrt{k_T t} \cdot 10^{-4}$ are equal to 4, 6 and 10 (curves a, b and c respectively). Such a selection of parameters may, as is seen from the nomogram (Fig. 37), correspond to the following values of the coefficients, if the interval of time taken for an example is equal to 200 days: $k_S = 5.8$ cm^2/sec; $k_T = 23$, 52 and 143 cm^2/sec respectively. In addition, in Fig. 38B are represented three T-S curves constructed at different values of the parameter, namely: $2\sqrt{k_S t} \cdot 10^{-4} = 4$; $2\sqrt{k_T t} \cdot 10^{-4} = 6$, 8 and 10 (curves a, b, c respectively). We see from Fig. 38 that with an increase in the coefficient of turbulent heat conduction in relation to the coefficient of diffusion, the T-S curves move ever further away from the straight T-S line I–II, which corresponds on the T-S plane to the case of equality of these coefficients. A brief study of this figure indicates how useful is the direct plotting of T-S curves from the formulae which represent the solutions of equations [39.5]. Indeed, as has already been said, the direct elimination of function $\Phi(z/2\sqrt{kt})$ from expressions [41.14] for the purpose of obtaining the analytical expression of the T-S curve, becomes difficult, provided that $k_T \neq k_S$; on the other hand, the study of the figure gives us an idea of the course of the process of mixing in its representation on the T-S plane, and moreover, in the majority of cases, so complete an idea that often the need for corresponding mathematical confirmation disappears. In the case considered, the parametric points z move from the initial position, fully corresponding to that considered earlier (given the condition of equality of the coefficients) – along the corresponding T-S curve to the middle of this T-S curve (to the point of symmetry or to the point equidistant from its ends). When $t = \infty$ all points z assemble in this central point, representing the thermohaline index of the resulting mixture. The positions of points with identical values of parameter z on different T-S curves are indicated by broken lines which in this case may be called *isobaths*.

Here the following important general observation should be made. In this example, and in all subsequent cases, the plotting of T-S relationships from

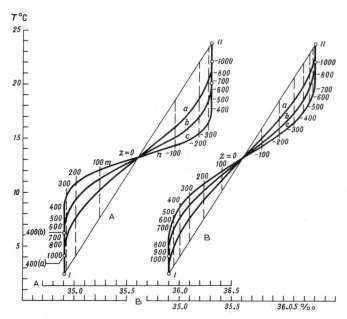

Fig. 38. *T-S* lines appearing on mixing of two water masses in an ocean of infinite depth when $k_T > k_S$. The broken lines are isobaths (or *T-S-t* curves). The thermohaline indexes are the same as in Fig. 36, and correspond to the deep and bottom water masses of the North Atlantic and the water mass of the Gulf Stream.

A. $2\sqrt{k_S t} = 2 \cdot 10^4$. (a) $2\sqrt{k_T t} = 4 \cdot 10^4$; (b) $2\sqrt{k_T t} = 6 \cdot 10^4$; (c) $2\sqrt{k_T t} = 10 \cdot 10^4$.

B. $2\sqrt{k_S t} = 4 \cdot 10^4$. (a) $2\sqrt{k_T t} = 6 \cdot 10^4$; (b) $2\sqrt{k_T t} = 8 \cdot 10^4$; (c) $2\sqrt{k_T t} = 10 \cdot 10^4$.

formulae [41.14] and others, more complex, leads to the picture represented by two families of parametric lines: *T-S* curves by parameter *z* when *t* = constant (these lines in Fig. 38 are represented by continuous lines and correspond to the "ordinary" *T-S* curves) and *T-S* curves by parameter *t* when *z* = constant (in Fig. 38 — broken lines). In the future, we will often call these families of *T-S* curves *T-S-z* curves and *T-S-t* curves. It is clear that, taking account of what has been said above, all equations of the type [40.9]–[40.11], with the analogous equations accompanying them with respect to salinity, are also *parametric equations*. The sole exception is the case considered in the preceding section when both sets of parametric *T-S* curves merge at any values of parameters *z* and *t* in one segment of the straight line.

In connection with what has been said, it is interesting to draw attention also to Fig. 39, which differs from Fig. 38 only by the fact that the three *T-S* curves represented in it are plotted not at a fixed value of parameter $2\sqrt{k_S t}$, but at a fixed value of parameter $2\sqrt{k_T t}$, namely at $2\sqrt{k_T t} \cdot 10^{-4} = 10$, and three different values of parameter $2\sqrt{k_S t} \cdot 10^{-4}$, namely 6, 4 and 2 (curves *a*, *b* and *c* in Fig. 39 respectively). We see that in this figure the *T-S-t* curves in principle are in no way different from the curves in Fig. 38, while the iso-

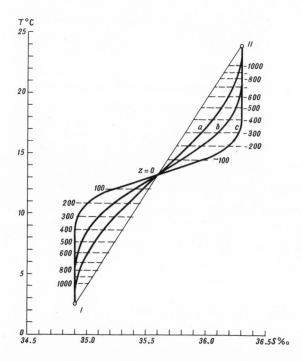

Fig. 39. *T-S* curves appearing on mixing of two water masses in an ocean of infinite depth when $k_T > k_S$. Compare with Fig. 38 (explanations in the text).
$2\sqrt{k_T t} = 10 \cdot 10^4$. (*a*) $2\sqrt{k_S t} = 6 \cdot 10^4$; (*b*) $2\sqrt{k_S t} = 4 \cdot 10^4$; (*c*) $2\sqrt{k_S t} = 2 \cdot 10^4$.

baths are arranged not vertically (parallel to the axis T), as in Fig. 38, but horizontally (parallel to the axis S). This is perfectly clear if we bear in mind that in the first case coefficient k_T/k_S changes and in the second case coefficient k_S/k_T. In case both coefficients change simultaneously, the *T-S-t* curves in the selection of a definite relationship between k_T and k_S, namely a relationship of the type $k_S/k_T = f(t)$, will have a more complex form, testifying to the distinctive "elusiveness" of the parameter z; this quantity, as it were, goes beyond the framework of the analytical theory of *T-S* curves, leaving us at the present stage of its development the possibility of only extremely approximate statements concerning parameter z.

The question of the effect of the uneven intensity of turbulent heat conduction and salt diffusion on the representation of mixing processes on the *T-S* plane was also considered by Stommel (1962a, b). Considering the results of an idealized logical experiment and not resorting to the solution of equations, Stommel, apparently unfamiliar with Ivanov's (1949) work, in one case comes to the same results which were considered by us above. He points out that, in the initial stage of mixing, in the presence of stable density stratification the *T-S* points, corresponding in our exposition to the points z, must move first of all from the points of the original thermohaline indexes in directions roughly parallel to the axis of temperature. After the temperature contrasts are smoothed out to a certain extent, diffusion becomes more effective, and the *T-S* points begin to move on the *T-S* diagram in a direction closer to the horizontal (the section *mn* of *T-S* curve *c* in Fig. 38A corresponds to this stage of Stommel's logical experiment). The final expression also corresponds to the median point. "The swarms of volumes" considered by Stommel in his logical experiment are nothing else but the parametric points *z* in the analytical theory; the logical considerations of the analytical theory lead, as we see, to the same results. Stommel develops a picture of the different versions of mixing and convection from the first case considered to an area of large generalizations, where analytical theories, at least in their present phase of development, are still unacceptable. Let us also indicate that Stommel, together with Turner (Turner and Stommel, 1964), carried out a real laboratory experiment to excite convection in a stably stratified medium. In order not to digress, we will not dwell on the results of these authors.

42. THE MIXING OF THREE WATER MASSES IN AN OCEAN OF INFINITE DEPTH

The solution of equation [39.5] in the case of mixing of three water masses in a sea of infinite depth is given by [40.10]; similarly for salinity we have:

$$S(z,t) = \frac{1}{2} \left[S_1 + S_3 + (S_1 - S_2) \, \Phi \left(\frac{z-h}{2\sqrt{kt}} \right) + (S_2 - S_3) \, \Phi \left(\frac{z+h}{2\sqrt{kt}} \right) \right] \qquad [42.1]$$

provided that $z = 0$ corresponds to the middle of the intermediate water mass. Before analyzing the properties of these solutions on the *T-S* plane it is useful to construct families of parametric *T-S-z* curves, and *T-S-t* curves, which we have done in Fig. 40 on the assumption that coefficients k_T and k_S are identical in the whole water layer. For the plotting of Fig. 40 we have selected the thermohaline indexes of real water masses observed in the Eastern North Pacific Ocean:

I (1.30°; 34.7‰) — North Pacific Deep (and Bottom) Water (Sverdrup et al., 1942);

II (9.0°; 33.6‰) — Subarctic Intermediate Water of the North Pacific;

III (20.0°; 35.1‰) — modification of the Eastern North Pacific Central Water (according to Sverdrup's terminology).

The type of real *T-S* curves, formed as a result of the vertical interaction of these water masses, was represented in Fig. 34 in the example of three stations of the research vessel "Carnegie"; these real *T-S* curves will be necessary below in order to compare them with theoretical curves.

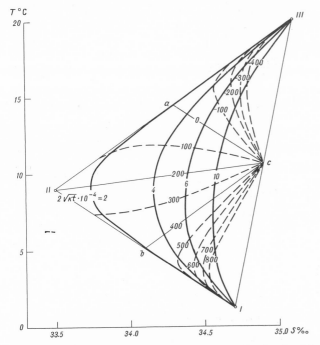

Fig. 40. T-S-z curves (solid lines) and T-S-t curves (broken lines) in the case of mixing of three water masses in an ocean of infinite depth. IIc is the principal, ac and bc are the secondary medians of the triangle of mixing (Mamayev, 1966b).

The T-S curves in Fig. 40 are plotted from formulae [40.10] and [42.1] at values of parameter $2\sqrt{k_T t} \cdot 10^{-4} = 2$, 4, 6 and 10 cm and at a thickness of the intermediate (subarctic) water equal to 400 m. This figure immediately gives us an idea of the course of the process of mixing of three water masses, which we will analyze below in more detail, paying most attention to the features of the transformation of the *intermediate water*, since it is precisely they which form the basis of the geometry of T-S curves of Shtokman (1943a), developed by him to apply to the mixing of three water masses in an infinite sea; Shtokman's theory is set forth below.

Thus, just as for the case of mixing of two water masses, let us trace the development of the process of mixing in its representation on the T-S plane. At the initial moment in time, when $t = 0$, all points z for region $+\infty > z > +h$ are grouped in point I; all points z for region $+h > z > -h$ are grouped in point II; all points z for the region $-h > z > -\infty$ are grouped in point III. Indeed, we have

for region $+\infty > z > +h$:

$$\Phi\left(\frac{z-h}{2\sqrt{kt}}\right) = \Phi(+\infty) = +1$$

$$\Phi\left(\frac{z+h}{2\sqrt{kt}}\right) = \Phi(+\infty) = +1 \; ; \qquad T(z,0) = T_1$$

for region $+h > z > -h$:

$$\Phi\left(\frac{z-h}{2\sqrt{kt}}\right) = \Phi(-\infty) = -1$$

$$\Phi\left(\frac{z+h}{2\sqrt{kt}}\right) = \Phi(+\infty) = +1 \; ; \qquad T(z,0) = T_2$$

for region $-h > z > -\infty$:

$$\Phi\left(\frac{-z-h}{2\sqrt{kt}}\right) = \Phi(-\infty) = -1$$

$$\Phi\left(\frac{-z+h}{2\sqrt{kt}}\right) = \Phi(-\infty) = -1 \; ; \qquad T(z,0) = T_3$$

Consequently, the initial moment of transformation is characterized by the presence on the *T-S* plane of only three points — the thermohaline indexes *I, II* and *III*. On the expiration of the initial moment, parametric points *z* instantaneously distribute themselves along the straight lines *I–II* and *II–III*; moreover, in however small vicinities of points *I* and *III* remains, however, large a quantity of discrete points *z*; in the vicinity of point *II* only a finite set of them appear. Just as instantaneously on the interfaces between water masses, mean values of temperature and salinity are established — values equal to half the sum of the values of the temperature and salinity of the water masses adjacent to the boundary. Indeed, from formulae [40.10] and [42.1] when $z = +h$ we obtain:

$$T(+h, t) = \frac{1}{2}\left[T_1 + T_3 + (T_2 - T_3)\,\Phi\left(\frac{h}{\sqrt{kt}}\right)\right]$$

$$S(+h, t) = \frac{1}{2}\left[S_1 + S_3 + (S_2 - S_3)\,\Phi\left(\frac{h}{\sqrt{kt}}\right)\right]$$

[42.2]

while when $z = -h$ we obtain:

$$T(-h, t) = \frac{1}{2} \left[T_1 + T_3 - (T_1 - T_2)\, \Phi \left(\frac{h}{\sqrt{kt}} \right) \right]$$

$$S(-h, t) = \frac{1}{2} \left[S_1 + S_3 - (S_1 - S_2)\, \Phi \left(\frac{h}{\sqrt{kt}} \right) \right]$$

[42.3]

At the initial moment, when $t = 0$, since $\Phi(\infty) = 1$, from these formulae we obtain:

$$T|_{z=+h} = \frac{T_1 + T_2}{2} \; ; \qquad S|_{z=+h} = \frac{S_1 + S_2}{2}$$

$$T|_{z=-h} = \frac{T_2 + T_3}{2} \; ; \qquad S|_{z=-h} = \frac{S_2 + S_3}{2}$$

[42.4]

A peculiar uncertainty, arising *in statu nascendi* and consisting of the fact that at the initial moment on the interfaces we have both discontinuity of functions $T(z)$ and $S(z)$ and the presence of the half-sums [42.4], is genetically linked with the concept of the physical (heat) impulse and with the even more abstract notion of the delta function of Dirac, and should not therefore be the subject of any misunderstanding.

The subsequent stage of the transformation consists of the appearance of *a continuous T-S curve*, the branches (ends) of which continue to coincide with sides *II–I* and *II–III* of the triangle of mixing, while in the region of the thermohaline index *II* a rounding appears — this is precisely the sign of a *T-S* curve. To prove the continuity of the *T-S* curve, let us construct an expression for the tangent of the angle of slope of the tangent to the *T-S* curve at any point of it; for this, recalling the expression of the derivative of the integral of probabilities $\Phi_z'(z)$ — formula [41.4], we obtain as a result of the differentiation of formulae [40.10] and [42.1] and the division of the results (Shtokman, 1943a):

$$\tan \alpha = \frac{dT}{dS} = \frac{(T_1 - T_2) \exp\left[-(z-h)^2/4kt\right] + (T_2 - T_3) \exp\left[-(z+h)^2/4kt\right]}{(S_1 - S_2) \exp\left[-(z-h)^2/4kt\right] + (S_2 - S_3) \exp\left[-(z+h)^2/4kt\right]}$$

[42.5]

In this expression taking out the multiplier $\exp\left[-(z-h)^2/4kt\right]$ from parentheses in the numerator and the denominator and reducing it, we obtain:

$$\tan \alpha = \frac{T_1 - T_2 + (T_2 - T_3) \exp\left[\{(z-h)^2 - (z+h)^2\}/4kt\right]}{S_1 - S_2 + (S_2 - S_3) \exp\left[\{(z-h)^2 - (z+h)^2\}/4kt\right]}$$

or, transforming the exponent:

$$\tan \alpha = \frac{T_1 - T_2 + (T_2 - T_3) \exp(-zh/kt)}{S_1 - S_2 + (S_2 - S_3) \exp(-zh/kt)} \qquad [42.6]$$

On the other hand, reducing the numerator and denominator of expression [42.5] by the quantity $\exp[-(z+h)^2/4kt]$, in the same way we can come to the expression:

$$\tan \alpha = \frac{T_2 - T_3 + (T_1 - T_2) \exp(zh/kt)}{S_2 - S_3 + (S_1 - S_2) \exp(zh/kt)} \qquad [42.7]$$

It is natural that both functions [42.6] and [42.7] when t = constant $(t \neq 0)$ are continuous in the whole region $+\infty > z > -\infty$; at small values of t (when $t \to 0$), $\exp(-zh/kt) \to 0$, and we have from expressions [42.6] and [42.7] respectively:
for the region $-\infty > z > 0$:

$$\tan \alpha = \frac{T_1 - T_2}{S_1 - S_2} \qquad [42.8]$$

for the region $-\infty > z > 0$:

$$\tan \alpha = \frac{T_2 - T_3}{S_2 - S_3} \qquad [42.9]$$

These results bear witness to the coincidence of the branches of the *T-S* curves with the corresponding sides of the triangle of mixing and moreover this coincidence, always taking place when $z \to \pm\infty$ *, is observed at the lower the values of parameter z, the earlier the moment of transformation t we consider (see Fig. 40).

Further, assuming in formula [42.6] or in formula [42.7] $z = 0$, in both cases we will obtain:

$$\tan \alpha = \frac{T_1 - T_3}{S_1 - S_3} \qquad [42.10]$$

Consequently, the tangent to the *T-S* curve in that point of it which corresponds to the *core* of the intermediate water mass ($z = 0$), is parallel to side *I–III*, which Shtokman called the *base* of the triangle of mixing.

Thus, mentally shifting the point of contact along the *T-S* curve throughout the whole region of values of parameter z, from $+\infty$ to $-\infty$, we are satisfied that the tangent from position *I–II* gradually proceeds to position *II–III*, bypassing, when $z = 0$, the position in which it proves parallel to the base of the triangle of mixing *I–III*.

---

* The same results are valid when t = constant and when $z \to \pm 0$.

The analysis performed characterizes any stage of mixing, excluding the initial and final stages (the latter we shall consider below), i.e., that basic interval of time during which are formed and gradually transformed the T-S curves themselves in which we are interested.

Let us now consider the process of transformation of the *special points* of the T-S curves, which correspond on the T-S diagram to the boundaries and core of the intermediate water mass. For this, let us return to formulae [42.2] and [42.3], which are noteworthy for the fact that they describe the nature of the change in the thermohaline indexes characteristic of the interface between water masses; in other words, these are parametric equations of T-S-h curves. Thus, unlike the mixing process of two water masses, the values of temperature and salinity on the interfaces between three water masses change. Physically this is explained by the gradual degeneration of the intermediate layer, the heat and salt reserve of which, by reason of the finiteness of this layer along the vertical, is gradually "absorbed" by the neighboring water masses, infinite upwards and downwards. In the final analysis, as we shall be able to see below, the intermediate water mass will degenerate altogether.

Eliminating function $\Phi(h/\sqrt{kt})$ from equations [42.2] and [42.3], in the first case we will come to the expression:

$$T - \frac{T_1 + T_3}{2} = \frac{T_2 - T_3}{S_2 - S_3}\left(S - \frac{S_1 + S_3}{2}\right) \qquad [42.11]$$

and in the second to the expression:

$$T - \frac{T_1 + T_3}{2} = \frac{T_1 - T_2}{S_1 - S_2}\left(S - \frac{S_1 + S_3}{2}\right) \qquad [42.12]$$

The first of them represents the equation of the straight line passing through two points: the middle of side I–II (point a in Figs. 40 and 41) and the middle of side I–III (point c in Figs. 40 and 41); the second expression is the equation of the straight line passing respectively through points b and c. The coordinates of points a, b and c are:

$$\left(\frac{T_1 + T_2}{2}, \frac{S_1 + S_2}{2}\right); \qquad \left(\frac{T_2 + T_3}{2}, \frac{S_2 + S_3}{2}\right); \qquad \left(\frac{T_1 + T_3}{2}, \frac{S_1 + S_3}{2}\right)$$

Thus, the lines of transformation of the thermohaline indexes, corresponding to the interfaces between water masses, are *straight lines*.

Finally, let us see how in the T-S plane the transformation of the thermohaline index of the *core* of the intermediate water mass takes place, understanding by this term, as has already been said, the thermohaline index corresponding to points with the value $z = 0$. Obviously, the core must also degenerate by reason of the above-mentioned "absorption" of the intermediate

$T°C$

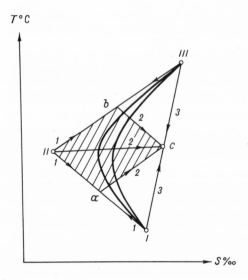

Fig. 41. Diagram of the successive transformation of three mixing water masses in an ocean of infinite depth (according to Shtokman, 1943a).
IIc is the principal, *ac* and *bc* are the secondary medians. The shaded area characterizes the intermediate water mass. The arrows show the direction of movement of parametric points z, while the numbering of these arrows corresponds to the sequence of events.

water mass by the surface and deep waters; assuming in formula [42.1] $z = 0$, we obtain:

$$T = \frac{T_1 + T_3}{2} + \left(T_2 - \frac{T_1 + T_3}{2}\right) \Phi \left(\frac{h}{2\sqrt{kt}}\right)$$

$$S = \frac{S_1 + S_3}{2} + \left(S_2 - \frac{S_1 + S_3}{2}\right) \Phi \left(\frac{h}{2\sqrt{kt}}\right)$$

[42.13]

Eliminating function $\Phi(h/2\sqrt{kt})$ from these expressions, we come to the equation of the $T-S-0$ straight line:

$$T - \frac{T_1 + T_3}{2} = \frac{T_2 - \frac{T_1 + T_3}{2}}{S_2 - \frac{S_1 + S_3}{2}} \left(S - \frac{S_1 + S_3}{2}\right)$$

[42.14]

passing through point *II* $(T_2; S_2)$ and point $c[(T_1 + T_3)/2, (S_1 + S_3)/2]$. Line *IIc* Shtokman (1943a) called the *principal*, and lines *ac* and *bc* — the *secondary medians* of the triangle of mixing.

Let us now consider how the final stage of the mixing process presents itself. Since when $t = \infty$ any of the functions $\Phi(z)$ turns into zero, from formulae [40.10] and [42.1], as well as from formulae [42.12] and [42.13], we obtain:

$$T|_{t=\infty} = \frac{T_1+T_3}{2}\;;\qquad S|_{t=\infty} = \frac{S_1+S_3}{2} \qquad\qquad [42.15]$$

Thus, the final product of the mixing is represented by the thermohaline index $c[(T_1+T_3)/2, (S_1+S_3)/2]$, corresponding to the middle of side $I–III$; we would have the same picture if only two water masses were mixed in an ocean of infinite depth, namely — water masses I and III. The identical final result, occurring on the mixing of two and three water masses in an infinite sea, testifies once again to the complete dissolution of the intermediate water mass when $t = \infty$.

Accordingly, the whole mixing process takes place schematically in the following way. The initial moment, characterized by the presence in the T-S plane of only three thermohaline indexes — I, II and III, is instantaneously replaced, by virtue of the movement of discrete points z, by the appearance of sides $I–II$ and $II–III$ of the triangle of mixing; reference has already been made to the distribution of parameter z along these sides. Because of the transformation of the intermediate water mass and its core, the angle $I–II–III$ is replaced by the rounded T-S curve. This T-S curve gradually tightens, occupying the successive positions represented in Fig. 40, to the base of the triangle of mixing — side $I–III$. At the same time, the points with the same values of parameter z slide along the broken lines represented on Fig. 40, gathering in point c. The state of the "two water masses", advancing when $t = \infty$, instantaneously gives way to the final picture in the form of thermohaline index c; moreover, all discrete points z of the entire region $+\infty > z > -\infty$ turn up in point c when $t = \infty$. This picture of the transformation, explained by Fig. 40, is also illustrated by diagram 41, borrowed from Shtokman's work (1943a, fig. 8). In Fig. 41 the shaded parallelogram characterizes the region of transformation of the intermediate water mass; the arcs of the curves, designated by the heavy lines, belong to this intermediate water. Comparing Fig. 41 with the percentage nomogram known to us (Fig. 25), we see that the boundaries of the intermediate water mass (the secondary medians) correspond to the lines of 50% content of the surface and deep water masses I and III. So far as the percentage amount of the intermediate water mass is concerned, along the secondary medians, as may be seen from a comparison of Figs. 25 and 41, it decreases from 50% (when $t = 0$) to 0% (when $t = \infty$).

Special interest attaches to the question of the arrangement within the triangle of mixing of the points with identical values of parameter z, but of the opposite sign, belonging to the same T-S curve; this question is important in the theory of the method of determining by T-S curves the coefficients of mixing k_S and k_T (see Section 47).

These points, obviously, lie on both sides of the principal median of the triangle of mixing, i.e., belong to regions $+\infty > z > 0$ and $0 > z > -\infty$ respec-

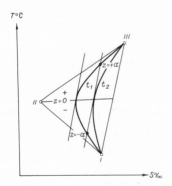

Fig. 42. Tangents to the *T-S* curves (according to Shtokman, 1943a). Explanations in the text.

tively. Let these points possess values of parameter $z = +a$ and $z = -a$ (Fig. 42). Then, on the basis of formulae [40.10] and [42.1], the values of temperature in these points are determined by the expressions:

$$T(+a,t) = \frac{1}{2}\left[T_1 + T_3 + (T_1 - T_2)\,\Phi\left(\frac{a-h}{2\sqrt{kt}}\right) + (T_2 - T_3)\,\Phi\left(\frac{a+h}{2\sqrt{kt}}\right)\right]$$

$$T(-a,t) = \frac{1}{2}\left[T_1 + T_3 + (T_1 - T_2)\,\Phi\left(\frac{-a-h}{2\sqrt{kt}}\right) + (T_2 - T_3)\,\Phi\left(\frac{-a+h}{2\sqrt{kt}}\right)\right] \qquad [42.16]$$

the values of salinity are determined by similar formulae.

Let us calculate the tangent of the angle of slope of the straight line joining these points; for this, apparently, it is necessary to construct differences $T_{+a} - T_{-a}$ and $S_{+a} - S_{-a}$ and divide them; bearing in mind oddness $\Phi(-z) = -\Phi(z)$, we obtain the following expression for the difference of temperatures:

$$T_{+a} - T_{-a} = \frac{1}{2}\left[(T_1 - T_2)\,\Phi\left(\frac{a-h}{2\sqrt{kt}}\right) + (T_2 - T_3)\,\Phi\left(\frac{a-h}{2\sqrt{kt}}\right) + \right.$$
$$\left. + (T_2 - T_3)\,\Phi\left(\frac{a+h}{2\sqrt{kt}}\right) + (T_1 - T_2)\,\Phi\left(\frac{a+h}{2\sqrt{kt}}\right)\right]$$

or, after the reduction of similar terms:

$$T_{+a} - T_{-a} = \frac{T_1 - T_3}{2}\left[\Phi\left(\frac{a-h}{2\sqrt{kt}}\right) + \Phi\left(\frac{a+h}{2\sqrt{kt}}\right)\right]$$

For salinity we obtain respectively:

$$S_{+a} - S_{-a} = \frac{S_1 - S_3}{2}\left[\Phi\left(\frac{a-h}{2\sqrt{kt}}\right) + \Phi\left(\frac{a+h}{2\sqrt{kt}}\right)\right]$$

Thus:

$$\tan \alpha = \frac{T_{+a} - T_{-a}}{S_{+a} - S_{-a}} = \frac{T_1 - T_3}{S_1 - S_3} \qquad [42.17]$$

i.e., the straight line connecting points $z = +a$ and $z = -a$, is parallel to the base of the triangle of mixing, $I-III$. Since it was proved above that the tangent to any T-S curve in the point $z = 0$ is also parallel to side $I-III$, it follows from what has been said that each such tangent to the given T-S curve is a secant passing through points $z = \pm a$ for some other T-S curve, preceding it in time; and, on the contrary, each secant to the given T-S curve is at the same time a tangent to some other T-S curve, subsequent to it in time. The relationship between the intervals of time Δt and the corresponding intervals Δz between points $+a$ and $-a$ will be brought out below (see Section 47).

With this we shall conclude the construction of the analytical theory of T-S curves for the case of mixing of three water masses in an ocean of infinite depth (provided the coefficients of heat and salt exchange are equal, $k_T = k_S$). The results of this theory, formulated by Shtokman (1943a) in the form of the following theorems, proof of which was demonstrated above, may be set forth thus:

(1) At the initial moment of mixing the T-S curve is a broken line, consisting of two straight lines which successively connect in the T-S plane the thermohaline indexes of the mixing water masses.

(2) In the points of the T-S curves, sufficiently far removed from the boundaries of the intermediate water mass, the tangents to the T-S curves practically coincide with the sides of the triangle of mixing.

(3) The points of the T-S curves, which correspond to the core of the intermediate water mass and possess the value of parameter $z = 0$, are simultaneously the points of the extreme of the T-S curves. The direction of the tangents in these points is parallel to the base of the triangle of mixing.

(4) The points locus with values of parameter $z = 0$, characterizing the transformation of the core of the intermediate water mass with time, represents the principal median of the triangle of mixing; in complete mixing of the water masses the latter are transformed into a new water mass, the thermohaline index of which corresponds to the coordinates of the midpoint of the base of the triangle of mixing.

(5) All points corresponding in the T-S coordinates to water on the boundaries of the intermediate layer lie on the secondary medians of the triangle of mixing. These straight lines cut off on the T-S curves the arcs characterizing water of the intermediate layer.

(6) The straight lines connecting on the T-S curve two points possessing equal values of parameter z, but of opposite sign, are parallel to the base of the triangle of mixing.

(7) A tangent drawn in a point of the extreme of the given curve, cuts off on another *T-S* curve, preceding in time, two points possessing equal values of parameter z, but of opposite sign.

The theorems formulated of the "geometry of *T-S* curves" constitute at the present time the basis for the analysis of curves of oceanographic stations, carried out for the study of the transformation of the water masses of the ocean. The question of applying analytical theory to the analysis of real *T-S* relationships of the water masses of the World Ocean is considered in Chapter 7.

The effect of inequality of the coefficients of turbulence

The theory of the mixing of three water masses in an ocean of infinite depth set forth above constitutes the basis for the practical analysis of the water masses of the ocean. However, it makes no allowance for the possible fact of inequality of the coefficients of turbulent heat conduction and turbulent salt diffusion. This inequality, as we shall now see, substantially distorts the results of the theory and leads to the most unexpected results.

As was already said in considering the inequality of the coefficients during the mixing of two water masses (Section 41), the elimination of function $\Phi(z)$ in this case becomes impossible in as simple a form as occurs when $k_T = k_S$; the theory becomes considerably more complicated, and its conclusions become not so obvious. Let us confine ourselves therefore to plotting in the *T-S* plane the solutions of equations [40.10] and [42.1] at various values of the coefficients of mixing. In Fig. 43 are represented three *T-S* curves, plotted for the same original water masses *I*, *II* and *III* as before (Fig. 40) with a fixed value of parameter $2\sqrt{k_S t} = 2 \cdot 10^4$ cm and different values of parameter $2\sqrt{k_T t}$. We see from the figure how much the form of the *T-S* curves deviates from the form which occurs provided the coefficients of exchange are equal. Fig. 44 proves even more significant, in which parameter $2\sqrt{k_T t} = 10^5$ is fixed and parameter $2\sqrt{k_S t}$ changes (from 2 to $8 \cdot 10^4$). We see that the greater the difference between the coefficients the more the *T-S* curves become complicated and the more they deviate from "classical" forms. The application of the geometry of *T-S* curves to the analysis of such curves would lead to the determination of false thermohaline indexes, not corresponding to the real original water masses; the example of curve *a* in Fig. 44 testifies to this in an especially striking way. It is curious that the appearance of false thermohaline indexes is tied in with the branches of the *T-S* curves; so far as the index of the intermediate water mass is concerned, it is maintained, and the application of the results of Shtokman's theory to such curves would not lead to such very great errors in the determination of the thermohaline index of the intermediate water mass. Therefore, bearing in

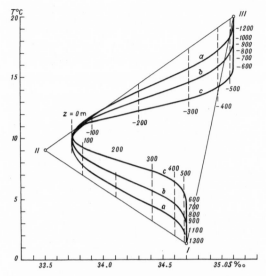

Fig. 43. T-S curves appearing on mixing of three water masses in an ocean of infinite depth when $k_T > k_S$, at value of parameter $2\sqrt{k_S t} = 2 \cdot 10^4$ cm. (a) $2\sqrt{k_T t} = 4 \cdot 10^4$; (b) $2\sqrt{k_T t} = 6 \cdot 10^4$; (c) $2\sqrt{k_T t} = 10 \cdot 10^4$.

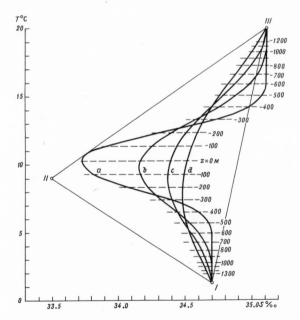

Fig. 44. T-S curves appearing on mixing of three water masses in an ocean of infinite depth when $k_T > k_S$, at value of parameter $2\sqrt{k_T t} = 10 \cdot 10^4$ cm. (a) $2\sqrt{k_S t} = 2 \cdot 10^4$; (b) $2\sqrt{k_S t} = 4 \cdot 10^4$; (c) $2\sqrt{k_S t} = 6 \cdot 10^4$; (d) $2\sqrt{k_S t} = 8 \cdot 10^4$.

mind the obvious fact of the inequality of the coefficients of turbulence, one must pay particular attention in the practical analysis of curves to the bends of such curves lying above and beneath a clearly marked extreme corresponding to the real intermediate water mass, remembering that these "secondary" extremes can be false. This observation applies especially to the surface branches of the T-S curves, since, precisely in the upper layer of the ocean, the distinction between the processes of heat and salt transfer is the most striking. The difference abates as we penetrate into the depth of the ocean, and in practical T-S analysis we do not, as a rule, notice any new or unexpected thermohaline indexes in the bottom parts of real T-S curves. In the surface part of the T-S curve, we repeat, false extremes may very well be observed, which do not correspond to a geographically real water mass (Section 63).

43. THE MIXING OF TWO WATER MASSES IN AN OCEAN OF SEMI-INFINITE DEPTH

In this section we will consider the problem of mixing in an ocean of semi-infinite depth, consisting at the initial moment of time of two layers, the upper of which has the finite thickness h, while the lower extends infinitely downward. This problem, as has already been indicated, will be considered given boundary condition [39.24], which corresponds to the situation when flows of heat and salts through the surface of the sea are absent. This condition assumes consideration of the purely internal mixing of water masses and conforms with the fact that the upper 100-m layer of the ocean, which interacts intensively with the atmosphere, is usually not considered in T-S analysis.

The problem to which we refer is extremely important from the point of view of the study of the origin of the tropospheric water masses of the oceans and of their zonal transformation. Without going into this question in detail, since it receives special consideration in Section 61, let us note that the application of the analytical theory of T-S curves (for the case of mixing of two water masses) to the study of the waters of the North Atlantic will enable us to draw the conclusion that the tropospheric waters of the North Atlantic are formed not in the region of Subtropical Convergence, but represent the result of the interaction of the water mass of the Gulf Stream itself (surface) A with the North Atlantic Deep and Bottom Water B underlying it.

Thus, the solution of the equations of vertical heat conduction and diffusion [39.5] at initial conditions [39.16] and boundary conditions [39.24] is determined by formula [40.33]; let us make special mention of the fact that here we will be considering the case of inequality of coefficients, $k_T > k_S$. In this case it becomes clear immediately that the T-S curves representing the mixing process will, as in the case of an ocean of infinite depth and provided $k_T \neq k_S$, deviate from the straight line of mixing.

In order to ascertain how temperature and salinity change on the sea surface in the process of mixing, let us assume in equation [40.33] that $z = 0$. Then, because of the oddness of $\Phi(-z) = -\Phi(z)$ we obtain for temperature and salinity:

$$T(0,t) = T_1 - (T_1 - T_2)\, \Phi\left(\frac{h}{2\sqrt{k_T t}}\right)$$

$$S(0,t) = S_1 - (S_1 - S_2)\, \Phi\left(\frac{h}{2\sqrt{k_S t}}\right)$$

[43.1]

These equations represent parametric equations (for the parameter t when z = constant) of the line which we shall call *the line of transformation of the surface T-S index*.

In order to elucidate how the process of mixing of two water masses in an ocean of semi-infinite depth is represented on the T-S diagram, it is useful to plot T-S curves and T-S lines of transformation from equations [40.33] and [43.1] respectively, which we shall do as applied to water masses A and B for three ratios of coefficients k_T and k_S:

(1) $k_T = 50$ cm^2/sec, $k_S = 5$ cm^2/sec.

(2) $k_T = 25$ cm^2/sec, $k_S = 5$ cm^2/sec.

(3) $k_T = 10$ cm^2/sec, $k_S = 5$ cm^2/sec.

The results of the corresponding calculations when $h = 500$ m $= 50 \times 10^3$ cm (the mean thickness of water mass A in the region of the Straits of Florida) are represented in Fig. 45, where the curves of transformation of the surface T-S index at various relationships of coefficients k_T and k_S are designated by roman numerals, while the corresponding T-S curves, plotted for one moment of time $t = 2000$ days (as an example) are shown by arabic numerals. The vertical solid and broken lines represent the isolines of parameter t (isochrones) for the lines of transformation and isolines of parameter z (isobaths) for the T-S curves respectively.

Considering Fig. 45, we see that in the process of mixing of these two water masses in an ocean of semi-infinite depth, the T-S curves are shortened and bent, while the apex of the T-S curve slides along the corresponding line of transformation of the surface T-S index in the direction of an increase of t; at the limit, when $t \to \infty$, the curve tightens in point B. In a particular case, when $k_T = k_S$, both the lines of transformation and the T-S curves coincide with the straight line of mixing AB; it is easy to see this by eliminating, when $k_T = k_S$, function $\Phi(z)$ both from equations [9.21] and from equations [43.1].

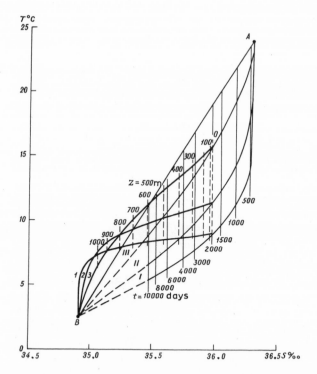

Fig. 45. Curves of transformation of the surface *T-S* index (*I, II, III*) and the corresponding vertical *T-S* curves (1, 2, 3) at *t* = 2000 days, appearing on mixing of water masses *A* and *B* in a semi-infinite sea (Mamayev, 1966a).

The curves are plotted respectively at the following values of the coefficients of mixing: (*1*) k_T = 50, k_S = 5; (*2*) k_T = 25, k_S = 5; (*3*) k_T = 10, k_S = 5. The vertical solid lines are isochrones (isolines of parameter *t*) for curves *I–III* (in days); the vertical straight broken lines are isobaths (isolines of parameter *z*) for curves *1–3* (in meters).

In both cases we will come to the equation of the straight line:

$$T - T_1 = \frac{T_1 - T_2}{S_1 - S_2}(S - S_1)$$

In the particular case mentioned, the *T-S* curve in the process of mixing will be reduced along the straight line *AB*, but again tightening in point *B* when *t* → ∞. In this case, the distribution of parameters *z* and *t* along the straight *T-S* line will depend on the values taken for coefficients k_T and k_S.

It should be pointed out that the solution of equations [39.5] for the case of mixing of two water masses in a sea of semi-infinite depth was considered by Ivanov (1949), but at mixed boundary conditions for temperature and salinity, namely:

$$T(z,t)\big|_{z=0} = T_2$$

$$\frac{\partial S}{\partial z}\bigg|_{z=0} = 0$$

[43.2]

It is obvious that in this particular case the line of transformation of the surface T-S index is a horizontal straight line, coinciding on the T-S diagram with the isotherm $T = T_2$; parameter t distributes itself along this line depending on the value taken for coefficient k_S *.

Having thus considered the general nature of the mixing of two water masses in a sea of semi-infinite depth at various values of the coefficients of mixing k_T and k_S, let us turn to the fundamental question of the formation of the T-S curves of the North Atlantic Central Water.

For this, let us consider Fig. 46, in which are represented two pairs of theoretical T-S curves, plotted from equations at the following values of parameter t and of coefficient k_T (coefficient k_S is taken everywhere as equal to 5 cm^2/sec): T-S curves 1 and 2 (Fig. 46, on the left): $t = 600$ and $1,000$ days respectively, $k_T = 25$ cm^2/sec; T-S curves 3 and 4 (Fig. 46, on the right): $t = 2,000$ and $4,000$ days, $k_T = 10$ cm^2/sec. These pairs of T-S curves can be considered as the extreme (enveloping) curves of two families ("clusters") of T-S curves, emanating from point B; the values of parameter t for the T-S curves lying between curves 1–2 and 3–4 are contained within the limits indicated for the two families: $600 < t < 1,000$ days and $2,000 < t < 4,000$ days, respectively. The isolines of parameter z for each of the two families within the limits from 0 to $1,000$ m are represented in Fig. 46 by thin broken lines. The figure also shows known T-S relationships of the waters of the North Atlantic, according to Sverdrup et al. (1942) and Jacobsen (1929 – compare Fig. 46 with Fig. 105).

Considering Fig. 46, we see that these T-S curves, obtained *as a result of the mixing of water masses A and B* (at selected values of t, k_T and k_S), in those parts of them which correspond to the distribution of parameter z from 0 to about 800–900 m, arrange themselves approximately along the straight line of mixing CS and the T-S relationships of the Central Water, according to Sverdrup. In their lower parts, when $z > 800$ days, the T-S curves have a

---

* Let us point out in passing the inaccuracy of the following assertion of Ivanov (1949, p. 69): "The process of mixing of two water masses is represented in the T-S plane by a segment of a straight line, strictly speaking, only when the sea is infinite in depth (the depths change from $+\infty$ to $-\infty$) and the coefficient of turbulent heat conduction is equal to the coefficient of turbulent diffusion ($k_T = k_S$)". We have seen, however, that the T-S curve can be rectilinear also in the case of mixing of two water masses in a semi-infinite sea. The point here is that, for a semi-infinite sea, to the condition of equality of the coefficients, $k_T = k_S$, it is necessary and adequate to add the condition of identity of boundary conditions for temperature and salinity (conditions [39.24] or conditions of the type of the first condition of [39.25]); Ivanov considered the problem only under mixed boundary conditions [43.2].

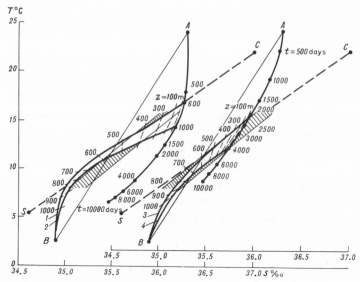

Fig. 46. *T-S* curves (*1–4*) appearing on mixing of water masses *A* and *B* at *t* = 600, 1000, 2000 and 4000 days respectively. For curves *1* and *2* k_T = 25, k_S = 5; for curves *3* and *4* k_T = 10, k_S = 5 (Mamayev, 1966a).

bend, characteristic for the *T-S* curves of the North Atlantic and distinctly demonstrated by the averaged real *T-S* curve represented in Fig. 105. Here a reservation should be made: this bend may very well be the result of the mixing of *three* water masses: the intermediate *S* and the extremes *C* and *B*; however, its formation, independent of water mass in the case of mixing of only *two* water masses, as considered in the present section, is beyond doubt.

Let us now turn to the consideration of Fig. 46 from a somewhat different angle; namely, let us assume that the left and right parts of the figure illustrate two successive stages of the mutual transformation of water masses *A* and *B*. These two stages are separated in time, as may be seen from the figure, by about 1,500 days ($\sim 130 \cdot 10^6$ sec). If one identifies the mixing of stationary water masses in time with their mixing in the process of movement with identical velocity *u*, then on the basis of the relationship $\Delta x = u \Delta t$ we can roughly evaluate distance *x*, which water mass *A* must travel over homogeneous water mass *B* so that the situation represented in the left part of Fig. 46 may be transformed into the situation shown on the right. For this, as a mean velocity of movement of water mass *A* let us take the value *u* = 5 cm/sec. Then for the corresponding distance *x* we will obtain:

$$x = 5 \times 130 \times 10^6 = 6.5 \times 10^8 \text{ cm} = 6,500 \text{ km} \approx 3,500 \text{ miles};$$

this distance, if it is taken along the axis of the Gulf Stream and the North Atlantic current, approximately corresponds to the distance from the Straits of Florida to the Irish coast.

Thus, along with the explanation of the form of T-S curves observed in the North Atlantic by the mutual transformation of water masses A and B, we come to the possibility of explaining this transformation as latitudinal, in the process of displacement of the water mass from the south to the north. It is obvious that in the given formulation of the problem the latitudinal transformation of water A is accompanied by a gradual shortening of the T-S curves along, say, the line CS, as well as by the gradual decrease of the ratio of the coefficients of mixing k_T/k_S; in other words, vertical turbulent heat exchange in the ocean decreases with an increase of latitude more rapidly than salt diffusion, which is perfectly understandable if one bears in mind the more considerable decrease with latitude in temperature (and its vertical gradients) as compared with salinity.

Let us remark that, in spite of all the approximateness of the calculation just quoted, it agrees with some other data. For example, the span of time between two stages of transformation (Fig. 46), taken as equal to about 1,500 days, agrees with the evaluations cited by Stommel (1966) of: (a) the time necessary for the full expenditure of the reserve of potential energy of the warm waters of the Sargasso Sea, i.e., of water mass A (1,700 days), and (b) of the time necessary for the removal of all the warm waters of the North Atlantic at full flow of the Gulf Stream (1,600 days).

Concluding these considerations, let us point out that in an analytical solution of the problem it is quite possible to set the condition of transformation of the surface T-S index strictly along any straight line, for example, the straight line CS. However, this is equivalent to the introduction, instead of [39.24], of boundary conditions dependent on time, which substantially complicates the solution of equations [39.5]; in particular, the numerical implementation of the solution proves to be connected in this case with the need to calculate improper integrals.

44. THE MIXING OF THREE WATER MASSES IN AN OCEAN OF SEMI-INFINITE DEPTH

Let us now consider in more detail the question of the mixing in an ocean of semi-infinite depth of three water masses under boundary condition [39.25], i.e., when constant values of temperature and salinity are maintained on the surface of the ocean (Mamayev, 1966b).

Fig. 47 shows a graph plotted from formula [40.26] and the similar formula for salinity at the same values of the thermohaline indexes I (T_1,S_1), II (T_2,S_2) and III (T_3,S_3) of the water masses of the Northeastern Pacific Ocean and at the same thickness of the intermediate layer (h_1 = 200 m, h_2 = 600 m, $h_2 - h_1$ = 400) as in Fig. 40. In addition, the coefficients of heat and salt exchange are also taken here as equal: $k = k_T = k_S$. The solid curves

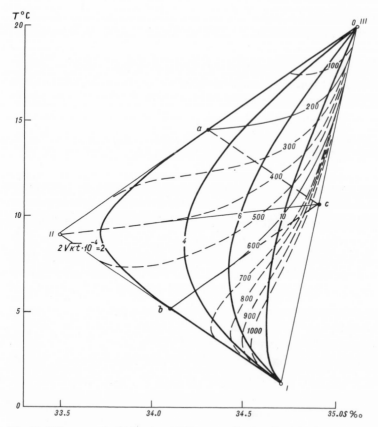

Fig. 47. *T-S-z* curves (solid lines) and *T-S-t* curves (broken lines) in the case of mixing of three water masses in an ocean of semi-infinite depth. The designation of the medians of the triangle is the same as in Fig. 40 (Mamayev, 1966b).

in Fig. 47 represent *T-S-z* curves at the same values as in Fig. 40 of parameter $2\sqrt{kt} \cdot 10^{-4} = 2, 4, 6$ and 10 cm, the broken lines: *T-S-t* curves *.

Comparing Figs. 40 and 47, we see immediately how substantially the theoretical picture of the transformation of three water masses in an ocean of semi-infinite depth differs from that for the infinite ocean. On the one hand, there is every reason for such a comparison, inasmuch as in both cases the triangle of mixing in the process of transformation of the water masses remains invariable and the *T-S* curves do not go beyond its limits (in particular, not one of the *T-S* indexes changes its coordinates, which cannot be said, for example, with respect to the surface index of *III* in the case of mixing of water masses at boundary condition [39.24]). Besides, the *T-S* curves (at the

---

* Using the terminology of thermodynamics, we may call Figs. 40 and 47 *diagrams of state* of water masses during their transformation.

same values of parameter $2\sqrt{kt}$) in both cases are extremely similar; therefore, the general geometric approach to the consideration of T-S curves, the principles of which were ascertained by Shtokman (1943a) for an infinite ocean, may also be extended to the case of the ocean of semi-infinite depth.

However, on the other hand, the picture of the distribution of T-S-t curves in the case of a semi-infinite ocean sharply differs from that for an infinite sea. The first thing which attracts attention is that the isolines, which correspond to the boundaries of the intermediate layer ($h_1 = 200$ m and $h_2 = 600$ m), do not coincide with the secondary medians of the triangle of mixing or, what is the same, with the lines of 50% content of water masses I and III (these lines ac and bc are also plotted in Fig. 47), but represent curves asymptotically approaching point III (T_3, S_3). Indeed, it follows from formula [40.26] when $z = h_1$ and $z = h_2$ that:

$$\lim_{t \to \infty} T(h_1, t) = T_3 ; \qquad \lim_{t \to \infty} T(h_2, t) = T_3$$

and similarly:

$$\lim_{t \to \infty} S(h_1, t) = S_3 ; \qquad \lim_{t \to \infty} S(h_2, t) = S_3$$

whereas on the basis of formulae [40.10] and [42.1] we have:

$$\lim_{t \to \infty} T(0, t) = \frac{T_1 + T_3}{2} ; \qquad \lim_{t \to \infty} T(h, t) = \frac{T_1 + T_3}{2} ;$$

$$\lim_{t \to \infty} S(0, t) = \frac{S_1 + S_3}{2} ; \qquad \lim_{t \to \infty} S(h, t) = \frac{S_1 + S_3}{2}$$

All the remaining T-S-t curves behave in the same way: when $t \to \infty$ they assemble asymptotically in point III (T_3, S_3) (Fig. 47), whereas, in the case of the infinite sea, as may be seen from Fig. 40, they assemble in point $c[(T_1+T_3)/2, (S_1+S_3)/2]$.

Let us consider the process of transformation of the intermediate water mass in a semi-infinite ocean: it is obvious that this water mass, as in the case of an infinite ocean, is also characterized on the T-S diagram (Fig. 47) by sections of T-S-z curves included within the parallelogram $IIacb$.

It will be seen from Fig. 47 that with time the intermediate water mass sinks deeper (the values of parameter z increase along median IIc), whereas in the case of the infinite sea the core of the intermediate water mass remains at the same depth.

Combining the T-S diagrams of Figs. 40 and 47 with the corresponding percentage nomogram, plotted on the same apexes, we can determine the percentage amount of each of the three water masses at any depth in the process of their mixing, which we have accordingly done for the intermediate water

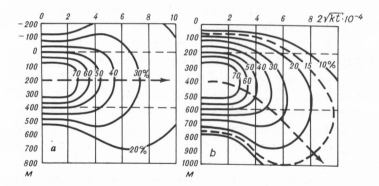

Fig. 48. Percentage content of intermediate water mass *II* in a section in the case of mixing of three water masses. a. In an infinite ocean. b. In a semi-infinite ocean. The thick broken line shows the direction of transformation of the core of the intermediate water mass, the thin broken lines show the original thickness of the intermediate layer (Mamayev, 1966b).

mass. Fig. 48 represents "sections", which show the content of this water mass at various depths in the case of transformation of three water masses in an infinite (a) and semi-infinite (b) ocean. Here the difference mentioned becomes even more evident.

We will not dwell any longer on the elucidation of the differences between these two versions of the theory of *T-S* curves, assuming that the main points have already been noted. Let us indicate in conclusion that everything said relates to the case when the coefficients of heat and salt exchange are taken as equal; it is only in this case that the *T-S* curves do not go beyond the framework of the triangle of mixing.

Drawing attention to the need for further development of the theory of *T-S* curves, in particular for an ocean of semi-infinite depth, we wish in conclusion to emphasize the following two points, which may be drawn as the main conclusions from this investigation:

(1) The question of the distribution of parameter *z* along the *T-S-z* curves is of extremely great importance in the theory of *T-S* curves. From the example of the comparison of Figs. 40 and 47 we saw that *T-S* curves close in form can differ substantially in this respect, and therefore the consideration only of the *form* alone of curves is insufficient for the analysis of interacting water masses and, in particular, of their distribution (percentagewise) by depth. Parameter *z* is distinguished by its peculiar "mobility" along the *T-S* curve, and the features of the transformation of water masses come out first of all in its redistribution along the curve and only later in the change of form of the latter.

(2) The calculation of special "stereotypes" for the analysis of water masses should be recognized as advisable. By this is understood a whole series of

theoretical *T-S* curves which can be calculated by computer for a number of different, to a considerable extent arbitrary, original thermohaline indexes and thicknesses of intermediate layers. By a selection of such theoretical *T-S* curves as would be similar to the real *T-S* curves investigated, not only in form but also by the distribution of parameter *z* along them, we can come to completely unexpected conclusions about the origin of the water masses of the ocean, which go beyond the framework of the usually accepted geometry of *T-S* curves. It is natural that the methods of such comparisons may be somewhat different; they require additional consideration, which should be carried out against the background of the further development of theory.

45. A SIMPLIFIED THEORY OF *T-S* CURVES FOR AN OCEAN OF FINITE DEPTH

The analytical theories of *T-S* curves considered above, constructed for cases of mixing of two, three and four water masses in an ocean of infinite and semi-infinite depth, suffer from one substantial shortcoming, which consists of the fact that they take no account of the *advection of water masses* with the exception of the particular case when the whole ocean may be considered as moving with identical velocity along the vertical, moreover in one direction. Meanwhile the water masses considered in the theory, move, as a rule, in *opposite directions*; this circumstance often makes the analytical theories unsuitable for practical application.

One may proceed in two ways in an attempt to overcome this contradiction. On the one hand, one can complicate the task by introducing convective terms into the equations of heat conduction and diffusion; in this case the analytical solutions also become substantially more complicated, which may deprive the theory of necessary clarity. On the other hand, considerable simplifications are possible which, without depriving the theory of effectiveness, can shed light on new circumstances unnoted in preceding theories.

A simplified theory of *T-S* curves for the case of moving water masses in an ocean of finite depth was constructed by the author (Mamayev, 1962a); an exposition of this theory follows below, from which it proves clear that the possibilities of the most elementary analytical theories of *T-S* curves are far from exhausted. We shall consider first the mixing of two and then three water masses of finite dimensions along the vertical.

The vertical mixing of two water masses

For the study of the vertical mixing of two water masses in a field of currents and of stationary distribution of temperatures along the vertical let us use equations:

$$u_A \frac{\partial T}{\partial x} = k_A \frac{\partial^2 T}{\partial z^2}$$

$$u_B \frac{\partial T}{\partial x} = k_B \frac{\partial^2 T}{\partial z^2}$$

[45.1]

in other words — let us consider stationary mixing along the vertical in a bi-dimensional current; let us direct the axis x along the current. The first of these equations is written for layer A, having thickness h, the second for layer B of thickness H, moreover the thicknesses of the layers are not equal: $h \neq H$. We will count the positive direction of axis z upward, the negative downward from the interface between water masses (Fig. 49). Wishing at the first stage to make the solution as simple as possible, let us assume in equations [45.1] $u =$ constant, $\partial T/\partial x =$ constant and $k =$ constant (in layers A and B these quantities may be different).

The assumption concerning the constancy of the horizontal temperature gradients in the direction of the current is justified for the greater part of the World Ocean (excluding boundary areas), as may be seen from the temperature distribution charts; this distribution is extremely monotonic. The assumption concerning the constant velocity of the current and the constant nature of the coefficient of turbulent heat conduction with depth is made for the purpose of simplifying the solution: the introduction of functions $u(z)$ and $k(z)$ must take place at the following stage, during the development of the elementary theory.

Bearing in mind the simplifications mentioned, we can rewrite equations [45.1] in the form:

$$\frac{d^2 T}{d z^2} = a \qquad (h > z > 0)$$

$$\frac{d^2 T}{d z^2} = b \qquad (0 > z > -H)$$

[45.2]

where a and b are constants. The integrals of equations [45.2] are:

$$T = \frac{az^2}{2} + C_1 z + C_2 \qquad (h > z > 0)$$

$$T = \frac{bz^2}{2} + C_3 z + C_4 \qquad (0 > z > -H)$$

[45.3]

The boundary conditions are as follows:

$$k_A \left. \frac{\partial T}{\partial z} \right|_{z=h} = 0$$

[45.4]

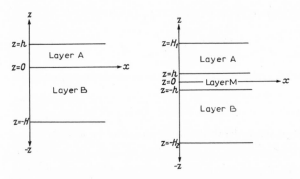

Fig. 49. Two and three water masses in an ocean of finite depth.

i.e., there is no heat flow from the atmosphere (we are considering a "pure" mixing of water masses among themselves):

$$k_B \left. \frac{\partial T}{\partial z} \right|_{z=-H} = 0 \qquad [45.5]$$

i.e., there is also no heat flow through the bottom. In addition:

$$T\big|_{z=+0} = T\big|_{z=-0} = \overline{\overline{T}} \qquad [45.6]$$

(on the interface some mean temperature is observed).

Determining the constants of integration $C_1 - C_4$ from boundary conditions [45.4]–[45.6], we obtain:

$$C_1 = -ah; \qquad C_3 = bH; \qquad C_2 = C_4 = \overline{\overline{T}} \qquad [45.7]$$

Substituting the values obtained of the constants [45.7] in equations [45.2], we obtain the following expressions:

$$T = \overline{\overline{T}} + a \left(\frac{z^2}{2} - hz \right) \qquad (h > z > 0)$$

$$[45.8]$$

$$T = \overline{\overline{T}} + b \left(\frac{z^2}{2} + Hz \right) \qquad (0 > z > -H)$$

For the determination of quantities a and b let us use the following conditions, which give constant T-S indexes on the outer boundaries of the water masses:

$$T\big|_{z=h} = T_A$$

$$[45.9]$$

$$T\big|_{z=-H} = T_B$$

These stationary temperature values form on the boundaries of the water masses in each point of the sea under the influence of constant climatic conditions. Substituting [45.9] respectively in [45.8], we obtain:

$$T_A = \bar{T} - \frac{ah^2}{2}$$

$$T_B = \bar{T} - \frac{bH^2}{2}$$

[45.10]

From these expressions it is possible to determine constants a and b:

$$a = \frac{2(\bar{T} - T_A)}{h^2} \qquad b = \frac{2(\bar{T} - T_B)}{H^2}$$

[45.11]

Expressions [45.11], where $a = u_A(\partial T/\partial x)$, $b = u_B(\partial T/\partial x)$ can correspond to the surface warm and the deep cold current, in Zubov's terminology, both in the same and in the opposite direction.

Finally, substituting expression [45.11] in equations [45.8], we obtain the final formulae which determine the distribution of temperature along the vertical:

$$T = \bar{T} + (T_A - \bar{T}) \left(2\frac{z}{h} - \frac{z^2}{h^2} \right) \qquad (h > z > 0)$$

$$T = \bar{T} - (\bar{T} - T_B) \left(2\frac{z}{H} - \frac{z^2}{H^2} \right) \qquad (0 > z > -H)$$

[45.12]

Formulae [45.12] have a similar appearance to formulae [41.2] and [41.3], with the difference that instead of the integral of errors we have the parabolic function appearing.

For the determination of mean temperature $\bar{T}$ let us use the condition of continuity of curve $T(z)$ at the interface ($z = 0$):

$$k_A \frac{\partial T}{\partial z} = -k_B \frac{\partial T}{\partial z} \qquad \text{when} \quad z = 0$$

[45.13]

Calculating derivatives $\partial T/\partial z$ of expressions [45.12] and satisfying condition [45.13], we obtain:

$$\frac{k_A(T_A - \bar{T})}{h} = \frac{k_B(\bar{T} - T_B)}{H}$$

[45.14]

whence it is easy to determine quantity $\bar{T}$:

$$\bar{T} = \frac{Hk_A T_A + hk_B T_B}{Hk_A + hk_B}$$

[45.15]

Let us note that in the particular case, when $H = h$:

$$\overline{T} = \frac{k_A T_A + k_B T_B}{k_A + k_B} \qquad\qquad [45.16]$$

Let us now apply, for the determination of vertical stationary distribution of salinity in the same two water masses, the equations of diffusion:

$$u_A \frac{\partial S}{\partial x} = k_A \frac{\partial^2 S}{\partial z^2}$$

$$u_B \frac{\partial S}{\partial x} = k_B \frac{\partial^2 S}{\partial z^2}$$

$\qquad\qquad [45.17]$

in which we will assume that the coefficients of vertical turbulent heat conduction and diffusion are equal.

Carrying on the same argument with respect to salinity, we come to the following equations, which are entirely analogous to equations [45.12]:

$$S = \overline{S} + (S_A - \overline{S}) \left(2\frac{z}{h} - \frac{z^2}{h^2} \right) \qquad (h > z > 0)$$

$$S = \overline{S} - (\overline{S} - S_B) \left(2\frac{z}{H} - \frac{z^2}{H^2} \right) \qquad (0 > z > -H)$$

$\qquad\qquad [45.18]$

where quantity $\overline{S}$ of mean salinity is determined from a formula entirely analogous to formula [45.15].

For the representation of the result in T-S coordinates let us eliminate parameter z from equations [45.12] and [45.18] respectively. As a result we will obtain:

$$\frac{T - \overline{T}}{T_A - \overline{T}} = \frac{S - \overline{S}}{S_A - \overline{S}} \qquad (h > z > 0)$$

$$\frac{\overline{T} - T}{\overline{T} - T_B} = \frac{\overline{S} - S}{\overline{S} - S_B} \qquad (0 > z > -H)$$

$\qquad\qquad [45.19]$

Equations [45.19] represent the equations of straight lines in the system of coordinates $T = f(S)$; the first of them is the equation of the straight line joining points (T_A, S_A) and $(\overline{T}, \overline{S})$; the second the equation of the straight line joining points $(\overline{T}, \overline{S})$ and (T_B, S_B). They correspond to equation [41.12] and also indicate that given these assumptions the process of vertical mixing of two water masses is represented on the T-S diagram by a *straight line of mixing*.

Vertical mixing of three water masses

For the study of the stationary distribution of temperature in the case of vertical mixing of three water masses (let us designate them by the letters A, M and B), superimposed over each other, let us make use of the same equations of heat conduction:

$$u_A \frac{\partial T}{\partial x} = k_A \frac{\partial^2 T}{\partial z^2}$$

$$u_M \frac{\partial T}{\partial x} = k_M \frac{\partial^2 T}{\partial z^2} \qquad\qquad [45.20]$$

$$u_B \frac{\partial T}{\partial x} = k_B \frac{\partial^2 T}{\partial z^2}$$

The respective arrangement of these three water masses and the method of reckoning the vertical coordinate are represented in Fig. 49. Assuming, as previously, $u_A(\partial T/\partial x) = a$, $u_M(\partial T/\partial x) = m$ and $u_B(\partial T/\partial x) = b$, let us write equations [45.20] and their integrals:

$$\frac{d^2 T}{dz^2} = a \qquad (H_1 > z > h)$$

$$\frac{d^2 T}{dz^2} = m \qquad (h > z > -h) \qquad\qquad [45.21]$$

$$\frac{d^2 T}{dz^2} = b \qquad (-h > z > -H_2)$$

$$T = \frac{az^2}{2} + C_1 z + C_2 \qquad (H_1 > z > h)$$

$$T = \frac{mz^2}{2} + C_3 z + C_4 \qquad (h > z > -h) \qquad\qquad [45.22]$$

$$T = \frac{bz^2}{2} + C_5 z + C_6 \qquad (-h > z > -H_2)$$

The boundary conditions:

$$k_A \frac{\partial T}{\partial z}\bigg|_{z=+H_1} = 0 \qquad\qquad k_B \frac{\partial T}{\partial z}\bigg|_{z=-H_2} = 0 \qquad\qquad [45.23]$$

indicate the absence of heat flows through the upper and lower boundaries of the sea, i.e., the internal nature of heat exchange. Further,

$$\frac{\partial T}{\partial z}\bigg|_{z=0} = 0 \qquad\qquad\qquad [45.24]$$

i.e., the extreme of curve $T(z)$ is given in the core of the intermediate water mass. Finally, on the boundaries between the water masses some mean temperatures are observed:

$$T\big|_{z=+h} = \overline{T}_1; \qquad T\big|_{z=-h} = \overline{T}_2 \qquad\qquad [45.25]$$

Here index "1" corresponds to the boundary between water mass A and M, index "2" to the boundary between water masses M and B.

In addition, by conditions:

$$T\big|_{z=+H_1} = T_A; \qquad T\big|_{z=-H_2} = T_B; \qquad T\big|_{z=0} = T_M \qquad [45.26]$$

let us set the stationary T-indexes on the outer boundaries of the surface and deep water masses and in the core of the intermediate water mass, and by conditions:

$$k_A \frac{\partial T}{\partial z} = k_M \frac{\partial T}{\partial z} \qquad \text{when} \quad z = +h$$

$$\left. \right| [45.27]$$

$$k_M \frac{\partial T}{\partial z} = k_B \frac{\partial T}{\partial z} \qquad \text{when} \quad z = -h$$

let us determine the continuity of curve $T(z)$ on the boundaries of the intermediate water mass.

On the basis of boundary conditions [45.23] and [45.24] respectively, we obtain:

$$C_1 = -aH_1; \qquad C_3 = 0; \qquad C_5 = bH_2 \qquad\qquad [45.28]$$

while on the basis of conditions [45.25] from equations [45.22] we obtain:

$$C_2 = \overline{T}_1 - \frac{ah^2}{2} + aH_1h \qquad (H_1 > z > h)$$

$$C_4 = \overline{T}_{1,2} - \frac{mh^2}{2} \qquad\quad (h > z > -h) \qquad\qquad [45.29]$$

$$C_6 = \overline{T}_2 - \frac{bh^2}{2} + bH_2h \qquad (-h > z > -H_2)$$

Substituting [45.28] and [45.29] in equations [45.22], we obtain:

$$T = \overline{T}_1 + \frac{a}{2}(z^2 - h^2) - aH_1(z - h) \qquad (H_1 > z > h)$$

$$T = \overline{T}_{1,2} + \frac{m}{2}(z^2 - h^2) \qquad (h > z > -h) \qquad [45.30]$$

$$T = \overline{T}_2 + \frac{b}{2}(z^2 - h^2) - bH_2(z - h) \qquad (-h > z > -H_2)$$

Having made use of conditions [45.26], we can determine quantities a, m and b from equations [45.30]; they equal:

$$a = \frac{2(\overline{T}_1 - T_A)}{(H_1 - h)^2}$$

$$m = \frac{2(\overline{T}_{1,2} - T_M)}{h^2} \qquad [45.31]$$

$$b = \frac{2(\overline{T}_2 - T_B)}{(H_2 - h)^2}$$

Finally, substituting [45.31] respectively in equations [45.30], we finally obtain:

$$T = \overline{T}_1 + \frac{\overline{T}_1 - T_A}{(H_1 - h)^2}[(z^2 - h^2) - 2H_1(z - h)] \qquad (H_1 > z > h)$$

$$T = \overline{T}_{1,2} + \frac{\overline{T}_{1,2} - T_M}{h^2}(z^2 - h^2) \qquad (h > z > -h) \qquad [45.32]$$

$$T = \overline{T}_2 + \frac{\overline{T}_2 - T_B}{(H_2 - h)^2}[(z^2 - h^2) - 2H_2(z - h)] \qquad (-h > z > -H_2)$$

It now remains to determine $\overline{T}_1$ and $\overline{T}_2$. For this it is necessary to differentiate equations [45.32] by z and to satisfy conditions [45.27]. As a result, on the basis of the first conditions of [45.27] we obtain:

$$k_A \frac{T_A - \overline{T}_1}{H_1 - h} = k_M \frac{\overline{T}_1 - T_M}{h} \qquad [45.33]$$

whence:

$$\overline{T}_1 = \frac{hk_A T_A + (H_1 - h)k_M T_M}{hk_A + (H_1 - h)k_M} \qquad [45.34]$$

In exactly the same way, on the basis of the second condition of [45.27] we obtain:

$$k_M \frac{T_M - \overline{T}_2}{h} = k_B \frac{\overline{T}_2 - T_B}{H_2 - h} \tag{45.35}$$

whence:

$$\overline{T}_2 = \frac{hk_B T_B + (H_2 - h)}{hk_B + (H_2 - h)k_M} \tag{45.36}$$

Stationary vertical distribution of temperature in a three-layer sea is accordingly determined by equations [45.32] simultaneously with [45.34] and [45.36].

For the determination of the vertical distribution of salinity, let us apply the equations of diffusion which are entirely analogous to the equations of heat conduction [45.20]. As before, let us assume that the coefficients of turbulent heat conduction and turbulent diffusion in each layer are equal $[(k_A)_T = (k_A)_S$ etc.]. Following the same arguments as for temperature, we come to the formulae determining the vertical distribution of salinity in a three-layer sea:

$$
\begin{aligned}
S &= \overline{S}_1 + \frac{\overline{S}_1 - S_A}{(H_1 - h)^2}\,[(z^2 - h^2) - 2H_1(z - h)] && (H_1 > z > h) \\[2ex]
S &= \overline{S}_{1,2} + \frac{\overline{S}_{1,2} - S_M}{h^2}\,(z^2 - h^2) && (h > z > -h) \\[2ex]
S &= \overline{S}_2 + \frac{\overline{S}_2 - S_B}{(H_2 - h)^2}\,[(z^2 - h^2) - 2H_2(z - h)] && (-h > z > -H_2)
\end{aligned}
\tag{45.37}
$$

where:

$$
\begin{aligned}
\overline{S}_1 &= \frac{hk_A S_A + (H_1 - h)k_M S_M}{hk_A + (H_1 - h)k_M} \\[2ex]
\overline{S}_2 &= \frac{hk_B S_B + (H_2 - h)k_M S_M}{hk_B + (H_2 - h)k_M}
\end{aligned}
\tag{45.38}
$$

Let us plot from formulae [45.32] and [45.37], taking account of the mean values of $\overline{T}$ and $\overline{S}$, as determined by formulae [45.34], [45.36] and [45.38], curves $T(z)$ and $S(z)$ for the vertical distribution of temperature and salinity within the following water masses:

Water mass	A	M	B
Temperature,°C	15	10	3
Salinity,‰	35.50	36.00	34.90
Thickness of layer, m	800 (from sea surface to depth of 800 m)	400 (from depth of 800–1,200 m)	1,300 (from depth of 1,200–2,500 m)

The graphs of vertical distribution of temperature and salinity, provided the coefficients of turbulent heat conduction and diffusion are equal in all three layers ($k_A = k_M = k_B$), are represented in Fig. 50 (curve *a* for temperature and curve *I* for salinity).

Let us now consider how the results obtained will be interpreted in *T-S* coordinates. For this, let us eliminate parameter z from equations [45.32] and [45.37] in pairs; as a result we will obtain expressions:

$$\frac{T-\bar{T}_1}{T_A-\bar{T}_1} = \frac{S-\bar{S}_1}{S_A-\bar{S}_1} \qquad (H_1 > z > h)$$

$$\frac{T-\bar{T}_{1,2}}{\bar{T}_{1,2}-T_M} = \frac{S-\bar{S}_{1,2}}{\bar{S}_{1,2}-S_M} \qquad (h > z > -h) \qquad \text{[45.39]}$$

$$\frac{\bar{T}_2-T}{\bar{T}_2-T_B} = \frac{\bar{S}_2-S}{\bar{S}_2-S_B} \qquad (-h > z > -H_2)$$

The first of these equations is the equation of the straight line joining points (T_A, S_A) and $(\bar{T}_1, \bar{S}_1)$, the second the equation of the straight line joining points $(\bar{T}_1, \bar{S}_1)$ and (T_M, S_M), as well as points (T_m, S_m) and $(\bar{T}_2, \bar{S}_2)$, and the third the equation of the straight line joining points $(\bar{T}_2, \bar{S}_2)$ and (T_B, S_B). Thus, the result has turned out the same as in the case of mixing of two water masses, while the *T-S* curve plotted from the data of vertical distribution $T(z)$ and $S(z)$ is represented in Fig. 51. From this the following conclusion should be drawn: in the case of vertical mixing of three water masses, characterized by identical equations of stationary heat conduction and diffusion with respect to parameter z, and given the equality of the coefficients of turbulent heat conduction and diffusion, the *T-S* curve coincides with two sides of the triangle of mixing.

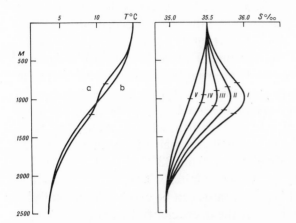

Fig. 50. Curves of vertical distribution of temperature in a three-layer (*a*) and a two-layer (*b*) ocean (on the left) and curves of vertical distribution of salinity in a two-layer sea (*V*) and in a three-layer sea (*I–IV*) (on the right) at different values of thickness of the intermediate layer and salinity extremes, determined by formulae [45.12] and [45.37] (Mamayev, 1962a).

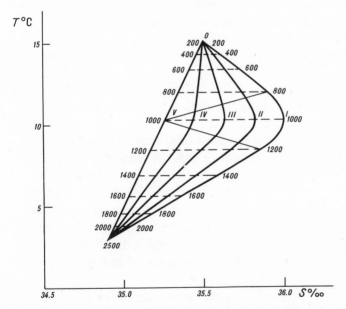

Fig. 51. *T-S* curves corresponding to the curves of vertical salinity distribution *I–V* (Fig. 50, on the right) and to the curve of vertical temperature distribution *b* (Fig. 50, on the left) (Mamayev, 1962a).

Combining the two- and three-layer models

In the real conditions of the World Ocean, in the majority of cases only one of the two properties — either temperature or salinity — is characterized by the presence of an extreme. This circumstance is indirect evidence of the fact that turbulent heat conduction and turbulent diffusion proceed with a different degree of intensity and, as a rule, the extreme of salinity is considerably more stable than the extreme of temperature. Thus, if we turn for an example to the T-S curves in the western part of the Atlantic Ocean, plotted by Defant and Wüst for various latitudes (Defant, 1961, fig. 98), we will see from them that the low temperature inversion, observed on the T-S curve for latitude 33°S., quickly disappears further to the north, whereas the two extremes of salinity on the T-S curves are distinctly maintained throughout the whole extent of the western meridian section from 33°S to 11°N.

This peculiarity of vertical mixing, which manifests itself in the considerable difference in the transfer of heat and salts, can be expressed indirectly, leaving aside the question of the ratio of the coefficients, by applying, say, to the temperature distribution a model of a two-layer sea and to the salinity distribution a model of a three-layer sea. In this case, the concept of the thickness of the layer, which is in T-S analysis in general a conventional concept (given the continuous distribution of temperature and salinity with depth), becomes somewhat more indefinite and characterizes rather the "sphere of influence" of the corresponding T-S index or extreme. Still another example may be cited: in studying the distribution of intermediate Mediterranean water in the Atlantic Ocean one may interpret with the three-layer model salinity alone, characterized by the present of an extreme; with respect to temperature it is perfectly possible to limit oneself to the two-layer model, for in its distribution along the vertical no factors determine the appearance of a temperature extreme.

Let us express, on the basis of what has been said, vertical temperature distribution by a two-layer model — equations [45.12], and vertical salinity distribution by a three-layer model — equations [45.37], having placed the boundary of the two "temperature" layers in the center of the intermediate "salinity" layer. The corresponding curves $T(z)$ and $S(z)$ are represented in Fig. 50 (when $T_A = 15°$, $T_B = 3°$), and moreover curve b corresponds to the two-layer model of temperature distribution. This vertical distribution of temperature and salinity leads to the T-S curve represented in Fig. 51 (curve I).

Looking at this T-S curve, we see that from the surface of the sea to 800 m and from 1,200 to 2,500 m, i.e. in layers A and B, the branches of the T-S curve are rectilinear. In the layer between the depths of 800 and 1,200 m it is curvilinear. The corresponding elimination of parameter z from equations [45.12] and [45.37] within the intermediate layer leads to the square dependence of T on S in the T-S coordinates.

Let us consider the question of the transformation of the salinity extreme in the direction of axis x, assuming for simplicity that over some distance L

the thickness of the intermediate layer decreases linearly from $h = 400$ m to zero:

$$h_x = \left(1 - \frac{x}{L}\right) h_M \qquad [45.40]$$

Let the decrease in salinity from $S_M = 36.00‰$ to $\bar{S} = 35.26‰$ (i.e., to the value of salinity corresponding to the two layer model when $S_A = 35.50‰$ and to the overall thickness of the two layers of 2,500 m) correspond to this degeneration of the intermediate layer. Then the decrease in the thickness of the intermediate layer and the decrease of salinity in the extreme may be linked by a simple linear dependence:

$$S_x = \bar{S} + (S_M - \bar{S}) \frac{h_x}{h_m} \qquad [45.41]$$

At the distances $x = 0.25\,L$, $0.50\,L$, $0.75\,L$ and L the thicknesses of the intermediate layer will equal 300, 200, 100 and 0 m, while the values of salinity in the extreme will correspondingly equal 35.82, 35.63, 35.44 and 35.26‰. The corresponding $S(z)$ and T-S curves (curves $II-V$) are also represented in Figs. 50 and 51 respectively. From the latter figure, in particular, it may be seen that when no change in temperature occurs in the core of the intermediate water mass, but when only salinity changes, the line of transformation of the core of the intermediate water mass is horizontal and coincides with the isoline $T = $ const, corresponding to constant temperature on the interface between the masses in a two-layer model.

The following tentative conclusions and assumptions may be drawn from the introduction to the elementary T-S analysis of moving water masses set forth above, which by no means claims to be complete.

(1) The nature of the T-S curves and the direction of the transformation of the extreme within the T-S triangle are determined by the relationship of the "spheres of influence" of the T-S indexes, or by the thicknesses of the corresponding layers. Apparently, this is determined in the last analysis by the ratio of the coefficients of turbulent heat conduction and turbulent diffusion (this question requires additional study). The transformation of the extreme along the principal median of the triangle of mixing (Shtokman, 1943a) is one of the possible paths. It is also quite possible that in the case of isopycnic propagation of the intermediate water mass, the line of transformation of the core will be an isopycnal (isostere) within the T-S triangle. In this case the change in temperature dT and the corresponding change in salinity dS in the process of transformation must satisfy the condition:

$$dT = dS \tan\varphi$$

where $\tan\varphi$ is the function of the equation of state of sea water (the thermohaline derivative), determined by formula [25.3]. In general, however;

$$dT = \tan \mu \, dS$$

where μ is the angle of slope of the line of transformation of the extreme to the abscissa axis, also determined by the relative intensities of the processes of turbulent heat conduction and diffusion.

(2) The question of the percentage ratio of the three water masses on the interfaces of the intermediate water mass requires further investigation. The determination of the boundaries of the intermediate water mass by its 50% content is, apparently, also a particular case; generally speaking, the percentage content of the intermediate water mass at its upper and lower boundaries may be different.

(3) An important question is that of the determination of the thermohaline index and the original thickness of the intermediate layer, inasmuch as in the stationary case we do not consider the initial moment of mixing with the homogeneous water masses which correspond to it. For the determination of the original thickness of the intermediate layer, it is necessary to study the curves $T(z)$ and $S(z)$, as well as the T-S curves in the regions of formation of the intermediate water masses. If for an example we turn once again to the Intermediate Mediterranean Water of the Atlantic Ocean, we may assume that the T-S curve for a station made in the Straits of Gibraltar, in that part which corresponds to the lower (outflowing) current from the Mediterranean Sea, must be similar to those parts of the T-S curves of the closest deep water stations in the region of the Atlantic near Gibraltar which are included within Mediterranean waters. Given this similarity of the T-S curves the value of the parameter z at the deep water stations must be "stretched" along the T-S curve as compared with a station in the Straits. Such a comparison can serve as a basis for the determination of the original thickness of the intermediate Mediterranean layer. So far as the thermohaline index of the intermediate water mass is concerned, it is necessary to consider within the framework of theory its horizontal transformation, which takes place in the same way as the zonal transformation of the indexes of the surface water masses (Section 61).

46. SOME PROBLEMS OF THERMOHALINE CONVECTION. AN ANALOGY: THE PROBLEM OF
    THE GROWTH AND DECAY OF AN ICE SHEET IN THE SEA

The problems of thermohaline convection, considered by Stommel (1961), represent a separate branch of the analytical theory of T-S curves. Although the process of thermohaline convection is considered by him under the conditions of an idealized experiment, these problems have great theoretical significance, since they demonstrate the essential difference between the stages of thermal and saline convection, showing that the process of heat exchange anticipates at the first stage of convection the process of salt exchange, and then, in the process of development of mixing, gives way to the latter. The simple models of Stommel can serve as a first theoretical explanation of the fact of difference in the processes of thermohaline convection in different geographical conditions, and in this sense are close to the theory of autumn—winter convection (winter vertical circulation), developed by Zubov (1938, 1945). Finally, one of Stommel's problems, namely the first, provides a highly interesting analogy with the problem of the growth and decay of the ice sheet; this question will be considered in the second half of the present section.

Let us consider an extremely simple model of convection: in a vessel (Fig. 52, upper part) by means of mixing (by the action of a mixer, say), identical values of temperature T and salinity S are maintained throughout the whole volume; though the porous walls of the vessel, heat and salt transfer is carried out from the external part of the reservoir, where temperature T and salinity S are maintained constant throughout the whole time of convection. In this case the equations of heat conduction and diffusion can be written in the following simple form:

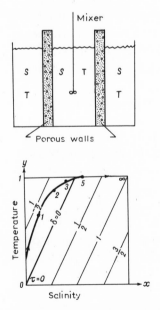

Fig. 52. Diagram of idealized experiment I (above) and dimensionless T-S diagram (the anomalies of density are plotted for the case $R = 2$) and dimensionless T-S curve for $\delta = 1/6$ and $\tau = 0, 1, 2, 3, 5, \infty$ (below). Explanations in the text. (Stommel, 1961).)

$$\frac{dT}{dt} = c(\mathrm{T} - T)$$

$$\frac{dS}{dt} = d(\mathrm{S} - S)$$

[46.1]

where c and d are the coefficients of heat and salt exchange respectively. Introducing the designations $\tau = ct$; $\delta = d/c$; $y = T/\mathrm{T}$; $x = S/\mathrm{S}$ equations [46.1] can be made dimensionless:

$$\frac{dy}{d\tau} = 1 - y$$

$$\frac{dx}{d\tau} = \delta(1 - x)$$

[46.2]

The solutions of equations [46.2] at initial conditions $x = x_0$, $y = y_0$ when $\tau = 0$ will be written in the form:

$$y = 1 + (y_0 - 1)\, e^{-\tau}$$

$$x = 1 + (x_0 - 1)\, e^{-\delta\tau}$$

[46.3]

It is obvious that when $\tau \to \infty$, $x = 1$ and $y = 1$. Stommel points out that inasmuch as $\delta < 1$ (since coefficient d of salt exchange by its physical nature is smaller than the coefficient of heat exchange c), temperature reaches its asymptotic value more rapidly than salinity. The solution [46.3] of system [46.2] can be represented parametrically on a dimensionless T-S diagram (x, y diagram) — Fig. 52, lower part; the corresponding curve is plotted for value $\delta = 1/6$.

Stommel brings into consideration the simplified equation of state:

$$\rho = \rho_0(1 - \alpha T + \beta S)$$

which can be expressed by dimensionless variables x and y in the following form:

$$\rho = \rho_0[1 + (\alpha \mathrm{T})(-y + Rx)]$$

where $R = \beta \mathrm{S}/\alpha \mathrm{T}$ represents the ratio of the temperature and salinity influences on the density of water in the final stage of the process. The velocity of change in density in the process of convection can in that case be expressed by the equation:

$$\frac{d\rho}{dt} = \rho_0[-1 + y + R\delta(1 - x)]$$

Stommel considers a particular case — at initial conditions $x_0 = y_0 = 0$ and

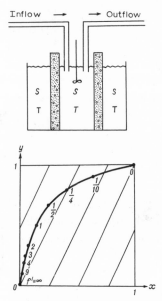

Fig. 53. Diagram of idealized experiment *II* (above) and corresponding dimensionless *T-S* diagram and dimensionless *T-S* curve for the values of parameter f = 0, 1/10, ..., 9, ∞ (below). Explanations in the text. (Stommel, 1961.)

values $R > 1$ and $R\delta < 1$; in this case in the initial stage of convection density decreases, then begins to increase, reaches its initial value at a definite moment of time, after which it still continues to increase, exceeding the initial value. The foregoing is well illustrated by the dimensionless *T-S* diagram (Fig. 52, lower part), on which are plotted the isolines of the dimensionless anomaly of density $\sigma = (\rho/\rho_0 - 1)/\alpha T$, calculated for the case $R = 2$. The asymptotic value of density, equal to zero at the initial moment, is equal to $+1$.

Let us now consider another idealized experiment, when water with the characteristics $T = 0$ and $S = 0$ flows into the vessel with velocity q; an evenly blended mixture flows out of the vessel with the same velocity (Fig. 53, upper part). The equations describing the process of heat conduction and diffusion prove even simpler:

$$\frac{dT}{dt} = 0 = c(T - T) - qT$$

$$\frac{dS}{dt} = 0 = d(S - S) - qS$$

[46.4]

In their dimensionless form they are written as follows:

$$1 - (1 + f')y = 0$$

$$\delta - (\delta + f')x = 0$$

$$[46.5]$$

where $f' = q/c$ is the dimensionless flow rate, whence:

$$y = \frac{1}{1 + f'}$$

$$x = \frac{1}{1 + \dfrac{f'}{\delta}}$$

$$[46.6]$$

It is clear that these dimensionless values of temperature and salinity are in equilibrium. Eliminating parameter f' from equations [46.6], we obtain the following equation for the *T-S* curve which describes the process of convection:

$$\frac{1 - y}{y} = \frac{\delta(1 - x)}{x}$$

$$[46.7]$$

This curve is plotted in Fig. 53, lower part, for the value $\delta = 1/6$ (the isolines of the anomalies of density are also plotted for $R = 2$). The parameter in the second problem, we repeat, is no longer time but flow rate of water; from Fig. 53 it is seen that it is precisely the latter which determines the density of the outflowing water: at $f' < 1/4$ the density of the outflowing water is greater than the density of the inflowing water, while when $1/4 < f' \leqslant \infty$ less dense water flows out. Stommel's second problem represents a highly interesting theoretical illustration (although also a highly tentative one) of the problem of the relation between convection and the water exchange of sea basins — a problem raised by Zubov in its broad geographical aspects.

Stommel considers still another more complex problem, when water exchange, as in the second problem, is not given as independent, but is determined itself in turn by the difference of densities. This problem leads to a system of two non-linear equations, which are investigated by methods of the theory of stability (on the corresponding *T-S* diagrams stable nodes and a stable focus appear). Here we will not consider Stommel's third problem, which is considerably more abstract in nature, but will proceed to consider the analogy of Stommel's first problem with the problem of the growth and decay of the ice sheet in the sea (Mamayev, 1964a).

In studying the situation of ice in the freezing regions of the oceans and seas, the question of simultaneous change in the process of growth or decay of the area of distribution of the ice sheet and the thickness of the ice is of great importance. It is known that a large number of investigations are devoted to questions of the formation and destruction of the ice sheet, but only one side of the problem is considered in the majority of them, namely the question of change in the thickness of ice under the influence of various conditions. Studies, however, devoted to the investigation of the simultaneous change of both factors of ice formation — the area of the ice sheet and the thickness of the ice, can literally be counted on the fingers of one hand. Meanwhile, this question is extremely important, in particular, from the point of view of forecasting. Thus, for example, if the analytical link is known between these factors of ice formation at different moments in time, it should be possible to judge the thickness of the ice flows and their passability from data on the area of the floes obtained as a result of aerial reconnaissance. The solution of the inverse problem would prove possible: from several measurements of the thickness of the ice to form an idea of the area of the ice sheet in the given basin.

Among the studies devoted to the investigation of the simultaneous change in the area of the ice sheet and its thickness, the article of Nazarov (1963) is highly interesting. In it, the graph showing the relationship between these factors is particularly worthy of note; this graph, obtained by Nazarov by an empirical method, is reproduced here in Fig. 54. The dependence between the area and the thickness of the ice in the process of growth of an ice sheet, according to Nazarov, is determined by the formula $F = \sqrt{h}$, where F is the area of the extent of the ice and h its thickness (both quantities in percentages of their maximum values).

Considering Nazarov's graph (Fig. 54), the author of the present work drew attention to the remarkable analogy of this problem with Stommel's first problem of the simultaneous change of temperature and salinity in the process of non-stationary thermohaline convection, when in the process of development of convection the change of temperature considerably anticipates, in any

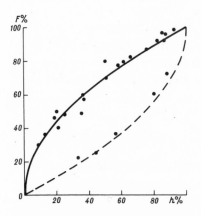

Fig. 54. Relationship between the area of ice $F\%$ and its thickness $h\%$ in the process of growth (solid line) and decay (broken line) of the ice sheet, according to Nazarov (1963).

case in the original stage of the process, the change of salinity, and, moreover, this takes place owing to the fact that the transfer of heat occurs with greater velocity than the diffusion of salts.

We see something similar in Nazarov's graph; in the initial stage the growth of the area of ice proceeds more quickly than the increase in its thickness; in the final stage, when the sea is covered by an almost solid sheet of ice, further growth takes place practically only because of an increase in its thickness.

The analogy noted led the author to the idea of applying Stommel's theory to the study of the process of growth of an ice sheet, extending it also to the inverse process – the process of decay. Below a corresponding elementary analytical theory of the growth and decay of an ice sheet is formulated, while its results are represented on the $s - h$ plane (s is the area of ice, h the thickness of the ice), which, thus, is considered an analog of the S-T plane.

The change in area s of a solid (unbroken) ice sheet and its thickness h in the process of growth may be expressed in a first approximation by the following obvious equations:

$$\frac{ds}{dt} = \alpha(s_L - s)$$

$$\frac{dh}{dt} = \beta(h_L - h)$$

[46.8]

where t is time, s_L the limiting area of the ice sheet (in case the sea freezes over completely, s_L represents its area), h_L the limiting thickness of the ice, α and β the coefficients of proportionality which can generally be called the coefficients of growth of the area and of the thickness of the ice respectively. Equations [46.8] are the most simple in studying the process considered: thus, for example, the first equation means that a change in the area of the unbroken ice is proportional to the area of clear water, $s_L - s$. These equations, as well as their solutions, are analogous to the corresponding simplified equations of heat conduction and salt diffusion of Stommel [46.1], with the difference that area s is used instead of temperature, and thickness h instead of salinity. Following Stommel, let us make equations [46.9] dimensionless, introducing the following designations:

$$\tau = \alpha t ; \qquad \delta = \frac{\beta}{\alpha} ; \qquad y = \frac{s}{s_L} ; \qquad x = \frac{h}{h_L}$$

Then equations [46.8] can be rewritten in the following form:

$$\frac{dy}{d\tau} = 1 - y$$

$$\frac{dx}{d\tau} = \delta(1 - x)$$

[46.9]

The solutions of equations [46.9] at initial condition $x = 0$, $y = 0$ when $\tau = 0$ are:

$$y = 1 - e^{-\tau}$$

$$x = 1 - e^{-\delta\tau}$$

[46.10]

From these equations it follows, in particular, that $y \to 1$, $x - 1$, when $\tau \to \infty$. We must consider the value of the dimensionless coefficient δ. The conditions observed in nature indicate to us that the area of the ice, at least in initial conditions, increases more rapidly than its thickness. Indeed, at the beginning of ice formation all the visible area of the calm sea (other conditions being equal) is rapidly covered by a thin crust of ice, and this process, under identical external conditions, proceeds the more quickly the larger the area of the sea; and only when the growth of the ice area begins to be hindered by the condition of its approaching the limiting value s_L, does the thickness begin to predominate in the growth. Therefore, it is obvious that we must usually take it that the coefficient of growth of area is greater than the coefficient of growth of thickness, $\alpha > \beta$, and, accordingly, that $\delta < 1$. In Fig. 55, in that part of it which lies between the bisector of the coordinate angle and the axis of the ordinates, are shown the "curves of growth" of the ice, plotted from equations [46.10] at various values of coefficient δ, namely 0.1; 0.2; and 0.4, as well as isochrones $\tau = 0.5$; 1; 2; 3; 4 and 5. In this part, the graph, represented in Fig. 55, is similar to Stommel's graph (Fig. 52).

The equation of the "curve of growth" in the x,y plane is easy to obtain, by eliminating parameter τ from equations [46.10]; as a result of this elimination we obtain:

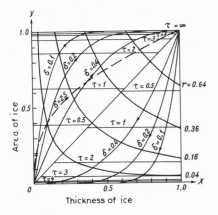

Fig. 55. Curves of growth and decay of an ice sheet at different values of the coefficient of growth (decay) δ. The graph also shows isochrones for $\tau = 0.5$; 1; 2; 3; 4; 5 and isolines of the equal volume of ice for $r = 0.04$; 0.16; 0.36; 0.64 (Mamayev, 1964d).

$$y = 1 - (1-x)^{1/\delta} \qquad\qquad\qquad\qquad\qquad\qquad\qquad [46.11]$$

Let us note that when $\alpha = \beta$ we have $\delta = 1$, and both equations [46.10] will become identical:

$$y = x = 1 - e^{-\tau} \qquad\qquad\qquad\qquad\qquad\qquad\qquad [46.12]$$

The "curve of growth" will coincide in this case with the bisector of the coordinate angle. It is obvious that for basins of different area the "curves of growth" will be different. For a basin of infinite area, both the area of the ice and its thickness will grow, apparently, with equal intensity (other conditions being equal); for this case $\delta = 1$. On the other hand, a basin of small area (lake) will be rapidly covered with ice and further growth will take place only owing to the thickness of the ice; in this case the "curve of growth" will reach its asymptotic value more rapidly and with a smaller thickness of ice. Thus, for example, when $\delta = 0.1$ and $\delta = 0.4$, to the same area of ice, let us say $y = 0.95$ reached at the moment of time $\tau = 3$, correspond, as may be seen from Fig. 55, various thicknesses of ice, namely $x = 0.26$ and $x = 0.70$ respectively.

Let us now consider the question of the change of volume of ice in the process of growth. Having designated volume by the letter q and introducing dimensionless volume r, we have:

$$q = hs = s_L h_L xy = q_L xy$$

$$r = \frac{q}{q_L} = xy \qquad\qquad\qquad\qquad\qquad\qquad\qquad\qquad [46.13]$$

where $q_L = s_L h_L$ is the limiting volume of ice. Substituting in [46.13] the results of [46.10] we obtain:

$$r = (1 - e^{-\delta\tau})(1 - e^{-\tau}) \qquad\qquad\qquad\qquad\qquad\qquad [46.14]$$

It is easy to obtain the differential equation of the growth of volume of ice, bearing in mind that:

$$\frac{dr}{d\tau} = x \frac{dy}{d\tau} + y \frac{dx}{d\tau} \qquad\qquad\qquad\qquad\qquad\qquad [46.15]$$

Substituting in [46.15] the values of the derivatives determined by equations [46.9], we obtain:

$$\frac{dr}{d\tau} = x(1-y) + \delta y(1-x)$$

or, bearing in mind [46.10]:

$$\frac{dr}{d\tau} = e^{-\tau}(1 - e^{-\delta\tau}) + \delta e^{-\delta\tau}(1 - e^{-\tau}) \qquad\qquad\qquad [46.16]$$

It is easy to verify by differentiation by τ of equation [46.14] that it is the solution of equation [46.16].

In Fig. 55 are plotted the isolines of the volume of ice $r = 0.04; 0.1; 0.36$ and 0.64. It is natural that the volume of ice in this figure will be formed for different moments of time by the rectangles inscribed in the hyperbolic isolines of volume, and in such a way that the free (not lying on axes x and y) summit of the rectangle lies on the corresponding "curve of growth" and slides along it with time, moving when $\tau \to \infty$ to its maximum value $r = 1$, i.e., to the area of the entire graph. In the case $\delta = 1$ we have again $x = y$, and the volume of ice will be represented by squares, the free summit of which lies on the bisector of the coordinate angle.

It is appropriate to point out that the family of isolines r is, in the given case, a kind of analog of the family of isolines of density (isopycnals) of water on the S-T diagram. The change in the density of water in the process of simultaneous change of temperature and salinity, considered by Stommel, represents, as compared with the change in the volume of ice, a considerably more complex process, even given the condition of the linearization of the equation of state of sea water carried out by him. The existence of so surprizing an analogy between two heterogeneous problems is hardly limited only to the foregoing, and probably further interesting comparisons are possible here.

The curves of growth of the volume of ice with time, i.e., the curves of function $r = r(\tau)$, represented in Fig. 56 by solid lines, are also of interest. These curves are plotted for various values of coefficient δ from equation [46.14], and we see from their consideration that the velocity of growth of ice changes substantially depending on the interrelationship (determined by coefficient δ) according to which the area of ice and its thickness change.

Let us now consider the *inverse* process – the process of decay of ice. It is

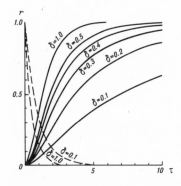

Fig. 56. Curves of the relationship of the volume of ice to time in the process of growth (solid lines) and decay (broken lines) of an ice sheet at different values of the coefficient of growth (decay) δ (Mamayev, 1964a).

obvious that it can be described by the following equations different from [46.8]:

$$\frac{ds}{dt} = -\alpha s$$
$$\frac{dh}{dt} = -\beta h$$

[46.17]

Indeed, the decrease in the area of ice on decay is no longer proportional to the area of clear water $s_L - s$, as was the case in the process of growth of ice, but to the area of the existing ice s. In equations [46.17] α and β are the coefficients of the decrease of area and thickness of ice respectively, which generally can be equal to the corresponding coefficients of growth (for this reason we keep the same letter designations for them). The first equation of [46.17] was solved by Zubov (1945), in a somewhat different form, he was considering the increase in the area of pure water in the process of ice decay.

Introducing the same dimensionless quantities, we reduce equations [46.17] to the form:

$$\frac{dy}{d\tau} = -y$$
$$\frac{dx}{d\tau} = -\delta x$$

[46.18]

The solutions of these equations at initial conditions $y = 1$, $x = 1$ when $\tau = 0$ are:

$$y = e^{-\tau}$$
$$x = e^{-\delta\tau}$$

[46.19]

The elimination of parameter τ from equations [46.19] leads to the following expression:

$$y = x^{1/\delta}$$

Finally, the change of dimensionless volume is determined, according to [46.15], by the equation:

$$\frac{dr}{d\tau} = -r(1 + \delta)$$

the solution of which is the product of equations [46.19]:

$$r = e^{-\tau(1+\delta)}$$

[46.20]

Fig. 55 also represents the "curves of decay" (they lie between the bisector of the coordinate angle and the abscissa axis) plotted for the same values of coefficient δ as the "curves of growth". The decrease in the volume of ice in the process of decay, given the same condition δ < 1, is shown by broken lines for the two values of δ, equal to 0.1 and 1.0, in Fig. 56. From a comparison of the curves of growth of volume of ice and the curves of its decrease, we may see that the destruction of the ice sheet proceeds, everything else being equal, considerably more intensively than its growth. This fact was pointed out by Zubov (1945), who drew attention to the "occasional exceptionally rapid disappearance of large masses of ice in the southern part of the Arctic Basin in the course of the polar summer, creating the impression of "melting before our eyes". Zubov explains this by the influence of the "internal decay" of the sea ice, which takes place thanks to the presence of brine in the ice; as we see, one may come to the same conclusion proceeding from the most general assumptions.

To draw some brief conclusions, let us note the main points: equations [46.9] and [46.18] are simultaneous in the sense that they describe the direct and inverse stages of a closed *circular process*. In the given specific case, under the physical assumption taken to the effect that the increase (decrease) in the area of ice proceeds more intensively than an increase (decrease) in the thickness of the ice (δ < 1), the circular process proceeds *clockwise* (Fig. 55). It is clear that from the physical point of view the case is also not ruled out when the process can proceed *counter-clockwise* *.

Recalling again Stommel's problem of thermohaline convection and comparing it with the problem just considered, we cannot help coming to the conclusion that they are both particular cases of a more general problem of cyclic processes. In this case, they are of broad significance and are the simplest for the description of reversible (but, apparently, not self-reversible) thermodynamic and physico-chemical processes.

47. THE DETERMINATION OF THE COEFFICIENT OF MIXING FROM *T-S* CURVES (JACOBSEN'S METHOD)

As early as 1927 Jacobsen had proposed a graphic method of determining the coefficient of mixing from two *T-S* curves successive in time. Jacobsen's method, extremely simple and elegant, won great popularity in the oceano-

---

* In water basins of small area (lakes), solidly covered by an ice sheet, the decrease in the thickness of ice in the process of decay can, as V.L. Tsurikov pointed out to the author, proceed more rapidly than a decrease in its area. In this case, obviously, it is necessary to take δ > 1. The corresponding "curve of decay" (for δ = 2.5) is represented in Fig. 55 by a broken line lying between the ordinate axis and the bisector of the coordinate angle.

graphic literature. Jacobsen's derivation is reproduced in the monographs of Shuleikin (1953) and Defant (1961), Jacobsen's formula was again obtained by a different, simpler method by Okada (1938) and Proudman (1953). The latter author extended Jacobsen's formula to various types of *T-S* curves, while Shtokman (1943) analyzed its correctness proceeding from the analytical theory of *T-S* curves. Ivanov (1944) considered the question of the determination of the coefficient of mixing from *T-S* curves in the case of mixing of four water masses.

Let us derive Jacobsen's formula by the method proposed by Okada. First of all, along any *T-S-z* curve the following relationships are valid:

$$\frac{\partial T}{\partial z} = \frac{\mathrm{d} T}{\mathrm{d} S} \frac{\partial S}{\partial z} \tag{47.1}$$

and:

$$\frac{\partial^2 T}{\partial z^2} = \frac{\mathrm{d}^2 T}{\mathrm{d} S^2} \left(\frac{\partial S}{\partial z}\right)^2 + \frac{\mathrm{d} T}{\mathrm{d} S} \left(\frac{\partial^2 S}{\partial z^2}\right) \tag{47.2}$$

(in obtaining relationship [47.2] it should be borne in mind that $\mathrm{d}T/\mathrm{d}S$ is a complex function: T is a function of S, while S in turn is a function of z, which must be taken into account in differentiation). Substituting [47.2] in the equation of heat conduction [39.5] and taking into account the equation of diffusion [39.5], we obtain:

$$\frac{\partial T}{\partial t} = k \frac{\mathrm{d}^2 T}{\mathrm{d} S^2} \left(\frac{\partial S}{\partial z}\right)^2 + \frac{\mathrm{d} T}{\mathrm{d} S} \frac{\partial S}{\partial t} \tag{47.3}$$

whence:

$$k = \left(\frac{\partial T}{\partial t} - \frac{\mathrm{d} T}{\mathrm{d} S} \frac{\partial S}{\partial t}\right) \bigg/ \left[\frac{\mathrm{d}^2 T}{\mathrm{d} S^2} \left(\frac{\partial S}{\partial z}\right)^2\right] \tag{47.4}$$

(here it is assumed that $k_T = k_S$). Let us now consider the *T-S-z* curve represented in Fig. 57 by the heavy line, and the three points on it, A, O and B with values of the parameters $(z_0 - \frac{1}{2}\Delta z, t_0)$, (z_0, t_0) and $(z_0 + \frac{1}{2}\Delta z, t_0)$, respectively, as well as point P of the intersection of the chord AB and line $z = z_0$; point P has the values of parameters $(z_0, t_0 + \Delta t)$.

Temperature and salinity in points A, B and P can be expressed approximately in the form of a sum of the first terms of a Taylor expansion in the vicinity of point O, namely:

$$T_A = T_0 - \frac{1}{2} \left(\frac{\partial T}{\partial z}\right)_0 \Delta z + \frac{1}{8} \left(\frac{\partial^2 T}{\partial z^2}\right)_0 (\Delta z)^2$$

$$S_A = S_0 - \frac{1}{2} \left(\frac{\partial S}{\partial z}\right)_0 \Delta z + \frac{1}{8} \left(\frac{\partial^2 S}{\partial z^2}\right)_0 (\Delta z)^2 \tag{47.5}$$

$$T_B = T_0 + \frac{1}{2}\left(\frac{\partial T}{\partial z}\right)_0 \Delta z + \frac{1}{8}\left(\frac{\partial^2 T}{\partial z^2}\right)_0 (\Delta z)^2$$

$$S_B = S_0 + \frac{1}{2}\left(\frac{\partial S}{\partial z}\right)_0 \Delta z + \frac{1}{8}\left(\frac{\partial^2 S}{\partial z^2}\right)_0 (\Delta z)^2$$

$$T_p = T_0 + \frac{\partial T}{\partial t}\Delta t$$

$$S_p = S_0 + \frac{\partial S}{\partial t}\Delta t$$

[47.5]

where the indexes designate the quantities in the corresponding points. Let us now take the equation of the chord AB:

$$T_p - T_A = \frac{T_B - T_A}{S_B - S_A}(S_p - S_A)$$

[47.6]

(instead of index "B" we can also write index "P", since point P lies on the chord) and let us substitute in it expressions [47.5]; as a result after some simple transformations we will obtain:

$$\left[\frac{\partial T}{\partial t} - \left(\frac{dT}{dS}\right)_0 \frac{\partial S}{\partial t}\right]\Delta t = -\frac{\Delta z}{2}\left[\left(\frac{\partial T}{\partial z}\right)_0 - \left(\frac{dT}{dS}\right)_0 \left(\frac{\partial S}{\partial z}\right)_0\right] +$$

$$+\frac{(\Delta z)^2}{8}\left[\left(\frac{\partial^2 T}{\partial z^2}\right)_0 - \left(\frac{dT}{dS}\right)_0 \left(\frac{\partial^2 S}{\partial z^2}\right)\right]$$

[47.7]

The expression appearing in this formula in the first term of the right-hand

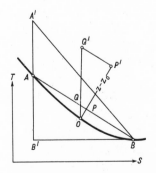

Fig. 57. Explanations in the text (according to Okada, 1938).

side in square brackets is equal to zero along the *T-S-z* curve, according to formula [47.1]; therefore, bearing in mind expression [47.2] also we can write this formula in the form:

$$\left[\frac{\partial T}{\partial t} - \left(\frac{dT}{dS}\right)_0 \frac{\partial S}{\partial t}\right]\Delta t = \frac{(\Delta z)^2}{8}\left(\frac{d^2T}{dS^2}\right)_0\left(\frac{\partial S}{\partial z}\right)_0^2 \qquad [47.8]$$

Substituting into [47.8] instead of the square bracket in the left-hand side the expression determined by formula [47.4], we finally obtain Jacobsen's formula:

$$k_z = \frac{(\Delta z)^2}{8\Delta t} \qquad [47.9]$$

Proudman (1953) had proposed an even simpler derivation of formula [47.9] and extended it to the case of vertical and horizontal stationary heat conduction (and diffusion) in the presence of advection, described by equations [39.6], as well as to the case of non-stationary horizontal heat conduction and diffusion. The corresponding formulae for the determination of the coefficient of mixing have the following form:

$$\frac{k_z}{u} = \frac{(\Delta z)^2}{8\Delta x} \qquad [47.10]$$

$$\frac{k_y}{u} = \frac{(\Delta z)^2}{8\Delta y} \qquad [47.11]$$

$$k_y = \frac{(\Delta y)^2}{8\Delta t} \qquad [47.12]$$

(the indexes in the last four formulae designate the direction of mixing).

In oceanographic literature there is a number of examples of the application of Jacobsen's formula; let us consider two of them. Fig. 58 represents the *T-S* curves of four stations of the R.V. "Meteor", made in the southern part of the Atlantic Ocean on a section along the western basin. Combining six pairs of *T-S* curves from the number of four mentioned, Defant (1954), applying formula [47.10], obtained the values of the quantities k_z/u as shown in Table IX.

The mean value of the ratio k_z/u amounts to 0.74 cm, and at mean velocity $u = 10$ cm sec^{-1} (the extremes of the *T-S* curves considered relate to the core of the Antarctic Intermediate Water, which spreads to the north with approximately the same velocity), the quantity of the coefficient of mixing amounts to 7.4 cm^2/sec, which is in full agreement with the order of magnitude of the coefficient according to determinations by many other methods.

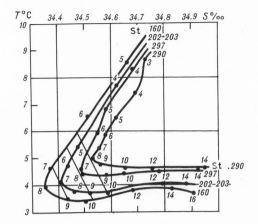

Fig. 58. Determination of the coefficient of mixing by pairs of T-S curves in the Atlantic Ocean from data of the R.V. "Meteor" (Defant, 1954).

TABLE IX

Values of the quantities k_z/u (After Defant, 1954)

Pairs of stations	Δx (km)	Δz (m)	k_z/u (cm)
160–202	600	215	0.96
160–297	3050	380	0.59
160–290	4150	485	0.71
202–297	2450	350	0.625
202–290	3550	510	0.92
297–290	1100	220	0.55

The second example relates to the extremes of T-S curves, corresponding to the sub-tropical subsurface maximum salinity in the equatorial part of the Atlantic Ocean; on the equator this maximum coincides with the core of the subsurface equatorial countercurrent — the Lomonosov Current. Let us select two pairs of T-S curves from the data of the stations of the R.V. "Crawford", made on a section along the equator (Fuglister, 1960). The quantities k_z/u, obtained on the basis of the application of the same formula [47.10], have the values which are shown in Table X (Mamayev, 1964b).

The ratios of k_z/u here, as we see, are approximately one order less than for the region of distribution of the Antarctic Intermediate Water; however, having taken the mean value of the coefficient $k_z = 10$ cm²/sec, for both pairs we will obtain values of velocity $u = 200$ and 222 cm/sec respectively. These values, to a greater or lesser extent, correspond to the velocity of the Lomonosov current in its core.

TABLE X

Values of the quantities k_z/u (after Mamayev, 1964b)

Pairs of stations	Δx (km)	Δz (m)	k_z/u (cm)
488–499	1600	80	0.05
480–488	1000	60	0.045

Formula [47.9] can be to a certain extent generalized to cover the case of inequality of the coefficients of heat and salt exchange. Thus, if $k_T \neq k_S$, instead of [47.3] we obtain:

$$\frac{\partial T}{\partial t} = k_T \frac{d^2 T}{dS^2} \left(\frac{\partial S}{\partial z}\right)^2 + \frac{k_T}{k_S} \frac{dT}{dS} \frac{\partial S}{\partial t} \qquad [47.13]$$

whence:

$$\frac{1}{k_T} \frac{\partial T}{\partial t} - \frac{1}{k_S} \frac{dT}{dS} \frac{\partial S}{\partial t} = \frac{\partial^2 T}{\partial S^2} \left(\frac{\partial S}{\partial z}\right)^2 \qquad [47.14]$$

and instead of [47.4] follows formula:

$$k_T k_S = \left(k_S \frac{\partial T}{\partial t} - k_T \frac{dT}{dS} \frac{\partial S}{\partial t}\right) \Big/ \left[\frac{d^2 T}{dS^2} \left(\frac{\partial S}{\partial z}\right)^2\right] \qquad [47.15]$$

Thus, instead of formula [47.8] we obtain, on the basis of [47.15], expression:

$$\left[\frac{\partial T}{\partial t} - \left(\frac{dT}{dS}\right)_0 \frac{\partial S}{\partial t}\right] \Delta t = \frac{(\Delta z)^2}{8 k_T k_S} \left[k_S \frac{\partial T}{\partial t} - k_T \left(\frac{dT}{dS}\right)_0 \frac{\partial S}{\partial t}\right] \qquad [47.16]$$

whence:

$$k_T k_S = \frac{(\Delta z)^2}{8 \Delta t} \frac{k_S \dfrac{\partial T}{\partial t} - k_T \left(\dfrac{dT}{dS}\right)_0 \dfrac{\partial S}{\partial t}}{\dfrac{\partial T}{\partial t} - \left(\dfrac{dT}{dS}\right)_0 \dfrac{\partial S}{\partial t}} \qquad [47.17]$$

When $k_T = k_S$, from [47.17] we have Jacobsen's formula.

Formula [47.17] by itself yields very little, but it can be somewhat simplified for the particular case when the *T-S-z* and *T-S-t* curves are perpendicular in any point; in practice such cases may be met with fairly often, and Fig. 57 corresponds precisely to such a version. Let us write the equation of the straight line *OP* (Fig. 57):

$$T_p - T_0 = \left(\frac{\widetilde{dT}}{dS}\right)_0 (S_p - S_0) \qquad [47.18]$$

where the tilda designates the tangent of the angle of slope of the tangent to the T-S-t curve (unlike the tangent of the angle of slope of the tangent to the T-S-z curve) in the same point O. If both of these curves are orthogonal in point O, then the condition of perpendicularity of two straight lines takes place:

$$\left(\frac{\widetilde{dT}}{dS}\right)_0 = -\frac{1}{\left(\dfrac{dT}{dS}\right)_0} \qquad [47.19]$$

Consequently:

$$T_p - T_0 = -\frac{1}{\left(\dfrac{dT}{dS}\right)_0} (S_p - S_0) \qquad [47.20]$$

From this formula on the basis of the last two formulae of [47.5] we obtain:

$$\frac{\partial T}{\partial t} = -\frac{\partial S}{\partial t} \bigg/ \left(\frac{dT}{dS}\right)_0 \qquad [47.21]$$

Let us substitute this expression in formula [47.17] and then, reducing the numerator and denominator of the right-hand side by $\partial S/\partial T$ and introducing (for brevity of notation) the designation $\alpha = (dT/dS)_0$ we obtain:

$$k_T k_S = \frac{(\Delta z)^2}{8\Delta t} \frac{k_S + k_T \alpha^2}{1 + \alpha^2} \qquad [47.22]$$

Finally, let us introduce the designation for the ratio of the coefficients $k_T/k_S = \beta \geqslant 1$; then, having divided [47.22] by k_S, we will obtain the formula for the determination of the coefficient of turbulent heat conduction:

$$k_T = \frac{(\Delta z)^2}{8\Delta t} \frac{1 + \beta \alpha^2}{1 + \alpha^2} \qquad [47.23]$$

and having divided [47.22] by k_T — the formula for the coefficient of turbulent diffusion:

$$k_S = \frac{(\Delta z)^2}{8\Delta t} = \frac{1 + \beta \alpha^2}{\beta(1 + \alpha^2)} \qquad [47.24]$$

The corrective multipliers to Jacobsen's formula [47.9], systematized in the form of the last two formulae, are > 1 in the first and < 1 in the second of the formulae; accordingly, Jacobsen's formula [47.9] gives a lower result for the coefficient of turbulent heat conduction and a higer one for the coefficient of diffusion. Consequently, it yields some average result, which, generally speaking, could have been expected in advance.

At first glance formulae [47.23] and [47.24] are not very promising, since it follows from them that for the separate determination of the coefficients of mixing it is necessary to know their ratio without knowing in advance the coefficients themselves; however, the point is that it is easier to determine this ratio than the coefficients themselves. This question is considered in detail from theoretical premises in a work of the author (Mamayev, 1958).

So far as the tangent of the angle of slope of the tangent to the *T-S* curve is concerned, it can be determined just as easily as all the remaining parameters which enter into formula [47.9]. For example, it may be indicated that the quantity α for all three tangents represented in Fig. 58 equals approximately -2.

Let us quote the example of the calculation of coefficient k_S from formula [47.24] for a pair of *T-S* curves of stations 160–290 (Fig. 58). Having taken as an example $\beta = 5$, we obtain (when $\alpha = -2$): $k_S = 0.71 \times 0.84 = 0.60$.

The investigation of the ratio of the coefficients on the basis of formulae [47.23] and [47.24] has, in our view, interesting prospects.

The question of the effect of the inequality of the coefficients of mixing on the dynamics of the waters of the ocean was considered from other premises by Shumilov (1964). Considering the divergence of the extremes of temperature and salinity in the process of mixing provided the coefficients of exchange are unequal, he came to an extremely interesting and important result, having plotted a chart of the divergence of the depths of the extremes; in the Southern Ocean this chart theoretically confirms the diagram of distribution of Antarctic Bottom Water, as obtained earlier from purely oceanographic considerations, as well as substantially improving the picture of the propagation of bottom waters.

CHAPTER 7

METHODS OF *T-S* ANALYSIS

The real *T-S* relationships of the waters of the ocean plotted on the *T-S* diagram serve as the immediate subject of thermohaline analysis. The "background" is also plotted on the *T-S* diagram, reflecting the equation of state of sea water; a family of isopycnals (or isosteres) usually serves as such a "background". The analysis of the *T-S* relationships, together with the field expressing the equation of state of sea water, allows us to take into account the most important factors which determine the nature of the transformation and interaction of different waters; in particular, the family of isopycnals allows us to take account, as was already said above, of the effect of contraction on mixing of sea waters. This question will be considered again in detail below (Sections 55–57). In addition, taking into account the nature of the field of isopycnals in the analysis of the *T-S* relationships of water masses is the basis for the *isopycnic analysis of water masses.*

Before turning directly to the methods of thermohaline analysis, we should draw up a classification of the types of *T-S* relations which can be plotted on the diagram. Among the many possible representations of the thermohaline field of the ocean on the *T-S* diagram, we should point out the main types of *T-S* relations; we shall briefly describe and illustrate these main types by specific examples.

(1) *T-S-z curves. Clusters of T-S-z curves.* The *T-S-z* curves represent on the *T-S* diagram the distribution of temperature and salinity along the vertical at an oceanographical station. For an oceanographical section, very often aggregates of *T-S* curves are plotted which form "clusters" or "fans". "Clusters" of *T-S* curves are characteristic for central ocean regions (Fig. 34), "fans" for so-called "transitional" regions characterized by a considerable horizontal transformation of water masses (Fig. 69). The question of the analysis of "clusters" and "fans" of *T-S* curves is considered in detail in Section 53.

Sometimes an aggregate of curves is not drawn, but an averaging curve is plotted (or not plotted) through a set of *T-S-z* points. Examples of computer-plotted clusters of *T-S-z* points may be found in the *Oceanographic Atlas of the International Indian Ocean Expedition* (Wyrtki, 1971).

Finally, we can outline on the *T-S* diagram the region in which all *T-S*

curves for a particular area of the World Ocean will lie; such regions on the
T-S diagram may be called fields of *T-S* curves. The fields of *T-S* curves for all
the oceans are represented on the generalized *T-S* diagram of Dietrich (1950,
1964), reproduced in Fig. 96.

(2) *T-S-t curves.* They characterize either a change in temperature and salinity in some point of the ocean due to periodic processes, or a change in the
process of transformation. In the first case, the *T-S-t* curves, as a rule, prove
to be closed; in the second case, the curves turn out to be interrupted, and for
a number of points (in particular, by depth) form families which combine in
a certain way with families of *T-S-z* curves. The question of the relationship of
curves by parameters *z* and *t* was considered in detail in the preceding chapter
in the example of analytical *T-S* curves.

(3) *T-S-l curves.* Relationships of this type are plotted for horizontal (or almost horizontal) distances (*l*), primarily in the following three cases:

(a) for the surface of the sea (or any other horizontal surface). Fig. 59
represents such *T-S* curves for the surface of the sea, averaged for the entire
World Ocean, according to Wüst et al. (1954).

(b) For a selected isopycnic surface; in this case the *T-S* relationship serves
as a basis for carrying out isopycnic analysis; set out in detail in Timofeev
and Panov (1962).

(c) For any characteristic quasi-horizontal surface. Most often a surface is
selected which characterizes the core of an intermediate water mass, for
example, the surface of maximum salinity in which the core of the Mediterranean water mass (Fig. 109) lies, or the surface of minimum salinity which
characterizes the position of the core of the Antarctic intermediate water

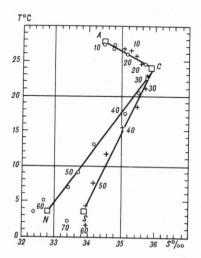

Fig. 59. Mean *T-S* relations of surface waters of the World Ocean. *A* = Equatorial; *C* = Central; *N* = Northern Subpolar; *S* = Southern Subpolar surface water masses of the World Ocean. (After Wüst et al., 1954.)

mass. The question of the *T-S* analysis of the transformation of the cores of intermediate water masses is considered in Section 62.

T-S-l relations can be plotted for quite obvious reasons for other surfaces too, say, for a surface of minimum oxygen content in sea water, the speed of sound (axis of the sound channel), etc.

(4) *T-S-z-l curves*. Such relations characterize a vertical (or any other) sur-face in the ocean, say, a contour formed by two verticals of oceanographic stations and two horizontals — on the surface of the sea and at some definite depth. It is clear that contours of such nature, having identical area but situ-ated in different regions of the World Ocean, will be characterized by con-tours on the *T-S* diagram which have a different area and a different location. The relation between the areas and location on the *T-S* diagram of the differ-ent *T-S-z-l* contours characterizes, conventionally speaking, the "thermohaline tension" of the section, which by its significance is similar to the tension of a solenoidal field on a cross-section through a geostrophic current. The example of closed *T-S-z-l* curves for two sections of the R.V.'s "Crawford" and "Atlan-tis", almost identical in area and located in the plane of circles of latitude 40°15′N and 32°30′S, is represented in Fig. 60. The study of such contours is connected with the application of Green's theorem for a potential *T-S* field (Section 37); this question is an interesting one for further development.

(5) *Statistical T-S relationships, or T-S-n relations*. In statistical *T-S* rela-

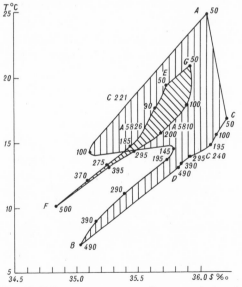

Fig. 60. *T-S* contours, corresponding to vertical sections in the ocean, which have approximately the same area and are located between pairs of stations: "Crawford" 221–240, along parallel 40°15′N (distance 1522 miles); "Atlantis" 5810–5826, along parallel 32°30′S (distance 1608 miles). Height of the contour in the sea: 500 m.

tionships a new parameter n appears — frequency (or p — probability). In this case, numbers attached to definite *T-S* points are plotted, which characterize the frequency (or probability) in percentages or in absolute quantities of the given *T-S* points in time or in space. For this reason, statistical *T-S* relations are divided into two main types: (a) temporal statistical *T-S* diagrams (see Fig. 90); (b) spatial and volumetric statistical *T-S* diagrams (see Figs. 91 and 92).

Statistical *T-S* analysis will be considered in detail below, in Section 58.

Such are the main types of real *T-S* relations of the waters of the World Ocean and its seas; there are, of course, and there may be further proposed other, intermediate, types as well. However, the possible representations of the thermohaline structure of the waters of the ocean on a *T-S* diagram considered are quite sufficient for fullfledged analysis of ocean waters. In the subsequent exposition we shall dwell in greater detail on ways and means of analyzing all these basic types, after having first considered the question of thermohaline indexes as they apply to the real water masses of the ocean.

49. WATER MASSES AND THERMOHALINE INDEXES

The concept of primary water masses homogeneous in temperature and salinity (which can be represented on the *T-S* diagram in the form of individual thermohaline indexes), which underlies the analytical theories of *T-S* curves and the methods of their analysis, does not correspond, as has already been said, to reality, for in the World Ocean we observe continuous distribution of temperature and salinity vertically and horizontally. Therefore, attention should be drawn to the relation of the real water masses of the ocean to their images on the *T-S* diagram, in particular to the thermohaline indexes.

The definition of a water mass of the ocean was given by Dobrovol'skii (1961); this definition is as follows: "The term water mass should be used to describe some relatively large volume of water, formed in a certain region of the World Ocean — the home or source of this mass, which possesses over a long time almost constant and continuous distribution of physical, chemical and biological characteristics, which make up a single system, and which displaces itself as one single whole". We see that the concept of a real water mass of the World Ocean is considerably wider than the concept which can be formed within thermohaline analysis; many characteristics of a water mass remain outside the scope of consideration in such analysis, although part of them can be represented on the *T-S* diagram parametrically.

On the other hand, a kind of generalization of the concept of the water mass is possible within the framework of *T-S* analysis; by water mass we do not necessarily have to refer to a discrete volume of water existing in the ocean, homogeneous or almost homogeneous in temperature and salinity. If, say, stationary values of temperature and salinity are observed on the surface of the ocean in some point, the corresponding thermohaline index may be considered an indicator of the water mass. One may imagine that such a fictitious water mass extends infinitely into the atmosphere, while alongside, in a neighboring point, another water mass also extends into the atmosphere. The steadiness of temperature and salinity in the point considered develops under the influence of climatic and other conditions; however, this constancy may conventionally be considered the consequence of the uninterrupted inflow of heat and salts from the fictitious water mass considered. Considerations of this nature were apparently expressed for the first time by Sverdrup et al. (1942), and he called such a water mass without real thickness a "water type". As an example of such "point" water masses one may cite the four main surface water masses of the World Ocean singled out by Wüst et al. (1954) and shown in Fig. 59. These water masses, obtained as a result of the consideration of T-S characteristics on the *surface* of the ocean, do not have any real volume; however, we are fully entitled to connect their thermohaline indexes by straight lines of mixing and to consider the surface waters of any latitude as the result of horizontal mixing of the corresponding water masses, although in point of fact the real case of the appearance of "mixtures" of primary waters is simply the climatic zonality of surface characteristics.

The foregoing applies to an even greater extent to waters mixing along the vertical; reference has already been made to this above, particularly in Chapter 6. Isolating the thermohaline indexes of the original "pure" water masses is necessary not only in order to define the boundaries between the water masses, but also to determine the percentage content of the waters and other numerical characteristics of interaction of water masses. The thermohaline indexes of the main primary water masses of the ocean are specified on the generalized T-S diagrams given in Chapter 8 (Figs. 97–100).

In contradistinction to the "water type", characterized by one T-S point, Sverdrup singles out the "water mass" as an object characterized on the T-S diagram by a rectilinear (or curvilinear) T-S relationship. An example of such water masses (determined also within the framework of thermohaline analysis) are the Central Water Masses of the oceans having almost rectilinear T-S relationships (Fig. 29). Such a view agrees to a certain extent with the determination of the boundaries of water masses on the basis of the geometry of T-S curves, with the difference in favor of the latter that makes it possible more clearly to determine the boundary of the water mass by the line of its 50% content in the triangle of mixing.

50. ANALYSIS OF WATER MASSES USING *T-S-z* CURVES

The most common case of mixing of water masses under the real conditions of the World Ocean is the vertical mixing of two, three and four superimposed water masses. The basis for the study of the vertical mixing of waters on the *T-S* diagram is the analytical theory of *T-S* curves for an ocean of infinite depth, considered in detail in the preceding chapter. It was already indicated there that the theory of *T-S* curves for a semi-infinite and finite sea introduces substantial corrections in the analysis of real *T-S* curves; however, the theory for the sea of infinite depth continues to remain the basis of analysis of curves, while the results of the latter two theories must be drawn upon to clarify the picture of the interaction and transformation of water masses obtained by applying the theory for the infinite sea as a first (and fairly close) approximation.

In chapter 6 it was indicated that the practical outcome of the analytical theory of *T-S* curves took the form of seven fundamental theorems of the "geometry of *T-S* curves" (p. 203). Inasmuch as part of the theorems concern not the stationary *T-S* curves, with which we usually have to deal, but *T-S* curves changing on the same vertical by time, in practical analysis it is more convenient to use *rules* resulting from the "geometry of *T-S* curves". These rules were proposed by Shtokman (1944) and they may be stated in the following way:

(1) The boundary between two water masses should be considered the depth at which the percentage content, as determined by the straight line of mixing or the triangle of mixing, amounts to 50% for each of the water masses.

(2) If the *T-S* curve is close to a straight line, then the straight line of mixing should be used for its analysis. In this case, the indexes of the two mixing water masses lie on the ends of the curve and correspond to the surface and deep water masses.

(3) If the *T-S* curve consists of two or more straight (or almost straight) sections, connected among each other, then there are three or more water masses. The quantity of water masses is equal to the quantity of extremes (or conjugations) plus two (in Fig. 61, *J, D* and plus *A* and *B*, see note on p. 253).

(4) The determination of the *T-S* indexes is made by drawing tangents to the straightened-out sections of the *T-S* curve. In this case the intersection of the tangents in the region of the extreme (conjugation) indicates the *T-S* index of the intermediate water mass (water masses *J* and *D* in Fig. 61); the ends of the branches of the *T-S* curve correspond to the surface and bottom water masses (water masses *A* and *B* in Fig. 61).

(5) For the determination of the boundaries and percentage content of the water masses at different depths, triangles of mixing (triangles *AJD* and *JDB* in Fig. 61) are plotted on the *T-S* indexes as apexes.

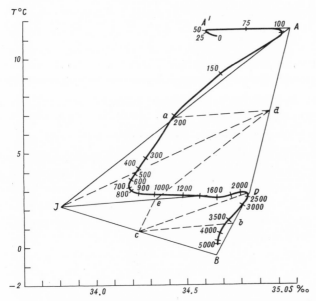

Fig. 61. Graphic analysis of the *T-S* curve of station No. 4 of the R.V. "Meteor".

 (6) The principal median of the triangle of mixing (*Jd* and *Dc* in Fig. 61) drawn from that apex which corresponds to the intermediate water mass to the middle of the opposite side (called the base of the triangle of mixing), intersects the *T-S* curve in that point where parameter *z* characterizes the position of the core of the intermediate water mass.

 (7) The secondary medians of the triangle of mixing (*ad* and *ae*; *ec* and *bc* in Fig. 61), drawn from the middle of the base of the triangle of mixing to the other two sides, intersect the *T-S* curve in those points of it where parameter *z* corresponds to the boundaries of the intermediate water mass. The part of the *T-S* curve contained between the secondary medians of the triangle of mixing corresponds to the intermediate water mass (see also Fig. 41).

 Let us consider the application of the "geometry of *T-S* curves" to the analysis of the *T-S* curve represented in Fig. 61 of station No. 4 of R.V. "Meteor", occupied on 7–8 June, 1925 in coordinates 41°27.0'S 52°47.0'W (Atlantic Ocean). It is characterized by the presence of three extremes (at depths of about 100, 700 and 2,500 m) * and two branches: surface (from 0 to 100 m) and deep (from 2,500 to 5,000 m). Thus, at the given oceanographical station the existence of five water masses should be assumed. Let us consider these water masses.

---

* The extreme at the depth of about 50 m is not considered, since it is not stationary: the section of the *T-S* curve from 0 to 50 m is characterized by an unstable stratification, temporary in nature.

(1) The water mass, characterized by the end of the surface branch (point A'), represents the *surface water mass*, lying in the layer from 0 to approximately 75–100 m. In the majority of cases, this surface water mass, subjected to the influence of the processes taking place on the surface (wind mixing, heat exchange with the atmosphere, etc.), is excluded from stationary *T-S* analysis (Sverdrup et al., 1942). In any case, the surface branches of the *T-S* curves from 0 to 100 m of the majority of stations of the World Ocean are characterized by uncertainty caused by the unsteadiness of the processes mentioned above.

(2) Water mass A, characterized by maximum salinity, is the *South Atlantic Central Water* (Sverdrup et al., 1942), formed in the region of the Subtropical Convergence (between 30 and 40°S) and lying in the layer between approximately 100 and 300 m. This water mass does not have any single definite *T-S* index, since it is zonally transformed (see Section 61); consequently, it has an infinite set of *T-S* indexes, determined by the latitude of the place, and the locus of all these *T-S* indexes is a straight line joining points (6°C; 34.5‰) and (18°C; 36.0‰). *T-S* index A, determined by the intersection of tangents $A'A$ and JA to the branches of the *T-S* curve, lies almost on the straight line mentioned and represents a zonal modification of the South Atlantic Central Water for the region of "Meteor" station No. 4.

(3) Water mass J, characterized by minimum salinity is the *Antarctic Intermediate Water*, formed in the region of Antarctic Convergence. The *T-S* index of this water mass, determined by Wüst, is: $T = 2.2°C$; $S = 33.80‰$ (Sverdrup et al., 1942). We observe this temperature and salinity on the surface of the sea in those places where the formation of the intermediate water mass takes place.

(4) Water mass D is the *South Atlantic Deep Water*. The question of the origin of this water mass is not quite clear. On the one hand, it is assumed that it is the result of the mutual transformation of the Antarctic Intermediate and the Antarctic Deep Water masses, i.e., according to Dobrovol'skii's (1961) terminology it is a secondary ocean water mass. On the other hand, the opinion is expressed that North Atlantic Deep and Bottom Water plays the main role in its formation. It is also quite likely that the South Atlantic Deep Water has its center of formation in the Southern Ocean and that it is thus a main ocean water mass (according to the terminology of Dobrovol'skii, 1961). The *T-S* index of this water mass ($T = 2.9°C$; $S = 34.85‰$) is determined by the intersection of the tangents to the branches of the *T-S* curve of "Meteor" station No. 4.

(5) *Antarctic Bottom Water B,* observed in the South Atlantic, is formed mainly in the Weddell Sea. Its thermohaline index, according to Sverdrup et al. (1942), is $T = -0.4°C$; $S = 34.66‰$.

Such are the water masses and their thermohaline indexes in the area of

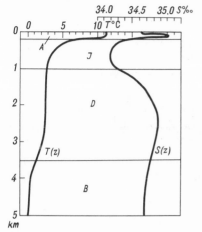

Fig. 62. Curves of vertical temperature and salinity distribution, as well as the position of water masses at "Meteor" station No. 4.

"Meteor" station No. 4, i.e., in the southwestern part of the South Atlantic. In Fig. 61 triangles of mixing AJD and JDB are plotted on these T-S indexes. From the intersection of the T-S curve by the secondary medians da, de, ce and cb we determine that the boundaries between water masses A and J, J and D, D and B lie at depths of 200, 1,000 and 3,500 m respectively. The boundaries between the water masses, as well as the curves of vertical distribution of temperature and salinity, are represented in Fig. 62.

Such is the process of analysis of a T-S curve of an oceanographic station. Having determined as a result the boundaries between water masses for a number of stations, we can characterize the distribution of the water masses of the region studied in sections or in charts. It goes without saying that the analysis of T-S curves is far from being always as simple and unambiguous as it turned out for "Meteor" station No. 4, selected as highly characteristic and fairly simple. Often T-S curves of neighboring stations are dissimilar, and sometimes it is extremely difficult to determine the T-S indexes, etc.

As a result it is necessary to consider an aggregate (clusters) of T-S curves, or averaged T-S curves, or, finally, idealized T-S curves, but even this does not always lead to one unquestionable value: every time the analysis of T-S curves represents a *research problem* and not merely a phase of standard processing.

51. DETERMINATION OF PERCENTAGE CONTENT OF WATER MASSES

Let us now consider the use of the triangle of mixing to determine the percentage content of water masses at different depths both at an individual

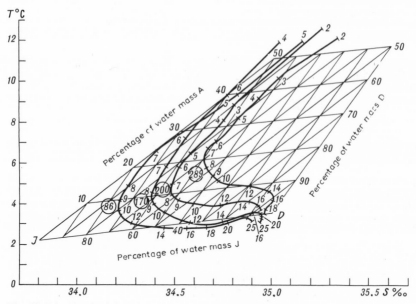

Fig. 63. Triangle of mixing of water masses *A*, *J* and *D* (lower part).

oceanographic station and in a section, i.e., the use of the graphic method which was mentioned in Section 30. Fig. 63 represents the lower part of the triangle of mixing plotted on the indexes of the following water masses as apexes:

A – South Atlantic Central Water. Modified index $T = 20°C$ and $S = 36.2\%_0$ is determined from the surface branches of eight "Meteor" stations (No. 86, 170, 157, 200, 210, 258, 299 and 289) composing a section in the South Atlantic along the eastern coast of South America. The *T-S* curves of these stations were chosen by Defant and Wüst (1930) for a description of the spread of the Antarctic Intermediate Water Mass. The *T-S* index of water mass *A* in Fig. 63 is cut off together with the upper part of the triangle of mixing;

J – Antarctic Intermediate Water ($T = 2.2°C$; $S = 33.8\%_0$);

D – South Atlantic Deep Water. Its *T-S* index ($T = 3.5°C$; $S = 35.0\%_0$) is also modified and determined from the *T-S* curves of the eight stations mentioned.

Within the triangle of mixing, indicated in Fig. 63, parts of the *T-S* curves of four of the eight stations mentioned (No. 86, 170, 200 and 289) are plotted. For the determination of percentage content it is most convenient to find the value of parameter *z*, corresponding to the percentage content, a multiple of 10%, of any of the three water masses; in other words, finding the point of intersection of the T-S curve with the corresponding isoline to determine the value of parameter z in this point. Thus, for example, at station No. 289 40% of water mass J is observed at depths of approximately 285 and 900 m. Thomsen (1935), who first applied percentage analysis of T-S curves using the triangles of mixing, proposes to determine the percentage content of each of the water masses in standard sections.

The result of the analysis carried out in respect of each of the water masses at all the oceanographic stations of the section (region) may conveniently be represented in the form of sections and charts. For example, in Fig. 64, representing a projection on the plane of the meridian of the section consisting of the eight "Meteor" stations mentioned, is shown the percentage content of the South Atlantic Central Water A (on top) and the Antarctic Intermediate Water J (below). In both figures the regions are shaded where the percentage content of the corresponding water mass exceeds 50%. In any point the sum must amount to 100% (for this triangle of mixing, water mass D must be added to water masses A and J).

The determination of the percentage content makes it possible to study more thoroughly the results, i.e., the determination of the boundaries of the water masses, since it gives an idea of the "density" of the corresponding water mass. Thus, it follows from Fig. 64 that the "density" of intermediate water mass J decreases from the south to the north (see isolines 70 and 60%). However, an even more important conclusion follows from both of the last figures, namely, that this water mass mixes more intensively with the underlying than with the overlying water mass (the thickness of the isolines above the core of the water mass and their rarefaction below). This fact was noted by the author earlier in the example of the Mediterranean Intermediate Water in the Atlantic (Mamayev, 1960b).

Thus, the determination of the percentage content of water masses makes it possible substantially to improve the picture of the distribution and transformation of the water masses of the ocean.

The example considered above relates to the case when mixing of water masses is observed along the vertical; however, the triangle of mixing is also applicable to the determination of the percentage content of waters in the case of horizontal mixing, as well as when two of the three water masses mix in a horizontal direction or along isopycnic surfaces. Let us refer to the highly significant example of "refined" T-S analysis of the interaction of water masses in the region of the coast of Central Brazil between Cabo Frio and Vittoria (Okuda, 1962), where in the process of horizontal mixing we find tropical, coastal and "deep" waters. Okuda plotted charts of distribution of these water masses on the surface of the sea as well as at levels of 25 and 50 m, from which it follows that the distribution of the water masses at these depths is spotty in nature. Thus, percentage analysis enables us to dis-

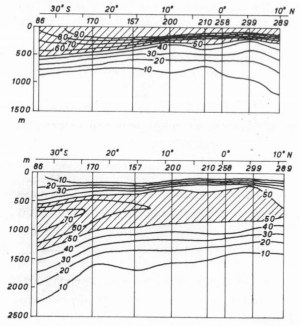

Fig. 64. Percentage content of Southern Atlantic Central Water mass *A* (above) and Antarctic Intermediate Water mass *J* (below) in a meridional section along the eastern coast of South America.

cover the extremely fine structure of the waters; its possibilities are very broad. Incidentally, this was demonstrated for the first time in the work of Thomsen (1935), devoted to the analysis of the interaction of waters in the southern parts of the Indian Ocean and Pacific Ocean.

Another example of when three water masses mix together both vertically and horizontally will be considered below in Section 53.

Finally, the case of mixing of four water masses is possible; if a minimum of two of the four water masses are in a process of horizontal mixing, the basis for the determination of the percentage content of the waters in the mixture is not the two triangles of mixing, plotted on the four *T-S* indexes as apexes and having one common side (such as triangles *AJD* and *DJB* in Fig. 61), but the *quadrangle of mixing*. A quadrangular nomogram was proposed by Miller (1950) and elaborated by the author; this question is considered in Section 54.

52. TYPIFICATION OF *T-S-z* CURVES

The determination of thermohaline indexes by means of analysis of the *T-S* curves of oceanographic stations in accordance with the analytical theories of *T-S* curves, with the subsequent determination of boundaries between water masses, their percentage ratio, etc., is not the only method of analysis of the water masses of the ocean. Apart from this, the *typification* of *T-S* curves is possible, i.e., their combination in groups following the principle of the similarity of form of curves, and the division into regions of the ocean — the determination of regions with identical types of *T-S* curves.

The analysis of water masses, based on the typification of *T-S* curves, was carried out for the first time by Jacobsen (1929) for the northern part of the Atlantic Ocean. The work of Mosby (1938) was very productive for an understanding of the structure of the water masses not only of the Atlantic but of the entire World Ocean as a whole.

On the basis of the typification of *T-S* curves Jacobsen broke the North Atlantic down into 23 regions *, each of which is characterized by a quite definite "cluster" of *T-S* curves. In the opinion of the author, the zoning of Jacobsen is too fractional; thus, the "clusters" of *T-S* curves for his regions XIV—XVII, embracing the Carribean Sea and the region adjacent to it on the east, whence the Carribean Sea is supplied with waters of the trade wind currents, are very similar. The *T-S* curves for these four regions are distinguished practically only by the length of the surface branches. Therefore, if we take into account the zonal (as well as latitudinal) transformation of the surface waters, the number of regions can be considerably reduced. Indeed, these four regions (XIV—XVII) characterize a genetically homogeneous province, the division of which crosswise into four regions is by no means in agreement with the principle of the continuity of ocean fields and oceanographic characteristics.

Proceeding from the foregoing, the author (Mamayev, 1960a), using the data of the IV cruise of the R.V. "Mikhail Lomonosov", drew up a typification of *T-S* curves with a view to a greater degree of generalization, identifying only five regions in the Central North Atlantic. Let us consider the corresponding results.

Concepts about the water masses of the North Atlantic in their most generalized form are set forth by Sverdrup et al. (1942). Sverdrup points out that the main feature of the North Atlantic is the presence of two basic water masses (the upper 100-m layer, subjected to the considerable influence of exchange with the atmosphere and seasonal fluctuations, is not considered by him):

---

* Tiuriakov (1964) and Tiuriakov and Zakharchenko (1965) repeated Jacobsen's typification on more modern and extensive material; they singled out 26 regions in the North Atlantic, the location of which improves the picture obtained earlier by Jacobsen. The map of Tiuriakov's and Zaharchenko's regions may prove useful for technical purposes.

(1) *North Atlantic Central Water* (*A*). Is characterized by a straight section of the *T-S* curve from the point $T = 8°C$, $S = 35.10‰$ to the point $T = 19°C$, $S = 36.70‰$. The correspondence between T and S is given more exactly in Table XI.

TABLE XI

Relation between T and S in North Atlantic Central Water
(According to Sverdrup et al., 1942)

T (°C)	S (‰)
8	35.12 ± 0.09
10	35.37 ± 0.09
12	35.63 ± 0.09
14	35.88 ± 0.09
16	35.12 ± 0.09

This water mass is formed in the region of the Subtropical Convergence, between 35 and 40°N, since in some seasons of the year the horizontal *T-S* curves for the surface of the sea are similar, as was noted for the first time by Iselin, to the vertical *T-S* curves of the Central Water (this region is shown on the map − Fig. 102 − by small squares). The North Atlantic Central Water lies below the 100-m depth, and the greatest depth of its lower boundary is equal to 900 m in the Sargasso Sea.

(2) *The Deep and Bottom Water* (*B*). Is characterized by temperatures between 2.2 and 3.5°C and salinities between 34.90 and 34.97‰. The thermohaline index of this water mass, according to Jacobsen (1929), is: $T_{in\ situ} = 2.5°C$; $S = 34.90‰$. The Deep and Bottom Water is formed, as Nansen had already pointed out, in the region to the southeast of the southern extremity of Greenland (as well as, according to Sverdrup, between the Labrador Peninsula and the southwestern coast of Greenland), sinking here to the bottom; it is observed at depths above 3,000 m in almost unmixed form throughout the whole North Atlantic − in regions where the features of the relief do not prevent its penetration.

Between these two basic types of water masses one may observe in different regions three types of intermediate water masses which influence the distribution of temperatures and salinities in the intermediate layers of the North Atlantic:

(3) *The Antarctic Intermediate Water* (*AnI*). Thermohaline index: $T = 2.2°C$; $S = 33.80‰$. This water mass is characterized by a considerable saline minimum; it is formed in the zone of Antarctic Convergence and extends between isopycnic surfaces $\sigma_T = 27.2$ and 27.4, primarily to the north.

In the Northern Hemisphere it is traced to 20°N. (It is possible that in the zone of the Antilles and Florida Currents, and further in the Gulf Stream too it may be traced somewhat more to the north.)

(4) *Arctic Intermediate Water* (*AI*). Formed to the east of the Great Newfoundland Bank in small quantities and has little influence outside the regions of its formation. Thermohaline index: $T = 3.5°C$; $S = 34.88‰$.

(5) *Mediterranean Intermediate Water* (*M*). When flowing out of the Straits of Gibraltar has a thermohaline index of $T = 11.9°C$; $S = 36.50‰$. Extends between the isopycnic surfaces $\sigma_T = 27.6$ and 27.8, reaching in the north the shelf of the British Isles.

Analyzing the water masses, we see an overall resemblance of the picture obtained with the generalized picture of the water masses of the North Atlantic according to Sverdrup. All the water area investigated by the "Mikhail Lomonosov", below the depths of 1,500–2,000 m, is underlayed the deep and bottom water mass of Arctic origin (*B*), having a thermohaline index of $T = 2.5°C$; $S = 34.90‰$. Above lie the surface and intermediate water masses, with the exception of the Antarctic Intermediate and Arctic Intermediate Waters, which were not observed in the region investigated in the IV cruise. The boundaries between the regions (structures) are shown in Fig. 65; they may be easily established after the typification of *T-S* curves, which is set forth briefly below, with an indication of the regions where individual types of *T-S* curves are encountered:

(1) *The region of the Subarctic Surface Water* (*SA*), under which the deep and bottom water mass stretches, lies to the south of the Straits of Denmark and Greenland; its southern boundary is situated approximately along 50°N. Typical *T-S* curves of this region are represented in Fig. 66 in the example of stations 234 and 243; they are almost vertical. The Subarctic Water is the subject of conversion into deep and bottom water masses, and from Fig. 106, which represents *T-S* points of the Subarctic Water for the surface of the sea at 24 stations: No. 234–244, 275–283, 321–324, it may be seen that, apart from the drop in temperature, for the formation of Deep and Bottom Water from Subarctic Water an insignificant change in salinity is also required, mainly salinization during ice-formation. Thus, the Subarctic Surface Water lies in the subarctic (or polar) region of convective mixing, according to Zubov (1947), and this mixing in the period of greatest spreading of ice-floes from the Baffin and Danish Straits creates the Deep and Bottom Water mass from the Subarctic Water mass.

(2) *The region of the North Atlantic Water* (*A*), *overlying the Deep and Bottom Water,* is characterized by the type of *T-S* curves represented in the example of stations 250 and 311 of the "Mikhail Lomonosov" (Fig. 66). This type of *T-S* curve has in the surface part a bend, which is a trace of the transformation of the more saline core of the North Atlantic Current together

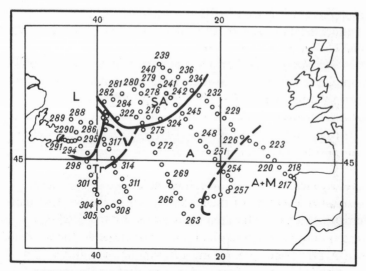

Fig. 65. Boundaries between the different water masses in the central North Atlantic (from data of the IV cruise of the R.V. "Mikhail Lomonosov"). *SA* = region of the Subarctic Water mass; *L* = region of the Labrador (arctic) Water mass; *A* = region of the North Atlantic Water mass; *M* = region of the Mediterrranean Intermediate Water mass; *Tr* = zone of horizontal transformation (Mamayev, 1960a).

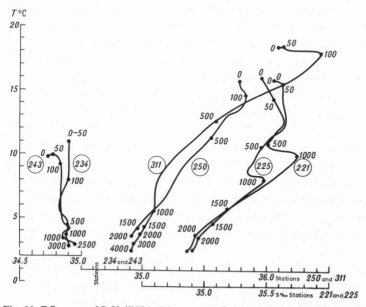

Fig. 66. *T-S* curves of R.V. "Mikhail Lomonosov" stations 234 and 243 (subarctic type), stations 250 and 311 (North Atlantic type) and stations 221 and 225 (North Atlantic type with Intermediate Mediterranean Water mass).

with the surface 100–150 m and underlying layers. Region A is a region of marked zonal transformation, downward along the currents of the Gulf Stream system, of the primary water mass of the Gulf Stream; passing through region A, these waters extend to Iceland and further, ending their journey as warm waters in the Arctic Basin. Precisely the zonal transformation of the surface waters makes it impossible to break down region A into parts; the thermohaline indexes of the different latitudinal modifications pass smoothly one into the other. The question of the nature of the zonal transformation will be considered again below (Section 61).

(3) *The region of the North Atlantic Water overlying the Mediterranean Intermediate Water* (*M*) is characterized by the *T-S* curves represented in Fig. 66 by the example of stations 221 and 225. The thermohaline index of the Mediterranean Water mass is taken at $T = 11°C$; $S = 36.50‰$; this core is situated approximately at a depth of 1,000–1,200 m.

In the nature of *T-S* curves of all stations at which Mediterranean Intermediate Water is present (region $A + M$, Fig. 65), an interesting feature is observed consisting of the asymmetry of the *T-S* curves as compared with that form which they should have in the case of mixing of three water masses, in accordance with the theorems of the "geometry of *T-S* curves". The asymmetry of *T-S* curves is explained by the influence of vertical stratification on the process of mutual transformation of water masses, as a result of which the Mediterranean Water M mixes mainly with Deep Water B. The unilateral direction of the transformation of the intermediate water mass is reinforced also by *contraction on mixing*. This question is considered in greater detail in Section 63 and in the work of the author (1960b).

(4) *Region of the water mass of the Labrador Current* (*L*). This water mass belongs to the arctic type (lies to the east of Newfoundland) and is characterized by the type of *T-S* curves represented in Fig. 67 (stations 290–293). Isopycnic mixing must exist between the water mass of the Labrador Current and the water mass of the Gulf Stream, presumably between the isopycnic surfaces $\sigma_T = 26.0$ and 26.5.

In the area of the zone of transformation contraction on mixing must take place, as was pointed out for this region by McLellan (1957). As an example, the water masses participating in isopycnic mixing along the surface $\sigma_T = 26.7$ must experience an increase in density up to the density of the waters lying on the straight line of mixing connecting points $(-1°C; 33.18‰)$ and $(15.1°C; 35.97‰)$. The contraction on mixing of the Labrador and North Atlantic Water masses does not lead to any considerable sinking of the water masses, since the analysis of many *T-S* curves does not reveal any deep points for the straight line of mixing noted above. Apparently, contraction on the isopycnic mixing of the Labrador and North Atlantic Water masses gives rise to the Intermediate Arctic Water mass; isopycnic contact leads to broad

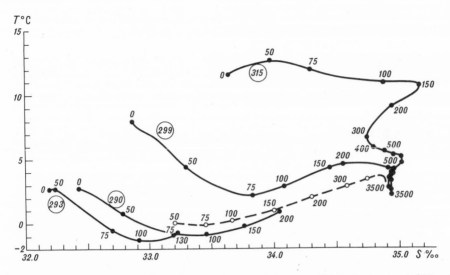

Fig. 67. *T-S* curves of R.V. "Mikhail Lomonosov" stations 290 and 293 (Labrador type), 299 and 315 (transitional type). The broken line shows the *T-S* curve of Labrador waters averaged for 1948–1958 (from data of the International Ice Patrol in the North Atlantic).

development of frontal activity and oceanic cyclogenesis, while isopycnic mixing in the zone of cyclogenesis leads to the creation of a zone of transformation, or a transitional zone, designated by the index "*Tr*" in Fig. 65.

(5) *The zone of horizontal transformation* (*Tr*) between the waters of the Gulf Stream and the Labrador Waters is characterized by the type of *T-S* curves represented in Fig. 67 by the example of stations 293 and 315. A curious feature of this zone is the presence of an isolated region of Gulf Stream Waters in the area of stations 284 and 285; this fact does not allow us simply to draw the northeastern part of the boundary between the zone of transformation and Atlantic Waters (between the third and fourth sections of the "Mikhail Lomonosov"). From Kashkin's data (personal communication, 1960) on the distribution of various forms (boreal and warm-water) of plankton, this boundary extends to the northwest, to stations 284 and 285. One may conjecture about the existence in the vicinity of these two stations of a temporary isolated heat eddy, detached from the main mass of Gulf Stream Waters in the process of meandering.

In conclusion, Fig. 68 gives model patterns of the interaction of water masses for two sections of the "Mikhail Lomonosov" (the first and the fifth). The nature of the interaction of the water masses in these two sections is extremely different. In Fig. 68 those boundaries between water masses which are more permeable are designated by dashed lines.

Division into regions on the basis of typification of the structures of the northwestern part of the Pacific Ocean was carried out by Dobrovol'skii

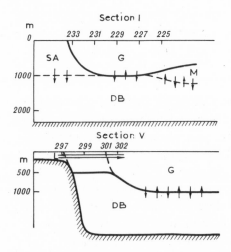

Fig. 68. Diagrams of the interaction of waters in the North Atlantic in sections *I* and *V* of the IV cruise of R.V. "Mikhail Lomonosov". *SA* = Subarctic Water mass; *G* = Gulf Stream (Atlantic surface) Water mass; *M* = Mediterranean Water mass; *DB* = Deep and Bottom Atlantic Water mass; *L* = Labrador Water mass.

(1962). He also identified only five regions (structures), in full agreement with the natural boundaries in the ocean, in particular, with convergences and fronts. Dobrovol'skii et al. (1960) extended such zoning to the central part of the Pacific Ocean also, while the definitive zoning of the structure of the entire Pacific Ocean has been drawn up by Radzikhovskaya and Leont'eva (Hydrology of the Pacific Ocean, 1968).

53. SPECIAL FEATURES OF *T-S* ANALYSIS IN REGIONS OF HORIZONTAL INTERACTION OF WATER MASSES

The study of the interaction of water masses in those regions of the World Ocean where the largest horizontal gradients of oceanographic characteristics are observed on the surface is of exceptionally great importance. In a number of cases this interaction is markedly frontal in nature, and in the process of horizontal mixing, as a rule, of two main water masses the formation of secondary water masses occurs, which often are quasi-stationary in nature; waters having the greatest biological productivity are tied in with these regions.

Regions of horizontal interaction of water masses are widespread in the ocean; the main such regions are the following:

(1) Regions of interaction of the Central Water masses of the northern parts of the Atlantic and the Pacific Oceans with Subarctic Waters. These regions are sometimes called: regions of the subarctic front. The subarctic frontal

zone in the North Atlantic stretches from the Gulf Stream front south of Newfoundland to the northeast, to the region of Iceland and Spitzbergen; in the Pacific Ocean it extends from the region of interaction of the waters of the Kuroshio and Oyashio in a latitudinal direction to the shores of British Columbia and California.

(2) An extensive region of the Southern Ocean lying between the subtropical and antarctic convergences. Here we observe latitudinal interaction of the South Central Water masses of three oceans with the surface circumpolar Antarctic Waters, while the product of mixing represents the Subantarctic Surface Water mass.

(3) Regions of intrusion of cold currents in the low latitudes. The most characteristic of these are the regions of interaction of the water of the California and Peru Currents with the Tropical and Equatorial Waters. These regions are distinctive "spurs" of the first two — namely, the subarctic and subantarctic regions of interaction of water masses, while the name given to them by Sverdrup — "transitional regions" — in a certain sense can be extended also to the regions of subarctic and subantarctic interaction of waters. A noteworthy feature of *T-S* curves plotted for oceanographic stations which are located in sections intersecting the "transitional regions" in the direction of the largest gradients of oceanographic characteristics is their "fan-shaped" nature in the surface layers (it is clear that we are referring to *T-S* curves plotted in the same scale). In this they are considerably different from the *T-S* curves for the central regions, which form distinctive "clusters" or "bunches". As an example, Fig. 69 gives the curves of six stations of the R.V. "Discovery", made in 1932 in the section between Australia and the Antarctic; the surface "fan" of *T-S* curves is formed as a result of horizontal mixing, which may be isopycnic in nature. One could multiply examples of *T-S* rela-

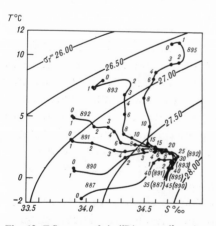

Fig. 69. *T-S* curves of six "Discovery" stations made in 1932 in a section between Australia and the Antarctic.

tions of this type; one of the most pronounced (with a large "opening of angle" between the extreme *T-S* curves) relates to the region of interaction of the water masses of the Gulf Stream and Labrador Currents (McLellan, 1957, see also Section 56).

The basis for the analysis of lateral interaction of water masses on a *T-S* diagram is a method apparently applied for the first time by Sverdrup and Fleming (1941) in studying the waters in the region of the Southern Californian coast. Fig. 70 is given as illustration of this method, and is borrowed from the work of these authors. The *T-S* points in this figure relate to the layer between 200 and 500 m in the vicinity of the coast of Southern California between Point Conception and San Diego (32.5–35°N) and are plotted on the basis of the data of sections made by the research vessels "Bluefin", "Scripps" and "Guide". The averaged *T-S* curves, plotted from these points, represent the "northern" and "southern" water masses, of which the first belongs to the California Current while the second belongs to the coastal countercurrent (Davidson's Current), which wedges itself in from the south to the north between the waters of the California Current and the shore. The *T-S* curves of both water masses together with the family of isopycnals (σ_T) plotted on the figure represent a *T-S* nomogram, which is highly convenient for the analysis of the percentage ratio of waters at different depths, isopycnic analysis, etc.

This method has subsequently been applied repeatedly and has been perfected. In particular, it was used in the works of Tibby (1941) and Rattray et al. (1962) for the study of the water masses of the northeastern part of the Pacific Ocean. In both these investigations the nomogram represented in Fig.

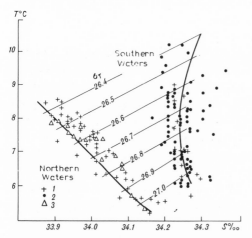

Fig. 70. *T-S* characteristics of "northern" and "southern" water masses in a layer of 200–500 m near the coast of Southern California (Sverdrup and Fleming, 1941). *1* = "Bluefin"; *2* = "Scripps", *3* = "Guide".

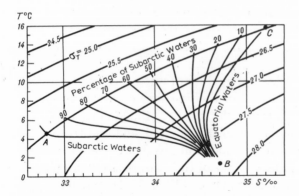

Fig. 71. *T-S* nomogram for the determination of percentage content in horizontal mixing of subarctic and equatorial structures of the Pacific Ocean (Rattray et al., 1962).

71 was used; in the work of Rattray et al. it is called an "analyzer" of water masses for the Northeastern Pacific Ocean". The nomogram represented in Fig. 71 makes it possible to determine the percentage ratio of the Subarctic and Equatorial Water masses of the Pacific Ocean, which are characterized by the extreme *T-S* curves of the nomogram (they are represented by heavy lines) in the mixed waters lying in the corresponding transitional zone.

In McLellan's work (1957) the method described was used to study the horizontal interaction of the waters of the Gulf Stream and the Labrador Current. McLellan supplemented the study of the transformation of water masses by taking account of the effect exerted on it by *contraction on mixing* (see Section 56); in the work of Sverdrup and Fleming (1941) this effect was neglected (the isopycnals in Fig. 70 have been straightened out for the sake of simplicity).

In spite of all the obviousness of the method of *T-S* nomograms similar to those represented in Figs. 70 and 71, it suffers from one shortcoming, which consists of the following: in studying the horizontal (lateral) interaction of two water masses it takes no account of the *vertical mixing* which takes place between each of these water masses and the Deep Water mass underlying them. Taking vertical mixing into account introduces into the corresponding methods of *T-S* analysis important adjustments, the gist of which can be explained from the example of the same Fig. 71, where we have plotted the following *T-S* indexes (the letter designations are arbitrary):

(1) *B* (1.30°C; 34.70‰) – the *T-S* index of the Deep Water mass of the equatorial Pacific. This index was determined by Sverdrup (Sverdrup et al., 1942).

(2) *A* (4.60°C; 32.80‰) – the surface *T-S* index of the Pacific Subarctic Water mass.

(3) C (15.70°C; 35.20‰) – the surface T-S index of the Pacific Equatorial Water mass.

Water mass B is observed in the northeastern part of the Pacific Ocean at depths ranging from approximately 1,000 m to the bottom (the T-S index determined by Sverdrup relates to the bottom part). The indexes of water masses A and C have been determined by us to a certain extent conventionally, within the framework of the figure (they both lie on the isopycnal $\sigma_T =$ 26.00), however, their presence makes it possible simultaneously with the nomogram to consider the triangle of mixing usual in the practice of T-S analysis, the apexes of which are T-S points A, B and C.

In this case the T-S curve of the Subarctic Waters (the extreme left in Fig. 71) may be considered as the result of vertical mixing of water masses A and B, while the T-S curve of the equatorial waters (the extreme right in Fig. 71) may be considered the result of vertical mixing of water masses C and B, applying to these waters the term "water structure" proposed by Dobrovol'skii (1961). Both these T-S curves deviate from straight lines AB and CB, but this fact is not so significant for it does not contradict the fact of mixing of precisely three water masses. The T-S points lying in the sector limited by these T-S curves may be considered as the result now of horizontal (isopycnic) mixing of the subarctic and equatorial structures, i.e., the mixing of those waters which are the product of vertical mixing of water masses A and C with water mass B. The foregoing becomes clear from study of Fig. 72, in which the usual triangle of mixing has been combined with a percentage nomogram of horizontal mixing similar to that shown in Fig. 71. Moreover, for purposes of general application this triangle is not attached to any specific indexes on the T-S diagram. Considering Fig. 72, we see that, say, point b represents the result of mixing of 42% of water A, 30% of water B and 28% of water C. On

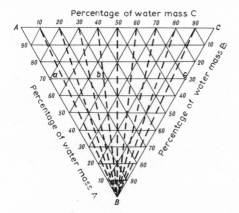

Fig. 72. Triangle of mixing. The broken lines plot the isolines of percentage content of waters CB on their mixing with waters AB.

the other hand, it may be viewed as the result of mixing of 60% of water *a* and 40% of water *c*; waters *a* and *c* are the products of mixing of 30% of water *B* with 70% of the waters of *A* and *C* respectively. The combination of the usual triangle of mixing with the percentage nomogram of horizontal mixing, it would seem at first sight, does not contribute anything new to the determination of the percentage composition of water masses. However, if we bear in mind that water masses *A* and *C* mix among themselves horizontally and that they also mix vertically with water mass *B* underlying them, and that any intermediate product of mixing is also in the same situation, we may draw some interesting conclusions therefrom if we consider the picture of the distribution of water masses in a section carried out across the region of interaction of the water masses.

Such a conventional section, or rather a section-nomogram, in which vertical and horizontal distances are conventionally reckoned in fractions of the percentage content, is represented in Fig. 73. Points *A* and *C* correspond to "pure" water masses interacting horizontally, line *BB'* — to the water mass underlying them. The content of the corresponding water masses in points *A* and *C* and on line *BB'* amounts to 100%. Such initial data make it possible, with the help of Fig. 72, to represent the percentage ratio of each of the water masses in the section, which has accordingly been done in Fig. 73. The percentage content in the section of water masses *C* and *A* is represented by two families of hyperbolae. The isolines of the percentage content of water mass *B* represent horizontal equidistant straight lines, with line *BB'*, as was already said, corresponding to 100%, and line *AC* to 0% of water *B*.

Thus, we see that, if in the case of vertical mixing of water masses the iso-

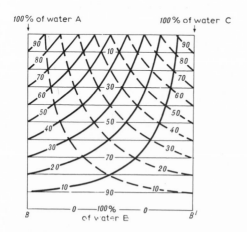

Fig. 73. Section-nomogram of the percentage content of water masses *A* (solid curves), *C* (broken curves) and *B* (horizontal straight lines) provided there is vertical mixing of water masses *A* and *C* with water mass *B* and horizontal mixing of waters *AB* with waters *CB*.

lines of the percentage ratio of the waters in the mixture extend in the section horizontally, and in the case of horizontal mixing vertically, then in the case of the combined effect, as may be seen from Fig. 73, they are completely different and indeed more complex in nature. It goes without saying that one can also imagine a more complicated case in which not two but three water masses are involved in the process of horizontal mixing, one of which is intermediate; in addition, one may consider the vertical mixing of these three water masses not with one, but, for example, with two underlying water masses, one of which is also intermediate, etc. Below one of these more complex versions will be considered.

The use of the nomogram shown in Fig. 73 in analyzing T-S curves taking account of the distribution of parameters σ_T (density) and z (depth) will make it possible not only more correctly to represent the percentage ratio of water masses in sections in cases of combination of horizontal and vertical mixing, but also substantially to improve and complete the methods of *isopycnic analysis of water masses*.

54. THE MIXING OF FOUR WATER MASSES OF THE OCEAN

Let us now proceed to consider the question of the mixing of four water masses of the ocean, drawing special attention subsequently to the case where at least two of the four water masses are involved in a process of horizontal mixing.

Let us consider how valid the construction of a triangle of mixing is for the study of the mixing of more than three water masses. It is clear that in such a situation triangles of mixing can be built on thermohaline indexes as apexes in such a way that, as a result, they have one or more common sides. Thus, in the case of the mixing of four water masses A, B, C and D it is possible to construct two triangles having one common side (Figs. 74 and 75). Underlying such constructions is the assumption (as a rule, not strictly provable), that in particular oceanographic conditions one of the four water masses plays — as compared with the three others — an insignificant part in the process of mutual mixing of the waters; this water mass is accordingly left outside the limits of the corresponding triangle of mixing. Let us consider several cases which arise in this situation.

Let us turn first to an example already known to us (Section 49) of the vertical mixing of four superimposed water masses of the ocean — the surface one, two intermediate (upper and lower) and deep (or bottom). This example is genetically connected with the development in Chapter 6 of the analytical theory of T-S curves for the case of vertical mixing of water masses. Thus, Fig. 74 (analogous to Fig. 61) can correspond to the case where surface water mass

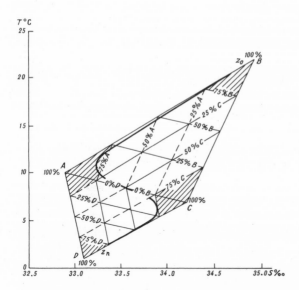

Fig. 74. Two triangles of mixing for the determination of the percentage content of four water masses mixing vertically. The heavy line shows the typical *T-S-z*.curve.

B, intermediate water masses *A* (upper) and *C* (lower) and deep water *D* mix under conditions of stable stratification. In such mixing there arises, in accordance with the conclusions of the theory of *T-S* curves (Chapter 6), the typical *T-S-z* curve, represented in the figure by a heavy line; parameter *z* is reckoned from $z = z_0$ to $z = z_n$. In this example, the two intermediate water masses form a kind of stratification screen which hinders direct vertical contact of water mass *B* with water mass *D*. This assumption accordingly constitutes the basis for the use of the two triangles of mixing: in the example represented in Fig. 74, the mixing of waters *A, B* and *C* is considered independently of water mass *D*, and the mixing of waters *A, C* and *D* in its turn independently of water *B*. Thus, the assumption referred to above in the case of vertical mixing of four water masses is to a certain extent justified (this statement is supported by the consideration of pictures of the percentage concentration of water similar to those shown in Fig. 64).

  Let us consider a second example when two different surface water masses *A* and *B* mix vertically with intermediate and deep water masses *D* and *C* respectively. Such a version of mixing, as well as typical *T-S-z* curves are shown in Fig. 75; given selected coordinates of thermohaline indexes under conditions of stable stratification it is fully possible. Two triangles with a common side of the type of Fig. 75 were used by Timofeev and Panov (1962) for the analysis of processes of mixing of water masses in the Norwegian, Greenland and Barents Seas.

  It should be pointed out that the use of triangles of mixing of the type in

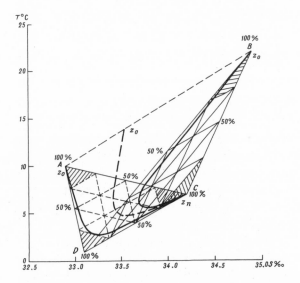

Fig. 75. Two triangles of mixing for the determination of percentage content on mixing of different surface water masses (A and B) with the same intermediate (D) and deep (C) water masses. The heavy lines show typical T-S-z curves: the "extreme" (solid lines) and the intermediate (broken lines).

Fig. 75 is valid only when waters A and B are not in contact with each other; however, if these waters can mix (horizontally), even if they are fairly distant from each other or if one of them is transformed into the other, for example, under the influence of climatic conditions, it is also necessary to take account of mixing along straight line AB. In other words, in the presence of horizontal (or isopycnic) mixing the intermediate T-S-z curve, represented in Fig. 75 by the heavy broken line, may appear — such that it cannot be inscribed in the triangles considered; consequently, the use of the "extreme" triangles of mixing ACD and BCD for the analysis of the mutual relationship of water masses becomes already less justified.

Finally, as the third and final example of interaction of four water masses, let us consider the case when two surface water masses A and B and two intermediate (or deep) water masses D and C are intermixing. The typical T-S-z curves corresponding to this version are represented in Fig. 79: the "extreme" by solid heavy lines, while the broken line shows the intermediate curve arising in the process of the mutual transformation of the first two. In this case the vertical mixing of waters A and D and waters B and C respectively predominates; this mixing forms structures AD and BC. Simultaneously horizontal mixing of the separate elements of these structures takes place, forming various intermediate structures. Thus, all the water masses are involved in a process of "equal" mixing, and the selection of triangles of mixing becomes invalid in general.

As one of the examples of this type of interaction of waters we may cite the region of the subarctic front of the northwestern part of the Pacific Ocean. Here we may identify four water masses mixing among themselves: surface subarctic water mass A (10.0°C; 32.9‰), water mass of the main axis of the Kuroshio B (22.0°C; 34.9‰), intermediate water mass in the Kuroshio Current, characterized by minimum salinity, C (7.0°C; 34.2‰) and intermediate subarctic water mass with minimum temperature, D (1.0°C; 33.1‰). The quadrangles shown in Figs. 74, 75, 78 and 79 as well as in Figs. 86 and 87, are constructed on summits which correspond to the thermohaline indexes of these water masses. The indexes themselves are determined in turn from the typical *T-S* curves represented in Fig. 86. These *T-S* curves are taken from Koizumi's work (1955) and will be considered in greater detail in Section 57 in connection with the question of the contraction on mixing of four water masses.

Another example of the interaction of four water masses relates to the region of continental slope of the Atlantic coast of North America, lying to the south and southeast of Long Island. The conditions of interaction of the waters are represented in Fig. 76 (Miller, 1950). Let us explain this figure. The aggregate of *T-S* points, stretching out in the form of the strip $C'CD$, represents the result of the horizontal mixing of waters $C'C$ with surface ocean waters D, and that part of the *T-S* points which lies in the sector $C'C$ characterizes the transformation of waters at shallow water and does not exert any visible effect on the mixing processes of the waters of the continental slope with ocean waters. Miller divides into three types the immediate ocean waters, lying in the vicinity of the edge of the shelf: surface slope waters D, intermediate slope waters A, lying directly under the strongly marked

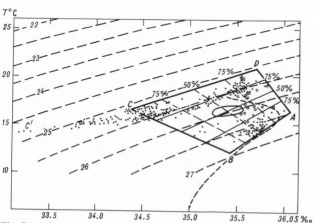

Fig. 76. *T-S* characteristics of water masses in the region of the continental slope of North America in the Atlantic Ocean, to the east of the Hudson submarine canyon, and *T-S* nomogram for the determination of the percentage content of the four main types of water masses. Further explanations in the text. (Miller, 1950)

thermocline, and the slope waters of medium depths B, observed at depths of approximately 200–600 m. The broken line in Fig. 76 represents the T-S curve for the slope waters, borrowed by Miller from Iselin's work (1936), and the little open circles designate waters directly belonging to the thermocline. Thus, the type of mixing considered by Miller also relates to our third example: two surface water masses, shelf and ocean, mix among themselves; in turn a process of vertical mixing takes place of the surface waters with the intermediate water mass of the continental slope and with the deeper water mass — the waters of the Gulf Stream at the slope.

A question arises: is it not possible for the case of mixing of four water masses, especially when they are in a state of vertical and horizontal mixing by pairs, to construct a quadrangular nomogram? In Section 30 it was indicated that the construction of a quadrangular plane nomogram (without any other additional limitation) is in principle impossible: owing to the fact that the problem is indeterminate, within the quadrangle of mixing it is possible, generally speaking, to construct an innumerable set of percentage grids. For the consideration of this question, i.e., the question of what can be the additional limitation imposed on the conditions of mixing of water masses so that the construction of a quadrangle of mixing might prove possible, let us turn to Miller's nomogram, constructed by him on the basis of intuitive considerations. Quadrangle of mixing $ABCD$ (Fig. 76) breaks down into 16 areas. In the four areas adjoining the angles of the quadrangle, the percentage content of each of the corresponding water masses must exceed 75%; in the closest areas, lying on both sides of the angles, more than 50% of the corresponding waters is observed. Finally, as Miller points out, in the four inner areas the

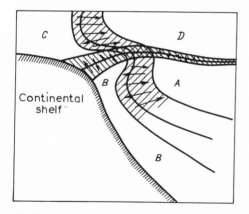

Fig. 77. Stylized diagram of the distribution and interaction of water masses in the vicinity of the continental slope of North America. The shaded regions correspond to zones of horizontal and vertical mixing of waters; the directions of movement of waters in the process of transformation are shown by arrows (Miller, 1950).

picture of percentage content is "more obscure", especially in the inner core of the nomogram lying within the oval in Fig. 76.

As a result of the use of the quadrangular nomogram described, a pattern of distribution of the water masses and of the nature of their interaction was obtained, and is shown in the form of a stylized section in Fig. 77. The "clear" regions in this figure correspond to the percentage content of the corresponding masses to the extent of 75% and more; the shaded areas of transformation, vertical and horizontal, and the curves lying within the shaded regions, correspond to 50% content of the water masses. It is clear that the boundaries of the zone of transformation correspond to 75% of the content of the "adjacent" and to 25% of the content of the "opposite" water masses.

The nomogram proposed by Miller is too approximate, since it does not even satisfy the basic requirement that in any of its points the sum of parts of the mixing waters should amount to 100%; this condition is fulfilled on the nomogram only on its sides. In order to correct this nomogram, let us formulate an additional limitation which was referred to above: let us assume that *equal* volumes of surface waters *A* and *B* are mixed with *equal* volumes of waters *C* and *D*; in turn, the elements of the structures *AD* and *BC*, which have arisen as a result of such vertical mixing, are mixed horizontally along the surfaces of their equal percentage content. Breaking down the sides of the quadrangle into equal parts (say into ten parts) and connecting the corresponding points on the opposite sides by straight lines, we obtain a grid (Fig. 78), which may be considered as an auxiliary nomogram, explaining this addi-

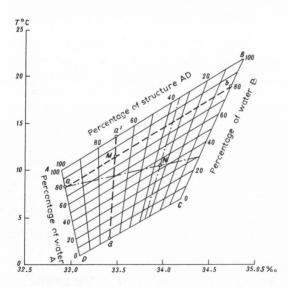

Fig. 78. Nomogram for the determination of percentage content on horizontal mixing of structures *AD* and *BC*; the latter are formed by the vertical mixing of types *A* and *D*, *B* and *C* respectively.

tional limitation. Such an approach makes it possible to construct one of the particular cases of a percentage grid, which may be shown by the example of the arbitrary point M. Through such a point it is always possible to draw two lines of mixing ab and $a'd$ — such that they have identical coordinates, expressed in percentages, on the sides of the quadrangle (in pairs). The structure can then be considered as the result of the mixing of 70% of the waters of structure AD and 30% of the waters of structure BC; in its turn the water in point a represents the product of mixing of 80% of water A and 20% of water D, while in point b — the product of mixing of 80% of water B and 20% of water C. Thus, for point M, expressing the concentration in fractions of a unit, we have:

$$0.7\,(0.8A + 0.2D) + 0.3\,(0.8B + 0.2C) = 1$$

The percentage content of waters A, B, C and D in point M amounts to 56, 24, 6 and 14% respectively.

Determining the percentage content of all four water masses, for example in the junctions of the grid in Fig. 78, we can construct the corresponding nomogram which is represented in Fig. 79. The isolines of equal percentage content represent the families of hyperbolae, plotted in the oblique-angled coordinates, while the sides of the quadrangle represent asymptotes. The foregoing will become obvious if we turn again to Fig. 73, extending the method used in constructing it to the case of mixing of four water masses. Assuming

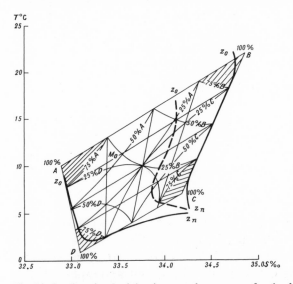

Fig. 79. Quadrangle of mixing (corrected nomogram for the determination of the percentage content of four water masses). The solid heavy lines represent typical T-S-z curves, corresponding to the "initial" structures AD and BC (compare with Fig. 78), the broken line represents the intermediate, appearing on mixing of the first two.

that each of the summits of the square corresponds to 100% content of each of the water masses, we will come to the conclusion that the isolines of percentage content must be represented by four perfectly identical families of hyperbolae, flowing around each of the right angles of the section-nomogram By means of a corresponding distortion on the *T-S* plane, i.e., the ties of the summits of the nomogram to specific *T-S* indexes (it is relevant to point out that Miller speaks of such a distortion of the nomogram in the transition from the *T-S* plane to the vertical section), we obtain the nomogram in Fig. 79. The sides of the nomogram are also isolines of zero percentage content in relation to the opposite indexes of the water masses: thus, the broken line *ABC* is the line of zero content of water *D*, line *BCD* is the line of zero content of water *A*, etc. In this way, Fig. 79 substantially improves Miller's nomogram.

As has already been pointed out, the percentage nomogram (Fig. 79) represents only one of the possible versions; any other version of mixing, for example, such as that shown in Fig. 78 by the dot-and-dash line, leads to other results: the percentage content of the waters in point *N* (28% of water *A*, 19.5% of water *B*, 45.5% of water *C* and 7% of water *D*) in no way corresponds to the nomogram in Fig. 79.

The limitation formulated above on the mixing of waters in equal proportions, $m_A/m_D = m_B/m_C$, is artificial, but from the oceanographic point of view looks authentic. Indeed, considering the position of quadrangle *ABCD* in the field of isopycnals (Fig. 86), we see that the position of the surfaces of equal percentage content of waters, along which "horizontal" mixing is assumed, corresponds fairly closely to the position of the isopycnic surfaces. Thereby the principle of isopycnic mixing is more or less observed.

The study of vertical and isopycnic mixing of four water masses is most promising in regions of substantial horizontal gradients of oceanographic characteristics, in particular in the "transitional" (according to Sverdrup) and frontal regions.

55. CONTRACTION ON VERTICAL MIXING OF WATER MASSES; TRIANGLES OF CONTRACTION ON MIXING

We have already touched upon the question of contraction on mixing from the point of view of the anomalous properties of sea water, as well as within the framework of the general theory of the *T-S* diagram (Section 36). There too we considered an accurate method for calculating contraction, based on computing the line integral along the straight *T-S* line. In this and the following sections we return again to this question, but now from a purely oceanographic point of view: namely, we shall consider the general methods of calculating contraction on mixing, as well as the question of how contraction on mixing affects the transformation of the water masses of the ocean.

Only a few studies devoted to the consideration of contraction on mixing of waters under the real conditions of the World Ocean are known. It has already been said that from the oceanographic point of view the phenomenon of contraction on mixing has been studied most thoroughly by Zubov (1938, 1947, 1957b). McLellan (1957) studied the effect of contraction on mixing on the interaction of the water of the Gulf Stream and the Labrador Current, Fofonoff (1956) the effect on the formation of the Antarctic Bottom Water in the region of the Weddell Sea, Bubnov (1960) the effect on the mixing of the waters of the Kuroshio and the Oyashio. The author (Mamayev, 1960b) considered the effect of contraction on mixing on the transformation of the intermediate water masses of the ocean using the example of the Mediterranean Intermediate Water in the Atlantic Ocean; he also proposed a "triangle of contraction" on mixing (Mamayev, 1963), which has proved very convenient for determining contraction on mixing from T-S curves. The questions touched upon in the above-mentioned works of the author, namely, the question of the use of triangles of contraction on mixing and the question of the effect of contraction on the transformation of the Mediterranean Intermediate Water in the Atlantic, were further developed in the works of Kin'diushev (1965) and Bubnov (1967). The triangle of contraction on mixing was used by Dubrovin to determine the lowering of the sea level as a result of contraction on mixing in the area of interaction of the waters of the Gulf Stream and the Labrador Current. Finally, the author (Mamayev, 1970) considered the question of contraction on mixing of four water masses of the ocean; it is taken up below in Section 57.

Methods of calculating contraction on mixing, mainly of two water masses, and its effect on the various oceanographic processes are considered in detail in the work of Zubov and Sabinin (1958), and therefore we will not deal with the details considered in that work, but will devote our attention to questions going beyond its scope, namely, as was already said above, mainly to questions of the contraction on mixing of three and four water masses, as well as the horizontal (isopycnic) mixing of stratified water masses in the ocean. The question of the mixing of two water masses, as set forth by Zubov and Sabinin, is a kind of preamble to the subsequent exposition.

The mixing of two water masses

In Section 21 we have considered an example of the mixing of equal parts of two water masses. In the case of mixing of two water masses in equal proportions, the mean temperature and salinity are calculated from formulae of mixing [30.1] and [30.2], then "densification" on mixing will be determined from formula:

$$\Delta\rho = \rho(\overline{T}, \overline{S}) - \overline{\rho} \qquad [55.1]$$

where $\rho(\overline{T}, \overline{S})$ is the density of the mixture, $\overline{\rho}$ the mean density, determined by a formula analogous to the formulae of mixing:

$$\overline{\rho} = \frac{V_1\rho_1 + V_2\rho_2}{V_1 + V_2} \qquad [55.2]$$

(V = volume). The quantity:

$$\Delta\alpha = \overline{\alpha} - \alpha(\overline{T}, \overline{S}) \qquad [55.3]$$

is actually the *contraction* on mixing (see p. 85); here $\overline{\alpha}$ is calculated from the formula:

$$\overline{\alpha} = \frac{m_1\alpha_1 + m_2\alpha_2}{m_1 + m_2} \qquad [55.4]$$

where m is mass.

The quantities of "densification" $\Delta\rho$ and contraction $\Delta\alpha$ are distinguished by sign and are extremely close to each other — see formula [23.6]:

$$\Delta\rho \approx -\Delta\alpha \qquad [55.5]$$

The distinction between "densification" on mixing and contraction on mixing was emphasized by Sabinin (Zubov and Sabinin, 1958); incidentally, the second term (contraction on mixing) is used in the English literature (Fofonoff, 1956). It was also demonstrated by Sabinin that in calculating the mean quantities $\overline{\rho}$ and $\overline{\alpha}$ from the formulae of mixing, in the first case volumetric proportions should appear, and in the second, mass proportions.

Continuing to consider the straight line of mixing, represented in Fig. 12, we see that, having broken it down into 100 parts, we can determine the densities (specific volumes) of the *T-S* points which correspond to a particular percentage content of both water masses; then, having calculated $\overline{\rho}$ or $\overline{\alpha}$, determine "densification", or contraction. A corresponding calculation for the water masses of the Gulf Stream and the Labrador Current is given in Table XII.

In the example cited, the densities of both original water masses were identical; the calculation of contraction, naturally, does not change if their densities are different. Fig. 80 represents the curves of density of two mixtures formed if the water mass *B* (its index is $T = 2.5°C$; $S = 34.90‰$) is mixed in different proportions either with water mass *A* (index $T = 12.5°C$; $S = 35.67‰$), or with water mass M (index $T = 11.9°C$; $S = 35.50‰$). This figure is the result of calculations completely analogous to those given in Table XII.

TABLE XII

Calculation of contraction on mixing in various proportions of the water masses of the Gulf Stream (A) and the Labrador Current (L)

Percentage of water A	Percentage of water L	$\bar{T}$	$\bar{S}$	$\sigma_T(\bar{T}, \bar{S})$	$\bar{\sigma}_T$	$\Delta\sigma_T$	Δv_T
(1)	(2)	(3)	(4)	(5)	(6)	(7)	(8)
100	0	30.00	35.36	22.02	22.02	0	0
90	10	26.85	34.57	22.46	22.02	0.44	−0.44
80	20	27.70	33.77	22.81	22.02	0.79	−0.79
70	30	20.55	32.97	23.09	22.02	1.07	−1.07
60	40	17.40	32.17	23.28	22.02	1.26	−1.26
50	50	14.25	31.38	23.38	22.02	1.55	−1.55
40	60	11.10	30.58	23.36	22.02	1.34	−1.34
30	70	7.95	29.78	23.22	22.02	1.20	−1.20
20	80	4.80	28.98	22.96	22.02	0.94	−0.94
10	90	1.65	28.18	22.56	22.02	0.54	−0.54
0	100	−1.50	27.38	22.02	22.02	0	0

Explanation of columns:
(1) − percentage content of water A;
(2) − percentage content of water L;
(3) and (4) − values of temperature and salinity on mixing of water masses A and L in the proportions indicated in columns 1 and 2. Calculated from formulae [30.1] and [30.2];
(5) − $\sigma_T(\bar{T}, \bar{S})$ is the density of the mixture. Is taken from the T-S diagram according to the corresponding T-S points;
(6) − $\bar{\sigma}_T$ is mean density. Is determined from formula [55.2];
(7) − "densification" on mixing, $\Delta\sigma_T = \sigma_T(\bar{T}, \bar{S}) − \bar{\sigma}_T$;
(8) − contraction on mixing, $\Delta v_T \approx −\Delta\sigma_T$.

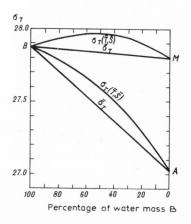

Fig. 80. Curves of contraction on mixing of water mass B with water masses A or M.

The mixing of three water masses

"Densification" $\Delta\rho$ or contraction $\Delta\alpha$ on mixing of three water masses is determined from the same formulae [55.1] and [55.3] with the only difference that by ρ and α are understood the mean weighted values of density or specific volume, determined from the formulae of mixing of three water masses:

$$\bar{\rho} = \frac{V_1\rho_1 + V_2\rho_2 + V_3\rho_3}{V_1 + V_2 + V_3}$$

$$\alpha = \frac{m_1\alpha_1 + m_2\alpha_2 + m_3\alpha_3}{m_1 + m_2 + m_3}$$

[55.6]

For the determination of contraction on mixing corresponding to any percentage ratio of the three mixing water masses, which is determined by the position of the corresponding figurative point within the triangle of mixing, the construction of *triangles of contraction* can be recommended. Let us explain the principle of such construction by the example of a triangle of mixing of three water masses which have the following indexes:

water mass A: $T = 12.5°C$; $S = 35.67‰$;
water mass M: $T = 11.9°C$; $S = 36.50‰$;
water mass B: $T = 2.5°C$; $S = 34.90‰$.

These three water masses correspond to the Atlantic Surface Water in the Azores–Pyrenean region, to the Intermediate Mediterranean Water and to the Atlantic Deep and Bottom Water (Mamayev, 1960b). The corresponding triangle of mixing is represented in Fig. 25. It is clear that the three water masses mentioned are in a state of vertical mixing.

For the construction of the triangle of mixing the following is necessary:

(1) To calculate from formulae [30.1] and [30.2], as well as from formula [55.6], the values of $\bar{T}, \bar{S}$ and $\bar{\alpha}$ * for all points in which the percentage content of each of the water masses is a multiple of 10%; in Fig. 25 these points are numbered from 1 to 66.

It is convenient to bring these computations together in a general table which makes it possible considerably to reduce calculations; inasmuch as the form of this table may be arbitrary, we will not provide it, confining ourselves to an example of a calculation for one of the points, namely, No. 16, the ratio of water masses A, M and B in which amounts to 10, 40 and 50% respectively (Fig. 25):

---

* Below we shall speak, for the sake of simplicity, of contraction on mixing in both cases, $\Delta\rho$ and $\Delta\alpha$.

$$\bar{T} = \frac{T_A m_A + T_M m_M + T_B m_B}{100} = \frac{12.5 \times 10 + 11.9 \times 40 + 2.5 \times 50}{100} = 7.26°$$

$$\bar{S} = \frac{S_A m_A + S_M m_M + S_B m_B}{100} = \frac{35.67 \times 10 + 36.50 \times 40 + 34.90 \times 50}{100} = 35.62‰$$

$$\bar{v}_T = \frac{v_A m_A + v_M m_M + v_B m_B}{100} = \frac{73.68 \times 10 + 72.96 \times 40 + 72.88 \times 50}{100} = 72.99$$

(2) From the values of the mean temperatures and salinities $\bar{T}$ and $\bar{S}$ for all 66 points of the triangle to determine from the T-S diagram the specific volumes $v_n(\bar{T}, \bar{S})$, where $n = 1, 2, ..., 66$.

For point No. 16 we have: $v_{16}(\bar{T}, \bar{S}) = 72.86$.

(3) To determine from formula [55.3] the values of contraction Δv also for all 66 points of the triangle. In our example $\Delta v_{16} = 72.86 - 72.99 = -0.13$. For the three sides of the triangle of mixing BA, AM and MB the values of contraction on mixing, determined in a similar way, are given in Table XIII.

TABLE XIII

Contraction on mixing, $-\Delta v \cdot 10^5$

Side of triangle BA		Side of triangle AM		Side of triangle MB	
No. of points	Δv	No. of points	Δv	No. of points	Δv
1	0	66	0	11	0
12	3	65	0	10	5
22	8	63	0	9	7
31	11	60	0	8	12
39	13	56	0	7	13
46	13	51	0	6	13
52	13	45	1	5	12
57	12	38	1	4	10
61	9	30	0	3	8
64	2	21	0	2	4
66	0	11	0	1	0

It is seen from this table that on side AM contraction on mixing is extremely insignificant; so far as sides BA and MB are concerned, it is much larger, and extremely close at the ends of the straight lines $2-12$, $3-22$, $4-31$, ..., $10-64$, parallel to side AB.

(4) To construct curves of contraction on mixing for the lines in all points of which an identical percentage content of one of the water masses is observed.

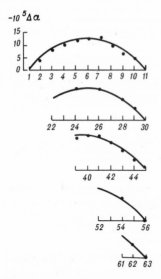

Fig. 81. Curves of contraction on mixing for lines corresponding to 0, 20, 40, 60 and 80% content of water mass *A* in the triangle of mixing represented in Fig. 25.

In the example considered, as such lines were selected line *BM* and the lines parallel to it on which 0, 10, 20, ..., 90% content of water mass *A* is observed.

Fig. 81 represents the curves of contraction on mixing for lines on which the percentage content of water mass *A* amounts to 0 (line *BM*), 20, 40, 60 and 80%. In constructing the curves of contraction the scatter of individual points, caused by the inaccuracy of calculation of contraction (this inaccuracy may be avoided by calculating v_T or σ_T not to the second but to the third decimal place), is smoothed out.

(5) To construct a triangle of contraction on mixing. For the case considered of vertical mixing of water masses *A*, *M* and *B* the triangle of contraction on mixing is represented in Fig. 82. In this example the construction of the triangle of contraction proved simple; indeed, returning to Table XIII, we see that contraction on mixing, on the ends of the lines parallel to side *AM*, differs by not more than $\Delta v = -2 \cdot 10^{-5}$. Disregarding this difference, and also assuming that contraction on mixing of water masses *A* and *M* is equal to zero, we have constructed the isolines of contraction on mixing parallel to side *AM*. In reality, the picture may be more complex.

With this the construction of the triangle of contraction on mixing is concluded. Plotting on this triangle the *T-S* curves of individual oceanographic stations, we can evaluate the extent of the effect of contraction on mixing, of an internal nature, on the process of vertical mixing of three water masses. For an example in the same Fig. 82 is represented a *T-S* curve of station 225, made by the R.V. "Mikhail Lomonosov", October 16, 1958, in coordinates

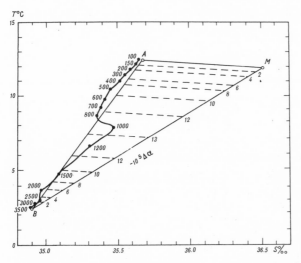

Fig. 82. Triangle of contraction on mixing of Atlantic surface (*A*), Mediterranean Intermediate (*M*) and Deep and Bottom North Atlantic (*B*) Water masses, as well as *T-S* curve of R.V. "Mikhail Lomonosov" station 225.

47° 46.2′ N 17° 14.3′ W. Looking at this *T-S* curve, we see that contraction on mixing reaches its greatest value ($-13 \cdot 10^{-5}$) on the lower boundary of the intermediate Mediterranean water mass *M* ($z \approx 1,100$ m). The effect of contraction on mixing causes uneven mixing of the three water masses, namely, greater intensity of mixing of water mass *M* with water mass *B* than with water mass *A*. This question is considered in more detail in the example of the transformation of the Mediterranean Water mass in the Atlantic, below in Section 63; this example is the one most studied, as has already been stated above, and therefore we devote a separate section to it

The application of the triangle of contraction on mixing to the analysis of individual *T-S* curves makes it possible to plot also the curves of vertical distribution of contraction on mixing, $\Delta\alpha = f(z)$, for each individual station. Such curves, for example, are given in the work of Kin'diushev (1965).

The triangle of contraction on mixing, represented in Fig. 82, is, as has already been said, extremely simple, and for other regions of the ocean the picture may prove more complicated. To illustrate the foregoing we may turn to Fig. 83, in which is shown the triangle of contraction on mixing of the Subarctic, Equatorial and Deep Water masses of the Pacific Ocean. The use of this triangle for the analysis of the mixing processes of water masses is considered in the following section.

A qualification should be made to the effect that the influence of the compressibility of sea water, which is not taken into account either in *T-S* analysis or in the analysis of contraction on vertical mixing of water masses, requires further study; it may introduce substantial corrections in what has been said.

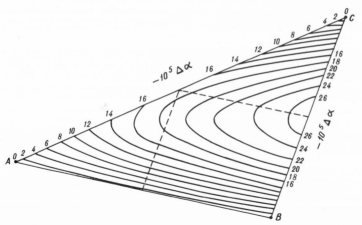

Fig. 83. Triangle of contraction on mixing $(-10^5 \Delta \alpha)$ of surface Subarctic (A), surface Equatorial (C) and Deep (B) Water masses of the North Pacific.

56. CONTRACTION ON HORIZONTAL MIXING OF WATER MASSES

Horizontal mixing of two (three) water masses, differing considerably in their *T-S* characteristics, is observed, as has already been said (Section 53), primarily in the frontal regions of the World Ocean. The most striking example of such mixing is the interaction of the water masses of the Gulf Stream and the Labrador Current; we have already dwelt on this example above (Section 53) and now we will return to it again.

It is known that the interaction of the two water masses indicated, A and L, leads to the formation of the product of their mixing — the specific *water mass of the continental slope* (*SL*). Contraction on mixing has a considerable effect on the formation of this water mass. It was studied by Zubov (1957b), as well as by McLellan (1957). Let us consider the effect of contraction on mixing on the horizontal mixing of the water masses of the Gulf Stream and the Labrador Current. Fig. 84 represents *T-S* curves of the waters of the Gulf Stream, the Labrador Current and of the product of their mixing — the water mass of the continental slope. McLellan, as well as Tibby (1941), makes an assumption that mixing proceeds between water masses of identical density: for example, the Labrador water, characterized by figurative point A, mixes with the water the the Gulf Stream b, while $(\sigma_T)_a = (\sigma_T)_b = 27.00$. As a consequence of the fact that the nature of the process of mixing deviates from the isopycnic (the product of mixing lies not on the isoline $\sigma_T = 27.00$, but on straight line ab), the effect of contraction becomes manifest. The products of the mixing of the two initial water masses are the points lying on the *T-S* curve of the water mass of the continental slope of the ocean and on the cor-

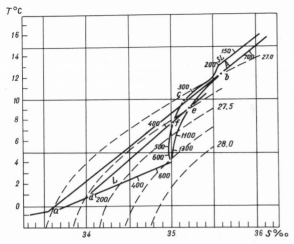

Fig. 84. Graphic determination of contraction on mixing from T-S curves of the water masses of the Gulf Stream (A) and the Labrador Current (L). The T-S curve of the continental slope water is designated by the index SL (McLellan, 1957).

responding straight lines of mixing: point c is the product of the mixing of waters a and b, point f the product of the mixing of waters d and e, etc. The calculation of contraction on mixing in this case is illustrated by Table XIV. We see from the table that contraction reaches its greatest value $\Delta\sigma_T = 0.17$ on the isopycnic surface $\sigma_T = 27.60$. In the overlying layers it keeps large values, but decreases with depth. In this way, as in the case of vertical mixing of water masses, contraction on mixing reaches its greatest values somewhere

TABLE XIV

Contraction on horizontal mixing of water masses of the Gulf Stream and Labrador Current, according to McLellan (1957)

Initial value of σ_T	Gulf Stream		Labrador Current		Slope water		σ_T	contraction $\Delta\sigma_T$	% of Gulf Stream Water
	T (°C)	S (‰)	T (°C)	S (‰)	T (°C)	S (‰)			
26.8	14.50	35.92	−0.80	33.30	10.48	35.23	27.07	0.09	74
26.9	13.30	35.72	−0.68	33.43	9.95	35.17	27.11	0.12	76
27.0	12.20	35.55	−0.42	33.57	9.42	35.12	27.17	0.17	78
27.1	11.08	35.41	0.06	33.73	9.00	35.09	27.21	0.11	81
27.2	9.93	35.27	0.65	33.90	8.43	35.06	27.26	0.06	84
27.3	8.90	35.18	1.20	34.07	7.81	35.03	27.33	0.03	86
27.4	7.88	35.12	1.78	34.24	7.10	35.01	27.43	0.03	87
27.5	7.02	35.08	2.36	34.42	6.40	34.99	27.52	0.02	87
27.6	6.10	35.04	2.90	34.61	5.60	34.98	27.61	0.01	86

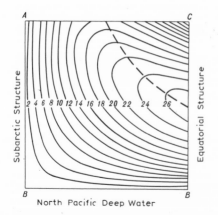

Fig. 85. Isolines of contraction on mixing ($-10^5 \Delta\alpha$) of surface Subarctic (*A*), surface Equatorial (*C*) and Deep (*B*) water masses of the northeastern part of the Pacific Ocean in a conventional section. 50% content of water mass *C* is plotted by a broken line.

in the intermediate depths. This phenomenon, a highly interesting one, requires additional careful study.

In considering the horizontal interaction of three water masses, the problem of the study of contraction on mixing is reduced once again to constructing the triangle of contraction on mixing; meanwhile, the formal side of the question, namely the construction itself of the triangle of contraction, does not change whether we are considering vertical or horizontal interaction of waters. The differences in the processes of interaction become manifest after we begin the consideration within the triangle of contraction of real *T-S* relations, plotted by using either parameter *z* or parameter *l*.

Considering the nature of contraction on mixing of waters both horizontally and vertically, let us return to the case of the interaction, considered in Section 53, of the Subarctic (*A*), Deep (*B*) and Equatorial (*C*) Waters of the North Pacific Ocean. Taking account of vertical mixing with the underlying water mass also introduces corrections in the picture of contraction, which may be seen from Fig. 83, which represents a corresponding triangle of contraction on mixing, and Fig. 85, in which the isolines of contraction on mixing are represented in a conventional section (cf. Fig. 73). The values of contraction, which served as a basis for the plotting of the isolines in Fig. 85, were calculated for all points of the square grid of Fig. 73 (excluding line *BB'*, we have 110 points) by the method described for the construction of "triangles of contraction".

Considering Figs. 83 and 85 we see that contraction on mixing of water masses *A* and *B* is practically insignificant; it reaches its greatest values, however, not on the horizontal mixing of waters *A* and *C*, but on the vertical mixing of waters *C* and *B*. The region of maximum values of contraction, as was to be expected, is close to the isoline of 50% content of water mass C, represented in Figs. 83 and 85 by a broken line. The considerable increase of contraction on mixing in the horizontal direction (at different depths) from the subarctic structure AB to the equatorial structure CB is, apparently, one of the reasons for the fact that the latter is more stable and reflects the result of mixing not only of water masses C and B themselves, but also the participation in mixing of water mass A. Thus, the bulging of the equatorial T-S curve in Fig. 71 to the left, in the direction of T-S index A, may very well be the result of precisely this process.

The reservation should be made that the type of mixing under consideration, strictly speaking, is not horizontal, inasmuch as the isopycnic surfaces, especially in frontal zones, are inclined, and therefore the analysis of the phenomenon of contraction itself can be usefully supplemented by the study of the question of the displacement of contracted waters from some levels to others (Zubov, 1957b).

57. CONTRACTION ON MIXING OF FOUR WATER MASSES

Let us consider the question of contraction on mixing of four water masses of the ocean homogeneous in temperature and salinity, and let us select such a case when a minimum of two primary water masses (or two pairs) are in a process of horizontal or isopycnic mixing. Such a type of mixing of four water masses was already noted above (Section 54) and shown in Fig. 79; it is of special interest since it is precisely in the regions of intensive horizontal exchange that the subarctic and subantarctic (intermediate) waters are formed, and the effect of contraction on mixing on their formation may be most appreciable.

The selection of the region of the subarctic front, lying immediately to the southeast and east of the shores of Japan, is supported by the fact that contraction on mixing for it was considered by Bubnov (1960); thus, there is a possibility of comparing the data on contraction obtained in different ways.

In Fig. 86 (which was in turn the basis for Fig. 79), three T-S curves are plotted, borrowed from Koizumi's work (1955), namely:

"T" – the mean annual curve for the weather ship "Tango" (29°N 135′E);

"X" – the mean annual curve for the weather ship "Extra" (39°N 153′E);

"Ca-119" – the curve of station No. 119 of the vessel "Carnegie", made July 7, 1929 (45° 24′N 159° 36′E).

These three T-S curves well reflect the thermohaline conditions in the region of interaction of the waters of the Kuroshio and Oyashio: station "T" is situated in the region of the main Kuroshio flow, somewhat to the south of it, and the T-S curve well reflects the structure of this current, agreeing with other averaged *T-S* curves of the Kuroshio region. *T-S* curve "*Ca-119*" is characteristic of subarctic waters, while the *T-S* curve of station "*X*", situated in the frontal region, is representative of the waters which have been formed as a result of the interaction of the water masses of the Kuroshio and subarctic waters. The line which joins the positions of these three stations on the map intersects the whole region of mixing of waters of different structure which is of interest to us.

In the same Fig. 86 the thermohaline indexes of four water masses are plotted, the contraction on mixing of which we will accordingly consider, namely:

A (10.0°C; 32.9‰) — the surface Subarctic Water mass;

B (22.0°C; 34.9‰) — the Water mass of the core of the Kuroshio current;

C (7.0°C; 34.2‰) — the Intermediate (of subarctic origin) Water mass, characterized by minimum salinity;

D (1.0°C; 33.1‰) — the Subarctic Subsurface Water mass (minimum temperature).

The positions of the indexes are determined by the extremes of the curves in accordance with the "geometry" of *T-S* curves, except for index *A*, the position of which is somewhat arbitrary. It is found on the tangent to the *T-S* curve of the station "*Ca-119*", while its temperature is corrected from Japanese climatic maps.

All four primary water masses in the zone of the subarctic front mix among

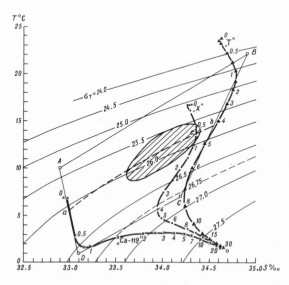

Fig. 86. *T-S* curves of "Carnegie" station 119 ("*Ca-119*"), of averaged stations of weather ships "Tango" ("*T*") and "Extra" ("*X*"), and the quadrangle *ABCD* of mixing of four water masses in the region of the subarctic front of the Pacific Ocean. Line *ab* is the line of maximum contraction on mixing; the shaded oval is the region of maximum contraction, $10^5 \Delta\rho \geqslant 30$. (Cf. Fig. 87.)

themselves: as has already been said, water masses A and D, as well as waters B and C, mix along the vertical; waters A and B — horizontally; waters C and D approximately along isopycnic surfaces. The products of the vertical mixing of waters A and D and B and C in turn also mix among themselves along the surfaces of equal percentage content. Thus, we may consider any T-S index lying within the quadrangle of mixing $ABCD$ as the result of the mixing of the four original water masses in different proportions.

The values of contraction on mixing of the four water masses within the quadrangle of mixing are calculated by the same method as in the construction of triangles of mixing (Section 55). In this, the grid represented in Fig. 78 is used. For each of the 121 junctions of this grid the values of density σ_T (or specific volume v_T) are taken from the T-S diagram; the mean density, or "density of mixing" $\bar{\sigma}$, is calculated from the condition of its linear change along the straight lines of the grid taking into account that the conventional densities in peaks A, B, C and D are equal to their true values, namely 25.33, 24.14, 26.80 and 26.54 respectively. Such an abbreviated method of calculating mean density is equivalent to calculating it from the formula of mixing for density:

$$\bar{\sigma}_M = \frac{m_A \sigma_A + m_B \sigma_B + m_C \sigma_C + m_D \sigma_D}{m_A + m_B + m_C + m_D} \qquad [57.1]$$

where $m_A, ..., m_D$ — are the amounts of percentage content of each of the four water masses in any point M, which may be determined from the percentage nomogram represented in Fig. 79.

The picture of contraction on mixing of the four water masses is represented in Fig. 87 in the form of the isolines of the smoothed-out values of contraction, $10^5 \Delta\rho = 10^2 \Delta\sigma = 10^2 [\sigma_T(S, T) - \bar{\sigma}]$; thus, we have *a quadrange of contraction on mixing*. It may be seen from this quadrangle that the value of the contraction of waters on mixing, $10^5 \Delta\rho$, fluctuates in the region of the subarctic front of the Pacific Ocean approximately between 5 and 30 · 10^{-5} g/cm^3, and that the maximum value amounts to 33. Line ab of maximum contraction extends from structure AD (the surface subarctic waters) to structure BC (the waters of the Kuroshio within the troposphere), while the region of maximum contraction, $10^5 \Delta\rho \geqslant 30$, is displaced towards the structure of the Kuroshio. Analyzing the contraction on horizontal mixing of winter subarctic waters and the waters of the Kuroshio by the method of Zubov and Sabinin, Bubnov obtained the same quantity, $10^5 \Delta\rho = 32$; he considered the mixing of the two waters along a straight line of mixing close to straight line BD, which, as may be seen from Fig. 86, also passes through the region of maximum contraction on the T-S diagram (shaded on the figure).

Let us now compare, in the most general form, the conditions of the con-

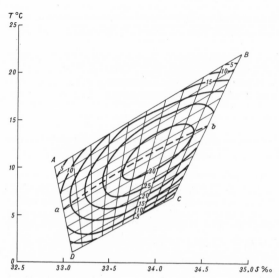

Fig. 87. Quadrangle of contraction on mixing (10^5 $\Delta\rho$) of water masses A, B, C and D. Line ab is the line of maximum contraction.

traction of the waters on mixing with the thermohaline conditions in the region considered. For this let us turn again to Fig. 86, on which line ab of maximum contraction is also plotted, as well as the region of maximum contraction, 10^5 $\Delta\rho \geqslant 30$ (the shaded oval). The line of maximum contraction lies approximately between isopycnals $\sigma_T = 25.8$ and 26.0; consequently, the region of maximum contraction in the ocean lies precisely in this isopycnic layer. This layer between stations "X" and "T", i.e., in the region of the most intensive interaction of the waters of the Kuroshio and the subarctic waters, gradually sinks from the northeast to the southwest from depths of 50–75 m to a depth of approximately 400 m (sector cb of line ab in Fig. 86); this may be seen from the distribution of depth marks along the T-S curve.

Line DC constitutes the line of transformation of intermediate cold waters D into the intermediate layer with minimum salinity C; this transformation takes place approximately along the isopycnic surface $\sigma_T = 26.75$, which was noted earlier by Hirano (1957). Contraction on mixing along line DC is the least in the whole region; however, the transforming waters also sink from the northeast to the southwest (from station "Ca-119" to station "T"), from a depth of 50–100 m to a depth of approximately 700–800 m (Fig. 86). Without going into details, one may express the surmise that this sinking is in some way assisted by the strong contraction taking place in the overlying layers.

The quantitative evaluation of the effect of contraction on the balance of the masses in the ocean cannot be obtained without an analysis of the content

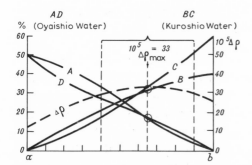

Fig. 88. Curves of percentage content of water masses A, B, C and D along line ab of maximum contraction on mixing. The values of contraction are shown by a broken curve. AD = waters of the Oyashio, BC = waters of the Kuroshio.

of each of the original waters in their mixture; some views can be expressed on the basis of a comparison of the quadrangle of contraction with the percentage nomogram (Fig. 79).

Let us consider, as an example, the change in the percentage content of each of the water masses along the line of maximum contraction on mixing ab, plotted also on the percentage nomogram. The curves of the percentage content of each of the waters have been entered in a separate Fig. 88 (here is also plotted the curve of contraction on mixing along line ab — the broken line). One may see, in particular, that maximum contraction, equal to $10^5 \Delta\rho$ = 33, is reached on the mixing of one third of the subarctic waters and two-thirds of the waters of the Kuroshio; precisely this explains the proximity of the zone of maximum contraction to the Kuroshio region as compared with the subarctic region. A more detailed picture can be obtained by an analysis of a sufficient quantity of T-S curves of stations lying between the area of station "Ca-119" and the weather ship "Tango".

As is already known (Section 36), when the mixing of three water masses is considered, the values of contraction on mixing represent the differences between density as determined by the true equation of state of sea water $\sigma_T = \sigma_T(S, T)$, and density corresponding to some linear equation of state. The latter is determined solely by the values of density of the apexes of the triangle of mixing, given by the true equation of state: the true and "mean" densities coincide in these three points (apexes) of the triangle. When two, as well as four or more water masses are mixing, the situation is different. On the mixing of two water masses the equation of state corresponding to the distribution of "mean" density, i.e., equation $\bar{\sigma} = \bar{\sigma}(\bar{S}, \bar{T})$, is linear along the straight line of mixing of these two water masses; outside this segment, generally speaking, it may also be non-linear. One may also set any linear equation of state corresponding to the linear distribution of density along a given straight line of mixing; these equations will be determined solely by the constant (arbitrary) values of the coefficients of thermal expansion α and saline contraction β in the expression for the linear equation:

$$\sigma = \sigma_0 - \alpha T + \beta S$$

(σ_0 is some constant value of conventional density).

If contraction on mixing of four water masses is considered, the true and "mean" densities coincide in the peaks of the quadrangle, the distribution of density is linear along the sides of the quadrangle as well as along the lines of the grid represented in Fig. 78. Within the quadrangle of mixing * the equation of state, however, will be non-linear, with the exception of one particular case, which will be referred to below.

Let us illustrate the foregoing in the following way. Let us imagine that the isopycnals on the *T-S* diagram represent the topographic lines of the relief of function $\sigma_T(S, T)$. In such a case, the true equation of state will appear in the form of a smooth, curvilinear surface rising in the region of high salinities and low temperatures and having a smooth bend along the line of temperatures of maximum density. The plane corresponding to the linear equation of state contains three points which correspond to the relief marks of the surface of the true equation of state in the apexes of the triangle of mixing, and intersects with the latter along some curves joining these points. Finally, the surface

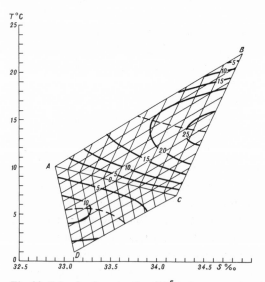

Fig. 89. Triangle of contraction ($10^5 \Delta\rho$) on separate mixing of water masses *A*, *B* and *C*, and *A*, *C* and *D*, respectively. The broken lines are the lines of maximum contraction.

* The possible generalized formal concept of "expansion on mixing" outside the triangle or quadrangle of mixing is not considered here.

plotted from the values of true density in the peaks of the quadrangle of mixing is curvilinear (a surface of the second order) and corresponds to the *nonlinear* equation of state for mean density $\bar{\sigma} = \bar{\sigma}(\bar{S}, \bar{T})$ as determined from formula [57.1]. This equation in one particular case can be linear, when four points (the angles of the quadrangle of mixing) lie in the same plane. The curvilinear surface mentioned "sags" with respect to the surface of the true equation of state; families of straight lines (the grid in Figs. 78 and 87) are its generatrices (within the quadrangle).

Fig. 89 represents (for comparison with Fig. 87) triangles of contraction on mixing *ABC* and *ACD*, constructed on the assumption that water masses *D* and *B* respectively do not participate in the mixing of the three other water masses. We see that the general picture within the two triangles differs from the picture of contraction in the quadrangle which is observed under conditions of mixing of all four water masses. These differences are fairly substantial, although the general nature of the picture of contraction is approximately similar (the region of maximum contraction is shifted in the direction of the straight line of mixing *BC* — the Kuroshio structure).

A comparison of Figs. 85 and 89 supports to a certain extent the hypothesis of "horizontal" mixing of water masses along the surfaces of equal percentage content, which served as a basis for the construction of the quadrangle of mixing (Section 54): within the region of interaction of waters there is no line of zero contraction on mixing (reference is made to line *AC* in Fig. 89), the nature of contraction on vertical mixing of waters *A* and *D* and waters *B* and *C* remains identical in both cases, but at the same time a region of maximum contraction appears inside the region of interaction of the four water masses.

58. STATISTICAL *T-S* ANALYSIS

The subject of statistical *T-S* analysis is the frequency of observations which correspond to definite points on the *T-S* diagram or definite squares on it. The study of the frequency of *T-S* relations, broken down in a certain way into classes by temperature and salinity, may be carried out by time, by space and finally, by volume. Statistical *T-S* analysis has considerably expanded the possibilities of *T-S* analysis as a whole; the latter was applied earlier, as we have seen, primarily to individual *T-S* curves.

Statistical *T-S* analysis was proposed by Montgomery (1955), whose work we shall consider first. Using 540 observations of temperature and salinity on the surface of the sea made in the course of 1948, 1949 and 1950 from the weather ship "J" in the North Atlantic (mean position: 53°N 20°W), Montgomery plotted the *statistical annual T-S diagram* represented in Fig. 90. Let us explain it. All observations of temperature and salinity were broken down into intervals of 0.5°C for temperature and 0.1‰ for salinity. In the corresponding squares of the *T-S* diagram was inscribed the frequency of the observations, expressed in parts per mille (number of cases per 1000, which we will designate here as ppm, so as to distinguish it from the designation of salinity (‰)). Thus, the frequency in the course of a year of temperature values between 9.5 and 10.0°C at simultaneous salinity values of between 35.3 and 35.4‰ amounts, as may be seen from Fig. 90, to 38 ppm. The isoline of frequency of *T-S* points, equal to 21.5 ppm, is plotted by a heavy line; this frequency covers 51% of all cases (the sum of the numbers included by the heavy line amounts to 516 ppm). The frequency of 5.5 ppm (thin line) covers 89% of all observations. Adding up the frequencies along the vertical and horizontal makes it possible to construct also one-dimensional distributions – histograms, separately for salinity and for temperature. These histograms are represented in the margins of the *T-S* diagram.

The smooth line, bending around the histogram, will be the curve of frequency; it is clear that in the margins of the *T-S* diagram cumulative frequency curves can also be constructed. Points have also been plotted on the statistical diagram (Fig. 90) which correspond to mean monthly values of temperature and salinity (these points are numbered by roman numerals *I–XII*); if these points are connected, we will obtain an example of a closed *T-S-t* curve.

Statistical analysis of *T-S* relations on the surface of the sea may be used for observations made not only in one point but on any water area. In this case frequency indicates not the quantity of cases in one point but the corre-

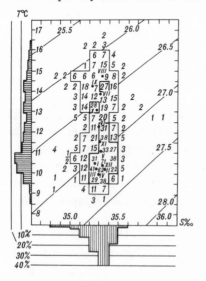

Fig. 90. Statistical annual *T-S* diagram for the surface of the sea. Position of weather ship "J" in the North Atlantic: 53°N 20°W. (Montgomery, 1955.)

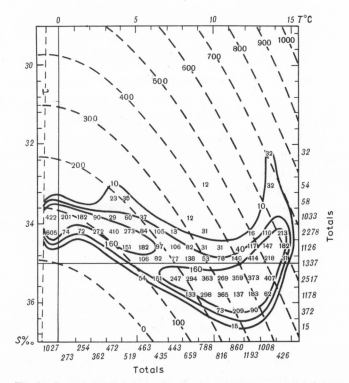

Fig. 91. Statistical *T-S* diagram showing frequency distribution of characteristics on the surface of the sea in the southern part of the Pacific Ocean (winter–August). The sum of frequencies amounts to 10^4; to convert to area the value of the frequency must be multiplied by 816 km^2. (Cochrane, 1956.)

sponding part of the total area in which temperature and salinity are observed simultaneously at given intervals.

Statistical *T-S* analysis for surface waters was carried out by Cochrane (1956) as applied to the entire Pacific Ocean including adjacent seas. Using G. Schott's maps of the distribution of temperature and salinity on the surface of the Pacific Ocean, Cochrane broke them down into squares (2.5° latitude by 5° longitude between 20 and 50°N and S; 5° latitude by 5° longitude in the other regions) and determined the areas occupied by *T-S* classes broken down into intervals of 2°C for temperature and 0.4‰ for salinity. Then he plotted statistical *T-S* diagrams for winter and summer, as well as mean annual diagrams for the northern and southern parts of the Pacific Ocean (six diagrams in all). One of Cochrane's diagrams is represented in Fig. 91 (the dashed isolines represent the anomaly of specific volume Δ_{ST}). The nature of these *T-S* diagrams is similar to the *T-S* diagram given in Fig. 90, with the difference indicated above, namely: each number in an individual class means that $n \cdot 10^{-4}$-part of the area of the ocean is occupied by water having *T-S* characteristics in the given limits. Cochrane points out that the numbers appearing in the classes of the statistical diagram satisfy the requirement:

$$N_{ij} = \frac{1}{\Delta S \, \Delta T} \int\limits_{S_i}^{S_i + \Delta S} \int\limits_{T_j}^{T_j + \Delta T} f(S, T) \, \mathrm{d}S \, \mathrm{d}T$$

where N_{ij} is the mean frequency for the characteristic class lying between S_i and $S_i + \Delta S$ and between T_j and $T_j + \Delta T$. The integrand function $z = f(T, S)$ represents (in the given case — empirically) the equation of "frequency sur⁺ face", i.e., the *T-S* relation for the waters of the Pacific Ocean. In Fig. 91 the contour lines of equal height of this surface (isolines of quantity N) have also been plotted, on the basis of which we come to a conclusion about the undoubted connection of statistical *T-S* diagrams with the question of the application of Green's theorem to the *T-S* diagram and the calculation of *T-S* areas. Analyzing the diagrams, Cochrane draws definite conclusions on the regional features of the thermohaline field, on which we will not dwell here.

Finally, statistical *T-S* analysis may also be applied to the *volumes* of seas and oceans. Volumetric statistical diagrams were published simultaneously for the Pacific Ocean (by Cochrane, 1958), the Indian Ocean (by Polak, 1958) and for the Atlantic as well as the World Ocean as a whole (by Montgomery, 1958). The three works mentioned, together with the two considered above, laid the foundation of statistical thermohaline analysis; they were followed by a number of other studies and, at the end of the section, we shall briefly take up the most interesting of them.

The method of plotting volumetric statistical *T-S* diagrams consists of the following. In the ocean (sea) under consideration, a uniform network of stations, made to the bottom, is selected. At the same time the condition must be satisfied that each station should be representative for some standard area (square). In the works of the authors mentioned such an area proved to be the area of 10^6 km^2 *. If one station covered a larger or smaller area, its statistical weight was correspondingly increased or decreased. A depth of 10 m was taken as the vertical unit of volume; thus, the volumetric unit proved to be equal to 10^4 km^3. Then each pair (temperature–salinity) for standard levels was assigned to a definite class; here, for analysis, two types of intervals were selected: a coarse scale (interval of temperature 2°C, of salinity − 0.5‰), and a fine scale (interval of temperature 0.5°C, of salinity − 0.1‰). The quantity of volumetric units, occupied by a *T-S* relation of a given class was determined in the following way: the *T-S* relationship was assigned to a layer, the boundaries of which along the vertical were situated in the middle between

---

* Given an area of the Atlantic Ocean of 106.2 · 10^6 km^2 105 representative stations were selected, i.e., one station just fitted an area of 10^6 km^2 (Montgomery, 1958).

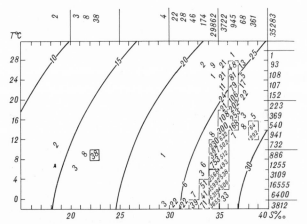

Fig. 92. Averaged volumetric *T-S* diagram for the Atlantic Ocean. Numbers in classes $2° \times 1‰$ signify volume in 10^4 km^3. The heavy line marks 75% of the entire volume of the ocean, the thin line 99%. Along the axes the sums of the numbers appearing in the classes of the diagram are plotted. (Montgomery, 1958.)

standard levels; the volumetric weight of the *T-S* relation was determined correspondingly in units of 10^4 km^3. Potential temperature was taken as the temperature.

One of the results — the coarse-scale *T-S* relationship for the Atlantic Ocean — is represented in Fig. 92. The numbers n in the unit classes $(2° \times 1‰)$ determine the volume times 10^4 km^3 of the water of the given class. The isopleth of frequency of 75% is plotted with a heavy line; consequently, more than 75% of the volume of the waters of the Atlantic Ocean is comprised between −1 and 3°C and between 34 and 35‰. The isopleth of 99% is plotted with a thin line. The sum of all the numbers appearing in the individual classes equals 70,566; a more accurate value of the volume of the Atlantic Ocean (Sverdrup et al., 1942, p. 15) amounts to 106,463 million km^3 — with adjacent seas and 82,441 million km^3 without adjacent seas.

The method of volumetric statistical *T-S* analysis was applied to the seas of Indonesia by Nefed'ev (1961), who improved the methods of plotting statistical *T-S* diagrams. This improvement consists of the fact that the vertical boundaries of the layer, assigned to the *T-S* characteristic, were determined from the curves of vertical distribution of temperature and salinity and not from the distances between standard levels. At large vertical temperature and salinity gradients, within one "standard" layer there prove to be several layers, assigned to the *T-S* pair within the given intervals of *T* and *S*; the result of the processing is made more precise, while the statistical *T-S* diagram proves close in its form to the "cluster" of *T-S* curves, conferring on the latter a quantitative (volumetric) characteristic.

Dubrovin (1965), studying the volumetric *T-S* characteristics of the Arabian and Red Seas, made a further improvement in methods: he determined the boundaries between layers along the vertical not by the curves of vertical distribution of temperature and salinity, but directly from the *T-S* curves. The use precisely of the *T-S* curves is not only more convenient but also more correct, since the identification of the layers may be carried out taking account of the vertical arrangement of the water masses and the boundaries between them. The same improvement was made in the works of Masuzawa (1964b, 1965).

As has already been said, a number of other works followed the first investigations considered. La Fond (1957) and Rama Sastry (1963) applied statistical analysis to the waters of the northern part of the Indian Ocean, particularly to the Bay of Bengal. Thus, in the latter of the works mentioned, Rama Sastry considers the characteristics of the surface waters in the coastal part of the Bay of Bengal near Madras. The material used for statistical *T-S* analysis was an eight-year series (1951–1957) of observations of temperature and salinity near Madras; as a result, a number of valuable conclusions was obtained concerning the seasonal interaction of water masses of different origins and concerning the effect on this interaction of various climatic factors: seasonal fluctuations in heat exchange with the atmosphere, river discharge, etc.

Let us mention several other works devoted to volumetric statistical analysis. In the articles of Karavaeva and Radzikhovskaia (1965) and Radzikhovskaia (1965), volumetric analysis was carried out for the waters of the northern and southern parts of the Pacific Ocean. In the work of Dubrovin (1965), which was already referred to above, a comparative analysis of the waters of the Arabian and Red Seas is carried out; finally, in Sturges' work (1965) the *T-S* characteristics of the waters of the Caribbean Sea are considered, while in the work of Yasui et al. (1967) characteristics of the waters of the Sea of Japan are given. All these studies contain a number of new and interesting conclusions, on which we will not dwell, noting only two for an example.

Thus, Dubrovin points out on the basis of statistical analysis that, in spite of the fact that the volume of the Red Sea is approximately 34 times less than the volume of the Arabian Sea, the reserve of salts in it is 29 times less, while the heat reserve is only ten times less. Sturges draws attention to an interesting feature of the waters of the Caribbean Sea, which manifests itself in the presence of strong "peaks" on the curve of frequency: almost half of all the Caribbean waters are comprised in the interval of only 0.1°C for temperature and 0.02‰ for salinity in relation to a mean value of 3.9°C and 34.98‰ respectively.

In addition, Sturges proposes a modernization of the method, consisting of the fact that the frequency of volumes in classes can be reckoned not only along vertical and horizontal strips, but also along isopycnic strips, which he accordingly does in the example of the Caribbean Sea.

Finally, volumetric distribution of the T-S properties for the entire Mediterranean Sea and its basins has been obtained by Miller (Miller et al., 1970)
and for the North Atlantic Ocean by Wright and Worthington (1970).

Apart from areas and volumes, the values of volume transports of water calculated, in particular, by the dynamic method, can be plotted in the corresponding classes of T-S diagram. For an illustration of the method, let us
consider the results obtained by Masuzawa (1964a, 1965) for the region of
the Kuroshio Current. From the data of a section made in May 1965 by the
R.V. "Riofu Maru" along the meridian 136.5°E, the geostrophic flows were
calculated between stations 2184 and 2189 (distance about 180 km) from
the reference surface of 2,000 dbar. The flows were calculated between each
pair of stations for layers of a thickness of 50 m in the upper 1,000-m layer
and for layers of a thickness of 100 m in the lower 1,000-m layer. The method
of combining geostrophic flows with the T-S diagram is as follows. For each
station the geostrophic flow is calculated as a quantity, which is the average
between the flows through neighboring pairs of stations, including the station
considered. Then, from the T-S curve the boundaries of the classes are determined and by interpolation — the corresponding transports within each of
the classes. Thus, the geostrophic transport is considered along the T-S curve
as a parameter.

In Fig. 93a are shown the results of a computation for the section mentioned above (the enveloping T-S curves belong to the extreme stations 2184
and 2189). It is seen from the figure that the maximum transport — 38.9
km³/h — belongs to the class (18.5°C; 34.85‰), which corresponds to waters
equivalent to the "18-degree waters of the Sargasso Sea" (Worthingon, 1959,
Istoshin, 1961). Full transport through the investigated part of the section
amounts to 247.2 km³/h. In Fig. 93b is shown for comparison the T-S diagram of the geostrophic transport of the Kuroshio through another section,
made further east (along meridian 144°E). One of the interesting results of
the comparison is the fact that the transport of "18-degree waters of the
Kuroshio" through the second of the sections considered is considerably
smaller. Finally, in Fig. 93c are shown the differences of geostrophic transports in corresponding classes between the two sections considered.

In Masuzawa's work (1965) the differences of T-S diagrams of geostrophic
transport through a section to the south of Cape Shiono — Misaki are also
considered — average for winter and summer, as well as the differences of
geostrophic T-S diagrams between the Kuroshio to the south of Japan (in the
summer) and the Northern Trade Wind Current (summer sections of 1956
along meridians 140 and 150°E). Let us note that as conventional dynamic
heights Masuzawa used integrated thermosteric anomalies Δ_{ST}.

Apart from frequency, area, volume, volumetric flow, other additive characteristics of water masses may also be represented on statistical T-S diagrams:

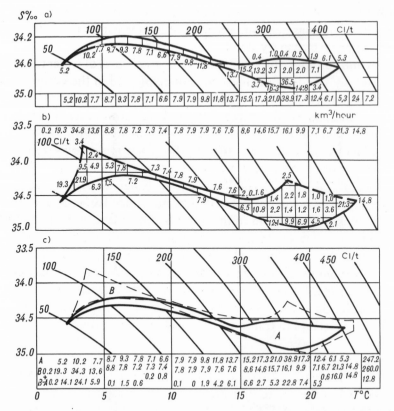

Fig. 93. *T-S* diagrams of geostrophic transport of the Kuroshio Current. (a) To the south of Japan, through a section made along meridian 136.5°E. (b) To the east of Japan, through a section along meridian 144°E. (c) Difference in geostrophic transports through the sections mentioned. (Masuzawa, 1964a; 1965.)

the oxygen, phosphate, or nitrate content, the biomasses of plankton, etc. As an example, a statistical *T-S* diagram is represented in Fig. 94, which shows the total content of dissolved oxygen in the plane of the section made by the scientific research vessel "Professor Deryugin" in the southeast part of the Pacific Ocean along meridian 109°W in May 1968 (part of the plane of the section was selected for calculation, namely, between the surface of the sea and the depth of 1,000 m and between latitudes 39.5 and 49.5°S, including ten stations made every other degree of latitude) (cf. also Mamayev, 1973b). The diagram represented in Fig. 94 is plotted in the following way. On the curves of vertical distribution of oxygen $O_2(z)$ at the oceanographical stations the thicknesses of the layers are laid out corresponding to the individual *T-S* classes (these thicknesses in turn were taken from *T-S* curves). Then, within each layer, the mean content of oxygen in the layer is graphically determined. Having multiplied the oxygen content in each square meter (or rather, in the

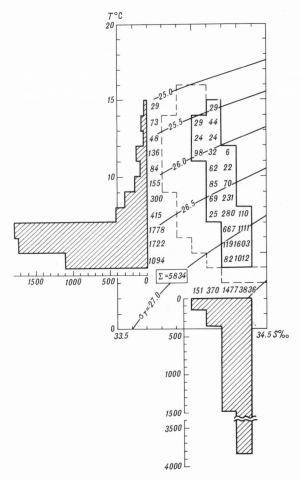

Fig. 94. Statistical *T-S* diagram of dissolved oxygen content (in ~$n \cdot 10^{-5}$ *l*) in a plane of a section of R.V. "Professor Deryugin", made at meridian 109°W in the Southeastern Pacific Ocean. The diagram relates to the part of the plane of the section lying between 39.5 and 49.5°S, in the 0–1,000-m layer.

volume 1 m × 1 m × 0.1 m), expressed in 10^2 ml/l by the thickness of the layer, we obtain the oxygen content in that part of the plane of the section which corresponds to the given *T-S* class; this content is presented in the classes of statistical diagram in liters per unit of latitude. In order to obtain the true oxygen content in the plane of the section (in a vertical layer of 0.1 m thickness) the sum of all the numbers of Fig. 94 must be multiplied by the distance between two stations, equal to $1.11 \cdot 10^5$ m. This content equals:

$$5833.85 \times 1.11 \cdot 10^5 = 6.5 \cdot 10^5 \text{ m}^3$$

Let us note that the statistical *T-S* diagram of oxygen content is very similar, by the nature of distribution, to the corresponding volumetric diagram which is not given here (cf. Mamayev, 1973b). This takes place because the oxygen content at the stations of the section of the "Professor Deryugin" along meridian 109°W is very uniform, changes little with depth and amounts in the layer 0–1,000 m on the average to about 6 ml/l. The latter quantity represents, as it were, the mean multiplier for the transition from the volumetric diagram to the diagram of dissolved oxygen content. The maximum content of dissolved oxygen belongs to the Antarctic Intermediate Water mass, the thermohaline index of which in the plane of the section considered amounts to (5.5°C; 34.34‰). By the way, the latter value was also determined from the statistical volumetric diagram as the "peak" of a mountainous country the relief of which can be plotted if the numbers appearing in the classes of the diagram are considered as heights (the foregoing is clarified by consideration of Fig. 91).

The possibilities of statistical *T-S* analysis, of course, are not confined to the examples quoted; in combination with "ordinary" *T-S* analysis it proves an extremely powerful means for the division of the World Ocean into regions by various factors, the study of the question of the seasonal and spatial interaction of water masses, the study of the interaction between the ocean and atmosphere, etc. Let us indicate in conclusion that the results of volumetric analysis were used by Bolin and Stommel (1961) for the study of the abyssal circulation of the World Ocean.

CHAPTER 8

WATERS OF THE WORLD OCEAN

59. GENERALIZED *T-S* RELATIONS OF THE WATERS OF THE OCEAN

In the present concluding chapter, the generalized T-S relations of the principal water masses of the World Ocean will be considered, their classification drawn up and their geographical distribution shown; the special features of the transformation of intermediate water masses will be considered in somewhat greater detail, as they are of the greatest interest in the vertical oceanographic structure of the ocean.

The generalized T-S relations, which are considered in this paragraph, represent the basic material for T-S analysis in the sense that they are the basis for the further study of the waters and the mapping of their characteristics: vertical and horizontal extent, paths of propagation, percentage ratio and other indicators of interaction; the quantitative comparison of the results of thermohaline analysis with other indicators of the dynamics of waters, as well as with the "secondary" characteristics of waters (say, with the distribution of various forms of plankton and its biomass). Finally, a highly important problem is the determination from generalized types of T-S curves of the coefficients of vertical and horizontal turbulent mixing, different for the regions of interaction of different waters, by means of Jacobsen's method (1927) and its possible modifications. Knowledge of the spectrum of averaged coefficients is absolutely necessary for the study of planetary processes of heat and salt exchange and for the subsequent solution of a number of the most important geophysical and geographical problems. Bearing in mind the connection between the T-S diagram and the coefficients of exchange, one may say that the coefficients of exchange are a distinctive parameter of state of the natural waters of the ocean, and to each point of the T-S diagram there can be placed in correspondence at least one value of the coefficient, which in turn is attached to a definite type of curve.

The question touched upon is not a new one; generalized T-S relations of the water masses of the World Ocean (excluding the 100-m surface layer) were constructed by Sverdrup et al. (1942) and Dietrich (1950, 1964). These diagrams constitute an introduction to what follows, and are shown in Figs. 95 and 96 respectively. The well-known T-S relations of Sverdrup show on the diagram the regions into which the main types of T-S curves of the World

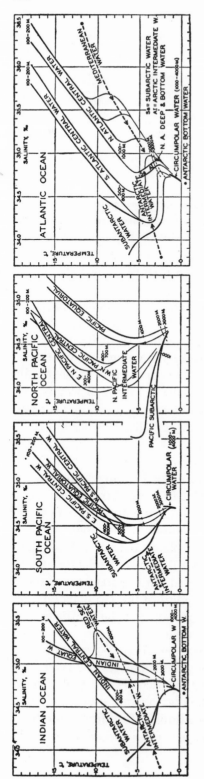

Fig. 95. Generalized *T-S* relations of the principal water masses of the World Ocean, according to Sverdrup (Sverdrup et al., 1942).

Ocean fit; looking at Fig. 95 we can easily imagine a particular form of a certain *T-S* curve in various regions of the ocean. However, Sverdrup's graph suffers from a certain shortcoming since it does not cover entire groups of *T-S* curves which lie *between* the main types and which correspond to regions which are transitional in relation to those referred to in the figure. To describe this shortcoming, let us cite the following example: the region of the Central waters of the eastern part of the North Pacific Ocean passes into the region of Equatorial waters if we follow along the meridian to the south; a gradual smooth change in the form of the *T-S* curve from one type to another corresponds to this "transition", even during transition through frontal regions. However, in Sverdrup's diagram, a considerable gap exists between the typical *T-S* bands, and it is not clear which form the "intermediate" *T-S* curves assume. In addition, as will be seen further on, Sverdrup's relations have become obsolete in some of their details and do not correspond to the true distribution of water masses.

The averaged diagram of Dietrich (Fig. 96) covers to a certain extent the shortcomings noted in Sverdrup's diagram, since all the possible *T-S* curves

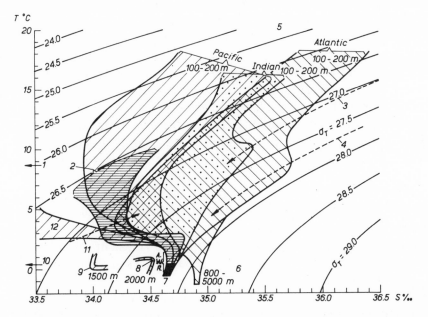

Fig. 96. Generalized *T-S* relations of the water masses of the World Ocean, according to Dietrich (1950).

1 = waters of the Black Sea; *2* = Subantarctic water ring; *3* = waters of the Red Sea; *4* = waters of the European Mediterranean Sea; *5* = oceans to the North of ~40°S; *6* = waters of the North Polar and North Seas; *7* = Antarctic bottom waters; *8* = waters of Baffin Bay (2,000 m); *9* = waters of the Sea of Japan (1,500 m); *10* = waters of the Sea of Okhotsk; *11* = Subantarctic intermediate waters; *12* = Subarctic Pacific waters; *A WR* = Antarctic water ring (1,000–4,000 m).

for a given T-S region fit within such a region, which corresponds to each of the oceans. However, the picture is too generalized, and we are denied the possibility of judging their form, since the main types ("clusters", if you will) are not defined here, and within each of the shaded T-S regions we can, generally speaking, imagine a T-S curve of any form. Of course, Dietrich's diagram also suggests to us the predominant configuration of T-S curves, but looking, for example, at the T-S region corresponding to the Indian Ocean, we can hardly imagine the probability of what is almost a straight T-S line, characteristic of the water masses of the southern part of the Bay of Bengal, where almost complete homohalinity is observed with depth.

Also, in Stepanov's work (1965), the main types of structures of the waters of the World Ocean are considered, their classification is proposed, and typical T-S curves are given corresponding to the main structures.

Thus, a clarification of the picture of the T-S relations of the main water masses appears necessary: on them, aside from the basic "fans", "clusters" and other aggregates of T-S curves, must also be represented the thermohaline indexes of the primary water masses (including the "point" masses too), as well as the main triangles of mixing, so that the analysis of T-S relations may be substantially supplemented by the conclusions which follow from the analytical theories of T-S curves. Furthermore, there is a need for a certain systematization of the T-S indexes of the main water masses.

The modified T-S relations of the main water masses of the Atlantic, Indian, Pacific and Southern Oceans, plotted by the author (Mamayev, 1969a), are represented in Figs. 97–100 respectively. Sverdrup's diagram (Fig. 95) was taken as a basis for these generalized T-S relations, and to a certain extent they may be considered as a modification of the former. In addition, in plotting these relations, apart from Sverdrup's T-S diagram, series of T-S curves plotted for selected sections in the oceans were used, as well as literary sources, of which the most important were the following:

On the Atlantic Ocean – the works of Defant and Wüst (1930), Jacobsen, (1929), Wüst (1935), Mamayev (1960a), Tiuriakov (1964), Tiuriakov and Zakharchenko (1965);

On the Indian Ocean – the works of Thomsen (1935), Rochford (1963, 1966a,b,c) the General Report (1966) by Uda about Japan's participation in the International Indian Ocean Expedition;

On the Pacific Ocean – the works of Dobrovol'skii (1962), Dobrovol'skii et al. (1960) and the monograph generalizing these works *The Hydrology of the Pacific Ocean* (Institute of Oceanology, Academy of Sciences, U.S.S.R., 1968), as well as the works of Robinson (1960) and Wyrtki (1963, 1967);

On the Soutern Ocean – the work of Yu.K. Gordienko (1964) (unpublished).

Apart from the thermohaline indexes of the primary water masses, Figs.

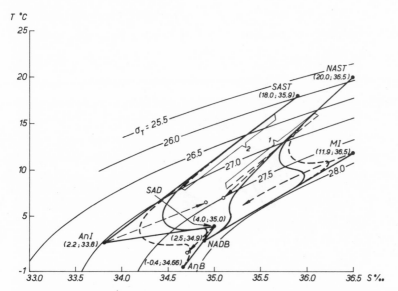

Fig. 97. Generalized *T-S* diagram of the water masses of the Atlantic Ocean.
1 = North Atlantic Central Water (according to Sverdrup), *2* = South Atlantic Central Water, AnI = Antarctic Intermediate Water, SAD = South Atlantic Deep Water, NADB = North Atlantic Deep and Bottom Water, AnB = Antarctic Bottom Water, NAST = North Atlantic Subtropical Troposheric Water, SAST = South Atlantic Subtropical Troposheric Water, MI = Mediterranean Intermediate Water.

97–100 show the positions of the main straight lines and triangles of mixing (although the *bases* of the triangles are not plotted in the majority of cases in order not to overload the diagrams) and the *types* of *T-S* curves (heavy lines – solid and dashed), formed as a result of the mixing of primary water masses (black dots). It is clear that the appearance on the *T-S* diagram of the thermohaline indexes of the primary water masses opens the way to greater scope in determining possible variations of *T-S* curves in accordance with the "geometry" of *T-S* curves.

The classification of types of water masses, based on generalized *T-S* diagrams, as well as their geographical distribution will be briefly considered in the following section; here we will give a characteristic example of the interpretation of generalized diagrams with the example of the eastern part of the Pacific Ocean, in explanation of the foregoing. For this purpose, the *T-S* relations for the eastern part of the Pacific Ocean have been singled out in a separate Fig. 101 in somewhat greater detail. The selection of this example is based on the fact that it is precisely in the Pacific Ocean that we observe in the most striking way (by comparison, say, with the Indian Ocean) the interesting effect of the interaction of two intermediate water masses which have extremes of the same sign, in particular salinity minima.

The main water masses of the central region of the North Pacific are

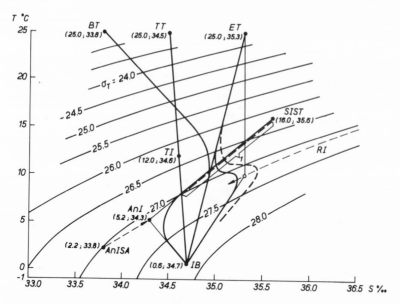

Fig. 98. Generalized *T-S* diagram of the water masses of the Indian Ocean. *1* = Indian Central Water (according to Sverdrup et al., 1942).

AnISA = Antarctic Intermediate Water of South Atlantic
AnI   = Antarctic Intermediate Water
TI   = Timor Sea Intermediate Water
BT   = Bengal Bay Tropospheric Water
TT   = Timor Sea Tropospheric Water
ET   = Equatorial Tropospheric Water
SIST = South Indian Subtropical Tropospheric Water
RI   = Red Sea Intermediate Water
IB   = Indian Deep and Bottom Water.

(Fig. 101): Surface (tropospheric) Subtropical Water mass C_1, Intermediate Subarctic Water mass A_2 and Deep Water mass B. Their interaction is illustrated by a T-S curve of type *1*, the most characteristic for the east of the North Pacific (Section 48).

For the central region of the South Pacific as a whole the following main masses are characteristic: tropospheric Subtropical Water mass C_2, Intermediate Antarctic Water mass A_1 and Deep Water mass B (the Bottom Antarctic Water mass, underlying Water B, is not considered in Fig. 101). The interaction of these water masses is represented by a T-S curve of type *3*.

In the equatorial part of the ocean the Intermediate Water masses, Subarctic and Antarctic, come into contact: the first of them, moving to the south, spreads over the second which is moving to the north. There occurs, as it were, mutual penetration of structures C_1A_2B and C_2A_1B, and a T-S curve of the transitional type *1–3* is formed; its upper part retains the features of T-S curve *1*, and the lower part of T-S curve *3*. The transitional

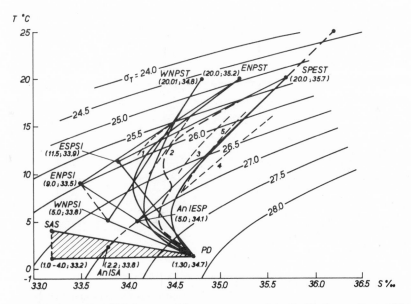

Fig. 99. Generalized *T-S* diagram of the water masses of the Pacific Ocean. Central Water masses (according to Sverdrup); *1* = Eastern North Pacific, *2* = Western North Pacific, *3* = Eastern South Pacific, *4* = Western South Pacific, *5* = Pacific Equatorial.

SAS   = Subarctic Surface Water
WNPSI   = Western North Pacific Subarctic Intermediate Water
ENPSI   = Eastern North Pacific Subarctic Intermediate Water
ESPSI   = Eastern South Pacific Subtropical Intermediate Water
WNPST   = Western North Pacific Subtropical Tropospheric Water
ENPST   = Eastern North Pacific Subtropical Tropospheric Water
SPEST   = South Pacific Equatorial and Subtropical Tropospheric Water
AnIESP   = Antarctic Intermediate Water of Eastern South Pacific
AnISA   = Antarctic Intermediate Water of South Atlantic
PD   = Pacific Deep and Bottom Water

curve *1–3* has two salinity minima – subarctic *a* and antarctic *b*, separated by salinity maximum *c*, *which represents the trace of the effect of water mass* C_2. The chief purpose of such discussion is to lay the way for a clarification of the origin of this kind of secondary maximum (of extremes in general), which are often observed on *T-S* curves – all the more since this clarification goes beyond the limits of the analytical theory of *T-S* curves developed up to the present time.

Thus, if we follow mentally from north to south along a meridional section in the eastern part of the Pacific Ocean, we can see how a *T-S* curve of type *1* gradually passes into type *3*, forming in the equatorial part of the ocean a transitional intermediate structure *1–3*. In the western part of the equatorial region of the Pacific Ocean a similar and even more striking picture is observed, since the Subarctic Intermediate Waters rise there to the surface.

A similar example is the formation of the *T-S* curve of the eastern-subtropi-

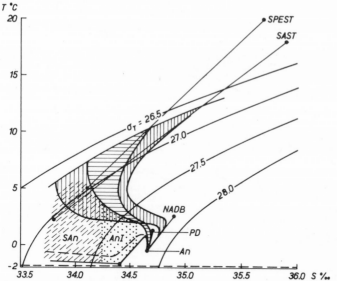

Fig. 100. Generalized *T-S* diagram of the water masses of the Southern Ocean (Atlantic and Pacific sectors).

SPEST = South Pacific and Equatorial Subtropical Tropospheric Water
SAST = South Atlantic Subtropical Tropospheric Water
NADB = North Atlantic Deep and Bottom Water
PD = Pacific Deep and Bottom Water
An = Antarctic Bottom Water
SAn = Surface Antarctic Water
AnI = Antarctic Intermediate Water

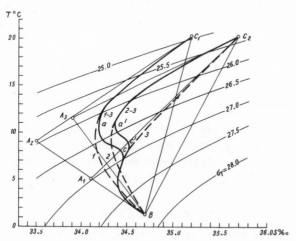

Fig. 101. Generalized *T-S* diagram of the water masses of the Eastern Pacific Ocean (explanations in the text).

Intermediate water masses: A_1 = Antarctic of the Eastern South Pacific; A_2 = subarctic of the Eastern North Pacific; A_3 = Subtropical of the Eastern South Pacific.
B = Pacific Deep and Bottom Water.
Subtropical Tropospheric waters: C_1 = of the Eastern North Pacific; C_2 = of the South Pacific Ocean.

cal structure *3* in the Peruvian–Chilean region of the southeast Pacific Ocean, *T-S* curve *3* of subtropical structure *2* is formed in the triangle of mixing C_2A_1B, whereas *T-S* curve *2* in triangle of mixing C_2A_3B. The "mutual penetration" of structures *2* and *3* leads to the formation of a transitional structure of the eastern–subtropical type (*T-S* curve *2–3* in Fig. 101). Intermediate secondary salinity maximum *c'* represents a survival of curve *3*, which reflects (at corresponding depths) the direct mixing of water masses C_2 and A_1.

Such is one of the examples of the application of *T-S* diagrams to the discovery of the genesis and transformation of water masses; other examples will be considered below.

60. THE CLASSIFICATION OF WATER MASSES

Thus, on generalized *T-S* diagrams (Figs. 97–100) *T-S* lines are considered not as images of individual water masses, but as lines of mixing between the primary water masses of the ocean. It may be seen from the *T-S* diagrams that these waters break down into three types: *tropospheric, intermediate* and *stratospheric* (*deep and bottom*) water masses. A return to Defant's terminology – "the troposphere of the ocean" and "the stratosphere of the ocean" – appears highly appropriate in the given context.

Let us consider briefly the basic features of these three types of water masses which are enumerated (together with their mean thermohaline indexes) in Table XV.

Tropospheric waters. These waters are situated in the near-surface layers of the oceans, at depths of approximately from 100 to 500–900 m and are linked basically to anticyclonic circulation of waters in the oceans. As a rule, tropospheric waters are characterized by a surface subtropical salinity maximum, which we have already seen in several figures (see for example, Figs. 61, 66); the *T-S* indexes of tropospheric waters are accordingly determined as the points of intersection of the tangents at the corresponding extreme. The surface water mass, very thin, with a mobile thermohaline index, is excluded in this (although it may serve as the subject of separate consideration). In the case where a subsurface salinity maximum does not occur, the index of the tropospheric mass "emerges on the surface". Moreover, the "absence of thickness" in such tropospheric water masses should not disturb us: precisely the stability of a *T-S* index in some surface point of the ocean may be treated as an indication of a "water mass" (such an approach was justified above, in Section 50). Tropospheric (central) waters and hypotheses of their formation are considered in detail in Section 61. The region of their distribution is near to what is shown in Fig. 102.

TABLE XV

Principal water masses of the World Ocean and their *T-S* indexes

Water masses	Atlantic Ocean	Indian Ocean	Pacific Ocean
Tropospheric[1]	North Atlantic Subtropical (20.0°C; 36.5°/oo)	Bengal (Bay) (25.0°C; 33.8°/oo)	Western North Pacific Subtropical (20.0°C; 34.8°/oo)
		Equatorial (25.0°C; 35.3°/oo)	Eastern North Pacific Subtropical (20.0°C; 35.2°/oo)
		Timor (Sea) (25.0°C; 34.5°/oo)	
	South Atlantic Subtropical (18.0°C; 35.9°/oo)	South Indian Subtropical (16.0°C; 35.6°/oo)	South Pacific Equatorial and Subtropical (25.0°C; 36.2°/oo)–(20.0°C; 35.7°/oo)
Intermediate	Subarctic (2.0°C; 34.9°/oo)	–	Subarctic (5.0°C; 33.8°/oo)–(9.0°C; 33.5°/oo)
	Mediterranean (11.9°C; 36.5°/oo)	Red Sea (23.0°C; 40.0°/oo)	Eastern South Pacific Subtropical (11.5°C; 33.9°/oo)
		Timor Sea (12.0°C; 34.6°/oo)	
	Antarctic (2.2°C; 33.8°/oo)	Antarctic (5.2°C; 34.3°/oo)	Antarctic (5.0°C; 34.1°/oo)
Stratospheric (Deep and Bottom)	North Atlantic Deep and Bottom (2.5°C; 34.9°/oo)	Deep and Bottom (0.6°C; 34.7°/oo)	Deep and Bottom (1.3°C; 34.7°/oo)
	South Atlantic Deep (4.0°C; 35.0°/oo)		
	Antarctic Bottom (−0.4°C; 34.66°/oo)		

[1] Not included in the table are surface (tropospheric) arctic, subarctic, antarctic and subantarctic waters with unstable *T-S* indexes.

Intermediate waters. These waters form a distinctive liquid boundary between the troposphere and the stratosphere of the ocean and are situated at depths of approximately from 600–800 to 1,200 m. Intermediate waters are determined on *T-S* curves by characteristic extremes, and are divided into three main types:

(1) Intermediate waters with a salinity minimum, formed in the subarctic and subantarctic latitudes; these are subarctic waters in the Atlantic Ocean, subarctic waters in the Pacific Ocean and antarctic waters in all three oceans (in their southern parts).

(2) Intermediate waters with salinity maximum, formed as a result of the

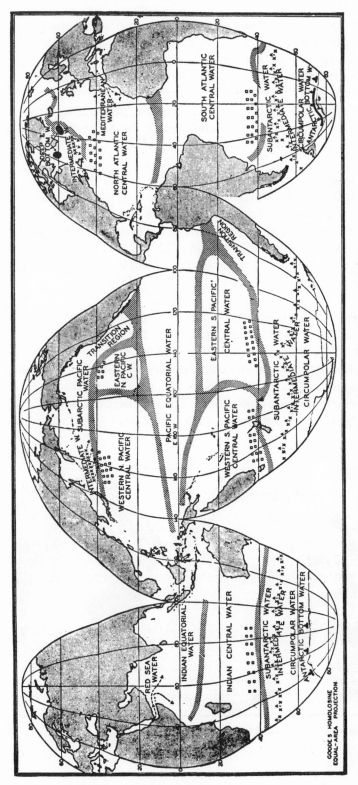

Fig. 102. Tropospheric (Central) Water masses of the World Ocean, according to Sverdrup (Sverdrup et al., 1942).

water exchange of the ocean with closed seas with a thermohaline structure different from the ocean. These are the Mediterranean Water mass in the Atlantic Ocean, the Red Sea Water mass in the Indian Ocean, the layers of increased salinity in the Arabian Sea, as well as the water mass of the Timor Sea in the Indian Ocean.

(3) Intermediate waters with a temperature maximum, penetrating into the high latitudes. These are the Atlantic Water mass in the North Polar Sea and the Antarctic Intermediate Water mass in the Southern Ocean, shown in Fig. 100.

The intermediate water mass in the southeast of the Pacific Ocean — the Eastern Subtropical Intermediate Water — stands somewhat by itself; its formation takes place in a similar way to the formation of the antarctic water mass, but in lower latitudes (it is also characterized by a salinity minimum, while its appearance was considered in discussing Fig. 101). The region of distribution of the main intermediate water masses of the World Ocean is shown in Fig. 103.

Stratospheric waters. The stratospheric waters may be divided into two principal types: waters formed in the high latitudes of the Northern Hemisphere and characterized by a salinity maximum, and waters formed in the high latitudes of the Southern Ocean and characterized by a salinity minimum ("maximum" and "minimum" must be understood here in a relative sense, when only stratospheric waters are considered). The stratospheric waters move from the high latitude regions toward each other, and where they come

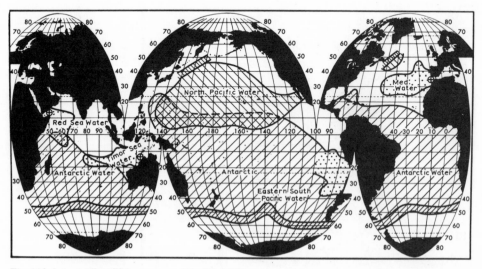

Fig. 103. Intermediate Water masses of the World Ocean.

into contact, the waters of the Northern Hemisphere prove to be deep, and the waters of the Southern Hemisphere bottom waters; the thermohaline characteristics are such that the antarctic waters always appear *lower* than the water masses of northern origin. The distribution of stratospheric waters is shown in Fig. 104.

A list of the principal water masses of the World Ocean, the indexes of which are represented in Figs. 97–100, is given in Table XV; here the water masses are classified by oceans, their vertical position (depth), as well as by their location from the north to the south.

The water masses listed embrace an area of water, excluding the regions of formation of stratospheric water masses, namely, the water area limited on the south by the line of antarctic convergence and on the north by the polar fronts of the Atlantic and Pacific Oceans. The water masses lying to the north and to the south are distinguished by greater "thermohaline complexity", which makes difficult the identification of individual thermohaline indexes. These waters (if the Arctic Basin is excluded) include: the subarctic (surface) waters of the North Atlantic; the subarctic waters of the North Pacific; and the antarctic waters in the Southern Ocean, surface and intermediate.

In spite of the difficulty (and even the impossibility) of isolating the thermohaline indexes of these waters, the *T-S* regions of their existence are indicated: for the Southern Ocean – in Fig. 100, for the Pacific Ocean – in Fig. 99 (the subarctic waters of the Atlantic are not indicated in Fig. 97; in

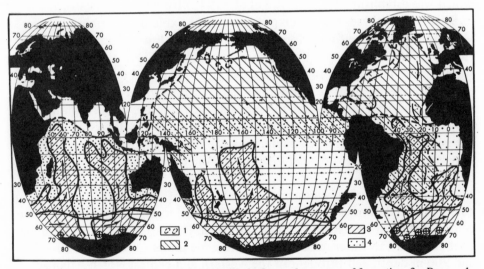

Fig. 104. Deep and Bottom Water masses of the World Ocean. *1* = centers of formation, *2* = Deep and Bottom Water mass of the North Atlantic and the North Pacific, *3* = Antarctic Bottom Water mass, *4* = Southern Deep Water masses of all three oceans.

order not to encumber the diagrams, they are shown in Fig. 106 and described below, in Section 61).

Regarding the subarctic waters, let us note in passing their difference for
the Atlantic and Pacific Oceans: the first are distinguished by pronounced
homohalinity along the vertical (mean salinity amounts to 34.9‰); the
second by pronounced homothermy (mean temperature along the vertical
1.5°C). Apparently, precisely this distinction explains the broad distribution
in the Pacific Ocean of Intermediate Subarctic Waters as compared with the
Atlantic Ocean.

61. THE TRANSFORMATION OF TROPOSPHERIC WATERS

One of the important questions in the study of the thermohaline structure
of the waters of the World Ocean is the question of the origin of the tropospheric, or central water masses of the World Ocean. As has already been
said, by this term we mean the waters belonging to anticyclonic gyres in the
oceans and observed both in the region of the currents themselves and in the
internal zones of circulation; in the vertical direction the Central Water masses
extend from a depth of 100 m to a maximum depth of 900 m in the Sargasso
Sea and a minimum depth of 200–300 m in the northern part of the Pacific
Ocean. Precisely in these horizontal and vertical limits (the upper 100 m
layer, interacting intensively with the atmosphere, is excluded) the Central
Water masses are characterized by a quite definite rectilinear T-S relationship,
which was established by Sverdrup et al. (1942) for all three oceans.

Several hypotheses have been put forward to explain the origin of these
water masses and the formation of the straight T-S lines which characterize
them. According to the first of these, formulated by Sverdrup, the central
water masses are formed in the regions of the subtropical convergences,
roughly speaking along the 40th north and south parallels as a result of convection, as well as of isopycnic mixing, and the fill in the entire region where
they exist. Justification for this hypothesis is provided by the coincidence of
the horizontal T-S curves, characterizing the regions of convergence in certain seasons, with the vertical T-S curves of the central water masses, noticed
by Iselin.

A second point of view concerning the waters of the North Atlantic had
been expressed even earlier by Jacobsen (1929). Jacobsen considers that the
central waters of the North Atlantic (although he does not use this term, he
does consider the water masses within the same limits — between 100–200
and 900 m depth and throughout the western and central North Atlantic)
are formed as a result of the mixing of water mass C (also central, but in
another terminology) and water mass S (southern). The first of these, accord-

ing to Jacobsen, represents a belt of warm and saline waters, extending lati-
tudinally from the Canary Islands to Puerto Rico at a depth of about 100 m;
the thermohaline index of this water mass C: $T = 22°C$; $S = 37‰$. The second
water mass S represents the intermediate layer, lying between the equator and
$10°N$ at a depth of 700–800 m; its thermohaline index: $T = 5.5°C$; $S = 34.6‰$.
It is obvious that water mass S represents nothing else but the transformed
Antarctic Intermediate Water mass, which, as is shown, spreads from the
south to precisely about $10°N$.

A shortcoming of the two hypotheses mentioned is the fact that they make
no allowance for the effect on the formation of the central water masses of
ocean currents, even such as the Gulf Stream and the Kuroshio, whose waters
are completely situated in the regions of distribution of the Central Water
masses. Therefore, another hypothesis was proposed by the author, based on
the example of the North Atlantic (Mamayev, 1960a), the gist of which is as
follows. The principal source of the Central Water mass of the North Atlantic
is the water mass of the Gulf Stream itself and the Antilles Current; moving
into the high latitudes, this water mass undergoes latitudinal (zonal) transformation, chiefly due to its continuous vertical mixing with the underlying deep
and bottom North Atlantic Water mass. So far as the rectilinear T-S relation
is concerned, characteristic for the central water mass, it is the locus of the
T-S points of the modifications of the "mother" water mass of the Gulf
Stream, observed at various latitudes within the entire thickness of the upper
layer.

Let us substantiate this point of view in somewhat greater detail. The
geographic layout of the problem under consideration is as follows. From
the region of confluence of the Florida Current and the Antilles Current,
initial water mass A spreads to the north; the thermohaline index of this
water mass may be taken as the following: $T = 24°C$, $S = 36.3‰$. This index
was obtained by Jacobsen as a result of averaging the data of oceanographical
stations, located precisely in the region of the Straits of Florida and the
Bahamas. As a result of vertical mixing with the Deep and Bottom Water
mass of the North Atlantic B ($T = 2.5°C$, $S = 34.9‰$) underlying it, the
characteristics of which remain invariable throughout the water area of the
North Atlantic, water mass A is transformed in such a way that its temperature and salinity decrease as the waters move to higher latitudes.

For the geometrical interpretation of this problem on the T-S plane, Fig.
105 is provided, on which are represented:

(1) The averaged T-S curve for the extensive region (zone XX according
to Jacobsen, 1929) lying between the coast of North America and meridian
$50°W$ and limited on the north and south by parallels 40 and $20°N$. This
region embraces almost all the west of the central part of the North Atlantic
(including the Sargasso Sea) and is directly fed by the waters of the Gulf

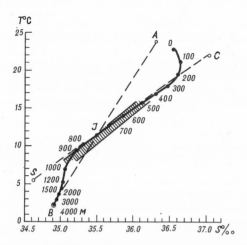

Fig. 105. Averaged *T-S* curve for the region lying between 20° and 40°N, 50°W and the coast of North America (western part of the North Atlantic). *AB* and *CS* are the straight lines of mixing of the corresponding water masses according to Jacobsen (1929). The shaded strip is the *T-S* relation for the North Atlantic Central Water according to Sverdrup (Sverdrup et al., 1942).

Stream. The *T-S* curve referred to was plotted by him from the data of 21 deep water stations.

(2) A linear *T-S* relation, characteristic for the North Atlantic Central Water according to Sverdrup — in the form of a shaded strip (all the *T-S* curves of the Central Water mass within its vertical extent fit into this strip).

(3) The straight lines of mixing, according to Jacobsen, of water masses *A* and *B*, as well as *C* and *S* (reference was made to these water masses above), as well as special *T-S* index *J*, which, as Jacobsen indicates, may be the result of mixing along either of the two straight lines.

A noteworthy feature of the *T-S* relations represented is the coincidence of the averaged *T-S* curve (within depths from 200 to 900 m), the straight line of mixing *CS* and the shaded strip; all these three relations accordingly characterize the Central Water mass of the North Atlantic in the range of depths indicated.

In order to prove the possibility of the formation of the central (or tropospheric; this term, which we will use below, is more general) waters by means of zonal transformation of the initial water mass (of the Gulf Stream and Antilles Current) it is necessary: (a) to demonstrate that the *T-S* curves lying in the shaded strip in Fig. 105 can be formed as a *result of the mixing of precisely water masses A and B*, and not necessarily, say, of water masses *C* and *S*, as Jacobsen assumes; (b) to demonstrate that in the shaded strip lie not only parametric points *z* of *T-S* curves in vertical limits corresponding to the thickness of the central water mass, but also the *thermohaline indexes of the subsurface waters, representing zonal modifications of the initial water mass of the Gulf Stream and the Antilles Current.*

Condition (a) was considered above, in Section 43; there it was demonstrated that by applying the analytical theory of T-S curves to the mixing of two water masses in an ocean of semi-infinite depth and provided that the coefficients of heat and salt exchange are not equal, the situation is precisely as stated in (a); let us now consider the second aspect of the problem, formulated in (b).

Considering the T-S curves for the region of North Atlantic Water mass A (Section 52), we see (Fig. 66) that this type of T-S curve has a bend in the surface part which is a trace of the transformation of the more saline core of the North Atlantic Current together with the 100–150-m surface layer and the underlying layers. The presence of this bend makes it possible to calculate the thermohaline indexes of the saline core of the North Atlantic Current, which we have accordingly done using the example of stations made in five sections of the IV cruise of the R.V. "Mikhail Lomonosov" (Fig. 65). These thermohaline indexes, determined from individual T-S curves and then averaged for each section, are shown in Table XVI, as well as being plotted in Fig. 106.

We see that the thermohaline indexes of the subsurface core of the North Atlantic Current fit well within the generalized T-S diagram (strip of central waters). The data of the IV cruise of the "Mikhail Lomonosov" relate to the northwestern part of the North Atlantic; in the eastern part of the North Atlantic a zonal transformation of the core of the North Atlantic Central Water mass is also observed (Kin'diushev, 1965).

Thus, the entire North Atlantic, or rather the entire region lying to the southeast of the left edge of the Gulf Stream and the North Atlantic Current is fed by the waters of the Gulf Stream system. These waters, as they move to the northeast undergo a zonal transformation, cooling off and experiencing decrease of salinity; therefore, each of the water masses lying on this path may be considered as a zonal modification of the "mother" surface Atlantic Water mass.

TABLE XVI

Mean termohaline indexes of the core of the North Atlantic Current for five sections of the IV cruise of the R.V. "Mikhail Lomonosov"

Section	T (°C)	S (°/oo)
I	12.49	35.67
II	13.17	35.75
III	15.36	36.02
IV	16.86	36.01
V	17.54	36.34

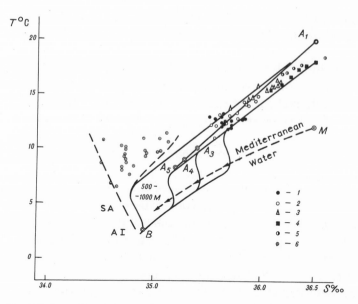

Fig. 106. Generalized *T-S* diagram of the water masses of the North Atlantic according to Sverdrup (Sverdrup et al., 1942), supplemented by data of the IV cruise of the R.V. "Mikhail Lomonosov". *1−5* = *T−S* indexes of the core of the North Atlantic Water mass at stations made in cross-sections *1−5* respectively; *6* = *T−S* points of the Subarctic Water mass on the surface of the sea; A_1A_5 = line of zonal transformation of the North Atlantic Water mass.

If the source of the North Atlantic Water mass were not the system of the Gulf Stream and the Antilles Current, the saline core of the North Atlantic Current would hardly occur; its existence is difficult to explain if one adopts the conception of the "central" formation of the North Atlantic Water mass.

For a clarification of the details let us turn to generalized Fig. 107, in

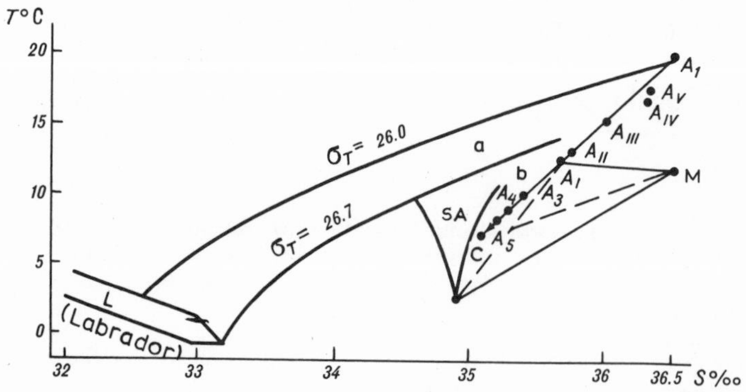

Fig. 107. Water masses of the North Atlantic and their interaction, shown on a *T-S* diagram (according to data of the I and IV cruises of the R.V. "Mikhail Lomonosov") (Mamayev, 1960a).
L = Labrador Water mass, *SA* = Subarctic Water mass, *a* = isopycnic mixing, *b* = zonal transformation. For further explanation see text.

which modifications of the initial Atlantic Water mass in five sections of the IV cruise of the "Mikhail Lomonosov" are plotted; these modifications are designated as A_I-A_V. In addition, the T-S indexes of the following water masses are plotted on Fig. 107:

A_1 – the water mass of the Gulf Stream itself in the region between $52°30'$ and $57°20'W$. The T-S index is determined from McLellan's work (1957) and equals: $T = 20.0°C$, $S = 36.5‰$. Depth is $0-150$ m;

A_3 – the North Atlantic Water mass in the section Iceland–Ireland, lying in the layer $0-500$ m and having index $T = 10.1°C$, $S = 35.40‰$;

A_4 – the North Atlantic Water mass in the section Hebrides–Iceland, lying in the layer $0-500-750$ m; thermohaline index: $T = 9.0°C$, $S = 35.28‰$;

A_5 – the North Atlantic Water mass in the Faeroe–Shetland section, lying in the layer $0-150$ m. Thermohaline index: $T = 8.3°C$, $S = 35.20‰$.

The thermohaline indexes A_3, A_4 and A_5 were determined by the author from the data of sections made in the I cruise of the "Mikhail Lomonosov" (October–November 1957), and were also shown in Fig. 106.

Finally, C is a water mass called by the author the *Eastern-Icelandic* (thermohaline index: $T = 7.2°C$, $S = 35.08‰$), representing the product of the zonal transformation of the North Atlantic Water mass in the region of Iceland and also determined from data of the I cruise of the "Mikhail Lomonosov". The final product of this zonal transformation is apparently the warm Atlantic intermediate * water mass in the Arctic Basin; its thermohaline index, according to the generalized T-S diagram of the waters of the Arctic Basin of Kusunoki (1962), is: $T = 0.3°C$, $S = 34.9‰$.

Thus, straight line A_1-C in Fig. 107 is the locus of all the modifications of the initial "mother" water mass of the Gulf Stream in the course of its transformation into water mass C. Naturally, there is an innumerable set of these modifications. Line A_1C may be called the line of zonal transformation of the North Atlantic Water mass. This transformation may take place for the following reasons:

(a) As a result of the consistent release of heat into the atmosphere as the warm waters of the Gulf Stream system progress into the high latitudes and of a certain degree of desalinization as a result of water exchange with the atmosphere.

(b) As a result of the interaction of the water mass of the Gulf Stream throughout its path from Cape Hatteras to Spitzbergen with the surface waters encountered by it on the left boundary – from the water mass of the continental shelf in the region of Chesapeake Bay and the water mass of the

---

* According to the terminology of Timofeev and Panov (1962), this water mass belongs to the main type of waters of the Arctic Basin and has the following thermohaline index from the generalized T-S diagram of the waters of the Arctic Basin: $T = 0.8°C$, $S = 34.9°/oo$.

Canadian continental slope in the area of the Great Newfoundland Bank to the polar waters in the region of the polar front near Jan Mayen Island. This interaction is frontal in nature and causes a broad frontal zone from Newfoundland to Iceland and further on to Jan Mayen.

(c) As a result of the mixing of the modifications of water mass A with water mass B underlying it.

In the process of horizontal transformation, the Atlantic Water mass wedges out to the north; its depth amounts to up to 1,000 m in the region of the Sargasso Sea, and only up to 150 m in the Faeroe–Shetland section. The increase in the mean density of the North Atlantic (surface) Water mass in the course of its transformation on the way from A_1 to C amounts to $\Delta\sigma_T =$ = 1.54. This increase owes its origin, in the first place, to the cooling of the Atlantic waters by 12.8°C (from 20 to 7.2°C) on the whole, as a consequence of their movement into the high latitudes and, in the second place, to their desalinization as a result of continuing contact with the less saline waters of the subarctic regions – on the whole, by 1.42‰ (from 36.50 to 35.80‰).

The nature of the zonal transformation of tropospheric waters must be similar in the Northern Atlantic and in the northern half of the Pacific Ocean; so far as the southern regions of the oceans are concerned, there it will be somewhat different in nature, since the final product may very well be the Antarctic Intermediate Water mass; formed as a result of convection in the region of the Antarctic Convergence.

Let us note that the T-S diagram represented in Fig. 107 supplements the generalized T-S diagrams of the main water masses of the Atlantic Ocean (Fig. 97).

62. THE TRANSFORMATION OF THE CORES OF WATER MASSES

Unlike the thermohaline indexes of the Tropospheric Water masses which, as we have seen, undergo continuous *zonal* transformation, the indexes of intermediate and deep water masses change for other reasons; this change may be called the *transformation of mixing*, inasmuch as the main factor which determines it is not the climatic factor, but the mixing processes of these water masses with the over- and underlying masses (see also the work of Khanaichenko, 1947). The primary effect of transformation of mixing is experienced by those water masses which, on the T-S curve, are characterized by extremes of salinity or (more rarely) temperature; the transformation of the cores of Intermediate and Deep Water masses may be zonal (meridional) in nature – for example, the Antarctic Intermediate Water mass in all three oceans. it may also be latitudinal in nature – for example, the Mediterranean Water mass in the Atlantic Ocean.

The essence of the "core method" of Wüst (1935) is extremely simple and consists of determining the characteristics which correspond to the core of the water mass and the subsequent charting of these characteristics. Thus, say, for the core of the Mediterranean Intermediate Water mass, one may determine the values of maximum salinity in the surface of the core, the temperature values, the dissolved oxygen content, the speed of sound, and so forth; the charting of all the characteristics peculiar to the core makes it possible to draw conclusions about the nature of the waters of the ocean, their geographical distribution and their paths of propagation.

The second phase of the "core method" consists of plotting the lines of transformation of the cores on the T-S diagram and determining the percentage content of the waters of the core, counting from the T-S point, which corresponds to the "pure" type of intermediate or deep water mass, to the T-S point which determines the practical disappearance of the core. Such a percentage nomogram as applied to the T-S line of transformation of the core allows us to determine, by comparing the line of transformation with the T-S curves of individual stations, the picture of the percentage content of intermediate or deep waters by space, and consequently to determine the nature and intensity of the mixing of the water masses. As a result of the analysis of the transformation of the cores of intermediate and deep water masses, in comparison with charts of salinity and temperature distribution, Wüst drew up a diagram of the meridional distribution of stratospheric water masses in the Atlantic Ocean, represented in Fig. 108 (Wüst, 1935).

To illustrate the "core method", let us cite a striking example of the transformation of the core of the Mediterranean Intermediate Water mass in the Atlantic Ocean. This question was also investigated for the first time by Wüst; however, he combined this water mass and the deep water mass of the South Atlantic of North Atlantic origin, also having a salinity maximum, in one "upper deep water mass"; as a result he obtained a formal picture differing from reality, namely: on his charts this "combined" water mass stretches

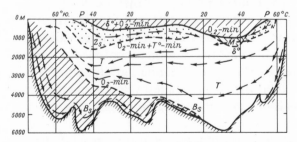

Fig. 108. Distribution of water masses on a middle section of the Atlantic Ocean (Wüst, 1935). B_S = Antarctic Bottom, T = Atlantic Deep, Z_S = Antarctic Intermediate, Z_N = Arctic Intermediate Water masses. Above: Tropospheric (Central) waters.

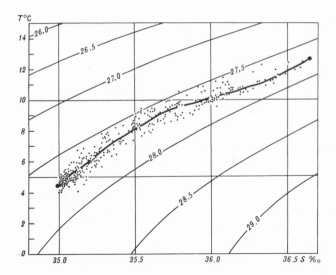

Fig. 109. Curve of transformation of the core of the Mediterranean Intermediate Water mass in the North Atlantic (Bubnov, 1971).

from the Straits of Gibraltar far into the southern hemisphere. It is clear that such a result does not correspond to geographic realities.

This question was taken up again from more realistic premises by Bubnov (1971). On the basis of the analysis of T-S curves of 385 oceanographic stations in the North Atlantic, he plotted a generalized T-S curve of the transformation of the core of the Mediterranean Water mass, represented in Fig. 109. The points in this figure are the cores determined from the corresponding intermediate salinity maxima of individual T-S curves. Having taken the T-S index of the "pure" Mediterranean Water mass ($T = 11.90°$C; $S = 36.50‰$ — Wüst, 1935; Sverdrup et al., 1942) as 100% and having determined the "final" T–S index of this T–S curve ($T = 4.5°$C; $S = 35.00‰$), Bubnov, taking the latter as 0%, plotted a chart of the percentage content of Mediterranean waters in the core with S_{max}.

It will be seen from this chart, as Bubnov points out, that the velocity of transformation of the Mediterranean waters is greatest in the northeast of the region outlined by the isoline of 0%; here the isolines of percentage content are most condensed. According to Bubnov, the propagation of Mediterranean waters to the northwest is limited by the North Atlantic Current, which represents a kind of liquid barrier on their path. The smallest velocity of transformation of Mediterranean waters is observed in the region lying to the south of parallel 30°N.

The lines of transformation of the cores of water masses on the T-S diagram make it possible to determine how much the nature of the transformation of intermediate waters differs from the isopycnic. Thus, for example, the transformation of the Antarctic Intermediate Water mass in the Southern Atlantic is not isopycnic; owing to the intensive nature of mixing, density in its core increases throughout the course of transformation from the value $\sigma_T = 27.0$ to $\sigma_T = 27.5$. On the other hand, the transformation of the Mediterranean Water mass is quasi-isopycnic in nature; the generalized T-S curve, shown in Fig. 109, is fairly well approximated by the isopycnic surface $\sigma_T = 27.70$. The difference in the nature of the transformation of the two water masses considered is explained by the different nature of their propagation. In the first case, when transformation takes place in a meridional direction, it entails a considerable redistribution of energy, the need for which is determined by climatic factors. In the second case, transformation proves to be isopycnic.

The investigation of the transformation of cores of water masses and of the boundaries of their propagation may also be carried out by means of the so-called *method of salinity anomalies*. This method was proposed by Helland-Hansen and Nansen (1926) and used for the determination of the boundaries of intermediate water masses in the North Atlantic by Iselin (1936) and Bubnov (1968). Let us consider briefly the gist of the method of salinity anomalies in the example of the work of the latter author, who dealt most thoroughly with this question.

The basis of the method is the selection of a standard or reference T-S curve, relating to that point in the ocean where the absence of intermediate water masses is most probable. As a reference standard for the North Atlantic a generalized T-S curve of the waters of the western part of the Sargasso Sea plotted by Iselin was selected. Comparing the T-S curves of individual stations with the standard T-S curve in the region of the cores of intermediate water masses, one may obtain the values of the salinity anomalies, wherein the positive anomalies will correspond to waters with high salinity, while the negative anomalies correspond to waters of low salinity. Having thus processed about three thousand stations in the North Atlantic, Bubnov constructed a chart of salinity isoanomalies, showing the distribution of intermediate water masses. Since in practice anomalies exceeding 0.05‰ are determined with reliability, precisely that quantity was accordingly taken as the boundary of distribution of intermediate water masses: +0.05‰ for the Mediterranean and −0.05‰ for the Antarctic and Subarctic Water masses. The location of the positive isoanomalies on Bubnov's chart is very similar to the picture of propagation of the Mediterranean waters obtained by the same author using the "core method"; however, the most interesting feature of the chart obtained is the region of negative isoanomalies, which correspond to the Subarctic Water mass. It turned out that this water mass has a wider propagation than was thought up to now, while the isoanomaly of −0.20‰ clearly shows the area of formation of the subarctic waters, situated to the south of Cape Farewell

(Greenland). The picture of the contact of the Mediterranean and Antarctic Intermediate Water masses is also interesting: these waters, as it were, "extinguish" each other approximately in the area of parallel 20°N, without spreading over each other, as occurs in the northern part of the Pacific Ocean. Bubnov's chart served to clarify the boundaries of the intermediate water masses shown in Fig. 103.

The method of standard T-S curves and salinity anomalies can be developed, and its further use for the study of the water masses of the ocean is necessary.

63. SOME SPECIAL FEATURES OF THE TRANSFORMATION OF INTERMEDIATE WATER MASSES

We have considered above the main features of the transformation of cores of intermediate and deep water masses, which we have called "the transformation of mixing", and although this transformation takes place mainly due to vertical mixing, we have considered the question of how the transformation of cores takes place in the horizontal direction and how it is connected with the isopycnic movement of water layers.

In the present section we shall consider some special feature of the transformation of intermediate waters, which arise first of all because of the differing intensity of vertical mixing upward and downward from the core of the intermediate water mass. If we turn once again to the analytical theory of T-S curves for the case of mixing of three water masses in a sea of infinite depth (this theory, as we know, continues to remain the basis for the practical analysis of T-S curves), we may note that the theory assumes symmetry of mixing upward and downward from the core of the intermediate water mass, which, in particular, follows from the regular distribution of parametric points z on the branches of the analytical T-S curve. However, some cases of practical analysis show that the fundamental principles of the geometry of T-S curves are often violated. In our opinion, this occurs for the reason that in analytical theory, even if we leave aside (in a first approximation) the question of the effect of horizontal currents, no allowance is made for the fact that triangles of mixing are constructed in the field of graph $\rho = \rho(T,S)$ and consequently, for the effect of vertical stability:

$$E \approx \frac{d\rho}{dz} = 10^{-3} \frac{d\sigma_T}{dz}$$

on the transformation of the intermediate water mass is not evaluated (Mamayev, 1960b). The differing values of criterion E on the upper and lower boundaries of the intermediate layer, as well as the appearance, in the majority of cases, of a second factor — contraction on mixing — violate the

TABLE XVII

Thermohaline indexes

Water mass	$T(^\circ C)$	$S(^o/oo)$	σ_T
North Atlantic surface (A)	12.5	35.67	27.04
Mediterranean (M)	11.9	36.50	27.78
North Atlantic Deep and Bottom (B)	2.5	34.90	27.86

symmetry predetermined by the geometry of T-S curves, cause the appearance of secondary and false T-S indexes on the T-S curves and change the percentage ratio of the water masses.

Let us explain the foregoing by the specific example of the mutual transformation of the Mediterranean Intermediate Water mass (M) with the overlying North Atlantic surface (A) and the underlying North Atlantic Deep and Bottom (B) Water masses; their thermohaline indexes are shown in Table XVII.

The T-S index of the first of these water masses was determined by the author and represents a zonal modification of the North Atlantic Water mass in the I section of the IV cruise of the R.V. "Mikhail Lomonosov" (Section 61); it fits into the generalized T-S diagram of the North Atlantic of Sverdrup (Sverdrup et al., 1942, p. 741). The second and third indexes are borrowed from Sverdrup and Jacobsen (1929).

If we consider the corresponding triangle of mixing (Figs. 25 and 110) in the field of isolines σ_T, it may be seen that the conditions of mixing of water mass A with water mass B on the one hand, and of water mass M with water mass B on the other, are not identical and, everything else being equal, depend on the degree of vertical stratification among the T-S indexes of the corresponding water masses. If it is assumed that the cores of water masses A, M and B lie respectively at depths of about 150, 1,200 and 2,500 m, which on the average corresponds to reality, the value of stability $10^8 E \approx 10^5 (d\sigma_T/dz)$ between water masses A and M will be equal to 70 CGS, and between water masses M and B to 6 CGS. Thus, vertical stability between masses A and M is almost twelve times more than between masses M and B. Consequently, mixing between water masses A and M, other things being equal (in particular, in the absence of horizontal advection), must also proceed twelve times less intensively than between water masses M and B.

The effect of stability on the form of the T-S curve is shown in Fig. 110, where, in the same triangle of mixing, a typical T-S curve is represented by a heavy line for the extensive region which lies to the southeast of the line connecting the southwestern extremity of Ireland with the Azores, and which has intermediate water mass M between masses A and B (see Fig. 65).

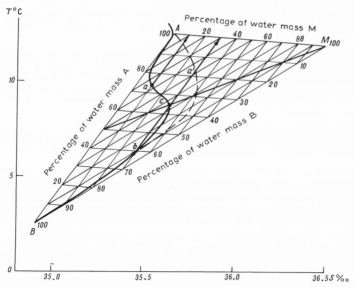

Fig. 110. Diagram of asymmetric T-S curve for the region where the transformation of the Mediterranean Intermediate Water mass takes place (Mamayev, 1960b).

The T-S curve in Fig. 110 is markedly asymmetrical as compared with the form which, within the given triangle, corresponds to the conditions of geometry of T-S curves (this form is shown by a dashed line). For this asymmetric curve, the theorems of the geometry of T-S curves are no longer valid. In particular, for points a and b, located at equal distance (in meters) from the core of intermediate water mass M, the percentage content of water mass M amounts respectively to 10 and 30%. In the case of a symmetric T-S curve these proportions for points of this type (a' and b) would be equal to 30 and 20%.

The mixing of water masses A and M is hindered by a kind of stratification screen ac, situated on the asymmetric T-S curve, between the salinity minimum in the area of point a and the salinity maximum in the area of point c. Branch Aa of the T-S curve gravitates towards straight line AB of mixing between the Atlantic surface and deep water masses, retaining the features of the T-S curves of those stations where the mixing of water masses A and B takes place without the participation of intermediate water mass M.

It follows from the foregoing that thermohaline index a is secondary (it may also be called false); it is not the index of any independent water mass as, for example, Jacobsen assumes, calling it "southern" water mass S (see Fig. 105). Fig. 111 * may serve as additional evidence of the "secondary" nature

---

* Fig. 111 was drawn up by V.A. Bubnov who kindly made it available to the author.

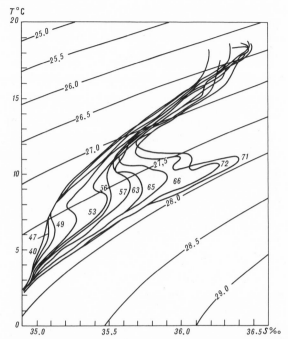

Fig. 111. *T-S* curves of R.V. "Chain" stations made in April–May 1959 in a section along 36°N (to the west of the Straits of Gibraltar).

of index a in Fig. 110, as Fig. 111 represents the T-S curves of stations of the R.V. "Chain", made in April–May 1959 in a section along 36°N lat. from the Straits of Gibraltar (station 71) to the west (station 40 on longitude 60°W). We see that as it moves away from the Straits of Gibraltar the extreme of the T-S curves, which corresponds to the core of the Mediterranean Water mass, is resolved, and together with it the secondary extreme (minimum) lying above it also disappears. The latter is the most pronounced in the immediate vicinity of the Straits of Gibraltar, in the same place where the Mediterranean maximum is also the most pronounced; it is clear that the Straits of Gibraltar cannot serve as the source of any other intermediate water mass than the Mediterranean. The salinity minimum is thus a secondary one. Let us note that the asymmetry of T-S curves with an intermediate water mass, considered above, is most marked in the whole World Ocean precisely in the region near Gibraltar.

On the other hand, the mixing of water masses M and B is assisted, apart from smaller vertical stability, by contraction on mixing also (this question was already considered in Section 55, and here we take it up again). Thus, for the region of the Straits of Gibraltar, where direct contact of undiluted Mediterranean water (of the "pure" type M) with water mass B takes place, the mixture of equal parts of both water masses, which is formed on the interface, has density $\sigma_T = 27.97$, which is 0.20 more than the density of water mass M and 0.12 more than the density of water mass B. Contraction on mixing reinforces the effect of stratification screen ac and stimulates a shift of the mixing reaction in the direction of water mass B. Apparently, it is precisely contraction on mixing which causes the unusually rapid transformation of core M into core c in the immediate proximity of the source of water mass M — the Straits of Gibraltar (to the nortwest of the boundary of Gibraltar region indicated above type M is no longer observed) (Mamayev, 1960b). Branch Bb of the T-S curve gravitates toward straight line MB of mixing between the Mediterranean and deep water masses (Fig. 110).

The question of the effect of contraction on mixing on the transformation of the Mediterranean Water mass studied by the author (his results were set forth in Section 55, as well as above in this section), was then examined at greater length by Kin'diushev (1965) and Bubnov (1967). The first of these authors constructed a number of triangles of contraction, using various zonal modifications of water mass A, and analyzed by means of these triangles the vertical distribution of contraction on mixing (from T-S curves of individual stations). His computations confirmed the conclusion drawn above (Section 55) from the consideration of Fig. 82: the maximum values $-10^5 \Delta\alpha = 10 \div 12$ are attained by contraction on mixing on the *lower* boundary of the Mediterranean Water mass.

Contraction on mixing of Mediterranean water with the deep waters of the North Atlantic was also considered by Bubnov (1967, 1971). Inasmuch as, and this was shown above, the maximum values ($\Delta\alpha = -14 \cdot 10^{-5}$) of contraction on mixing are observed on the lower boundary of the Mediterranean waters, while the contraction on mixing of the Mediterranean waters with the overlying surface Atlantic Water mass is extremely insignificant, Bubnov accordingly confined himself to studying contraction on mixing of the Mediterranean waters only with the waters underlying them.

Inasmuch as the main differences in T-S curves are observed in the surface parts, while below the core of the Mediterranean waters they are practically insignificant, in this case, as Bubnov showed, there is no need for division into zones for the purpose of constructing triangles of mixing (let us emphasize that this applies *only* to the region under consideration; due precisely to the foregoing, the triangle of contraction on mixing shown in Fig. 82 is extremely simple: the isolines of contraction on mixing are parallel to one of the sides). For the analysis of contraction on mixing Bubnov made use of real T-S curves *; for each T-S curve there was determined the T-S index of the modi-

---

* Real curves were processed for the region lying between 35 and 45°N 15°W and the coast; for the remaining part of the water area the T-S curves were averaged by 5° squares, while the value calculated for contraction referred to the center of the trapezium.

fied core of the Mediterranean Water mass, the T-S index of the core of water B and maximum contraction, corresponding to the mixing of the water masses in equal proportions (50% each), was calculated. Thus, contraction was determined from the straight line of mixing, while the point of the straight line corresponding to 50% content of each of the water masses, following the theory of T-S curves. accordingly corresponds to the lower boundary of the upper (Mediterranean) water mass. As a result, he constructed a chart of isolines of contraction on mixing, which in its main lines repeats the picture of salinity distribution in the core of the Mediterranean Water mass as well as the picture of the percentage content of the Mediterranean waters in their core.

The special features considered of the transformation of an intermediate water mass depend to a varying extent on the reciprocal position of the sides of the triangle of mixing and the isopycnals in the field of the T-S diagram and may also be observed in other cases. The example quoted of the effect of vertical stability and contraction on mixing on the mutual transformation of three water masses is the most striking. Generally speaking, however, a change in the sign of curvature of the T-S curve within the triangle of mixing may also fail to occur, but the effect of stability (with a correction introduced for contraction on mixing) causes a deviation of the T-S curve in the direction of one of the sides of the triangle of mixing, and also leads to uncertainty in finding the T-S index. This testifies to the more intensive interaction of the intermediate water mass with only one of the two water masses lying above or below it.

It follows from the phenomenon considered that it is necessary further to develop the theory of T-S curves taking into account the special features noted, as well as the inconstancy of the coefficients of turbulent exchange of momentum and turbulent diffusion along the vertical, particularly upward and downward from the core of the intermediate layer.

Above we have considered the vertical transformation of the intermediate water mass without allowing for horizontal advection. In particular, this was valid for a region of weak currents, lying near to the Iberian Peninsula and the Bay of Biscay. But the case of symmetric T-S curves in the field of graph $\sigma_T = f(T,S)$ is also quite possible, provided that symmetry "is restored" thanks to the existence of advection. In this case it is natural to assume that:

$$\bar{k}_{1-2} = \bar{k}_{2-3} = f(\overline{Ri}) \qquad [63.1]$$

where $\bar{k}$ is the mean coefficient of turbulence (for example, of diffusion) between surface 1 and intermediate 2, and between intermediate and deep 3 water masses:

$$Ri = \frac{g}{\rho} \frac{E}{\left(\dfrac{du}{dz}\right)^2}$$

is Richardson's number (u is the velocity of the mean current), whence:

$$\overline{Ri}_{1-2} = \overline{Ri}_{2-3} \qquad [63.2]$$

where the line signifies the averaging of the number Ri between the surface of the sea and the core of the intermediate water mass and the core of the intermediate water mass and the bottom of the sea. This assumption means that the different extent of the effect of stratification on mixing between the surface and intermediate water masses, on the one hand, and between the intermediate and the deep, on the other, is compensated by the effect of the velocity gradient. Let us assume that depth $z = H$ of surface of no motion coincides with the core of intermediate water mass 2. In this case expression [63.2] makes it possible to evaluate bottom ($z = B$) velocity u_B of an abyssal system of circulation. Assuming for simplicity that velocity and density are distributed with depth on a linear basis, we will obtain for velocity u_B from expression [63.2] the following simple expression:

$$u_B = u_0 \sqrt{\frac{\rho_B - \rho_H}{\rho_H - \rho_0} \left(\frac{B}{H} - 1\right)} \qquad [63.3]$$

where u_0 is the velocity of the current on the sea surface.

Thus, the nature of T-S curves must inevitably be linked also with circulation; it can provide a certain idea not only of the special features of the transformation of intermediate water masses but also, say, of the conditions of abyssal circulation of the World Ocean.

CONCLUSION

Our attention in the present monograph has been devoted mainly to the following questions of T-S analysis of the waters of the ocean: the *theoretical aspects*, which include the equation of state and questions of the thermodynamics of sea water, the analytical properties of the T-S diagram, including such an important phenomenon as contraction on mixing of waters; and finally, the analytical theories of T-S curves in an ocean of infinite and semi-infinite depth; the *practical aspects*, which include the most widespread methods of analysis of the interaction of water masses, wherein the methods of T-S analysis were considered as they apply to the main water masses of the ocean, which extend over the entire globe; the main emphasis was laid on the consideration of the intermediate water masses, which represent the most interesting and specific phenomenon in the thermohaline structure of the waters of the World Ocean.

T-S analysis, naturally, is not confined to this set of questions, and therefore it is interesting in conclusion briefly to consider the prospects for its development, and in so doing to touch upon those questions which are not reflected or are insufficiently reflected in the main body of the present work. In the author's opinion, the following questions present the greatest interest:

(1) *The theory of T-S curves and the geostrophic currents.* The first step in this direction was taken by Stommel (1962b), who examined the *cause* of the formation of a T-S curve, stable in form, in the field of geostrophic currents, provided the equation of state was linear, and who, in particular, made clear the connection between surface waters, characterized by a scatter of T-S points, and deep waters, determined by a single T-S curve. This theoretical aspect is very interesting from the point of view of the general theory of ocean currents. Even earlier, investigators had been attracted by the question of the connection between the stability of form of the T-S curve and the methods of calculating dynamic depths and heights. Here one may refer to the works of Thompson (1939), Stommel (1947) and Yasui (1955), devoted to the same question of the use of T-S curves for the calculation of dynamic heights. In particular, Yasui considers the question of the possibility of an approximation of the stable T-S curve in the Kuroshio region by a third-powered polynomial and of the subsequent elimination of salinity in dynamic computations. The

practical value of the recommendations proposed in these three works is not especially great at the present time, but this question is interesting from the point of view of principle.

(2) *The development of the theory of T-S curves for an ocean of finite depth taking account of advection.* The extension of the analytical theory of T-S curves to the case of an ocean of finite depth is of special interest for the analysis of waters in a comparatively shallow sea. As early as the work of Shtokman (1943a) and Ivanov (1949), a sea of finite depth was studied, but the results obtained by these authors were not carried to the point of practical application. In addition, taking account of the effect of currents on the possible forms of T-S relations are of exceptionally great importance, as was stated above at the end of the last chapter. The elementary theory of T-S curves for moving water masses in an ocean of finite depth, considered in Section 45, is naturally insufficient for a solution of this question, and therefore it is interesting to consider the solutions of equations [45.1] at different boundary conditions and variable (by depth) current velocity, and to present the interpretation of these solutions in the T-S plane.

(3) *T-S analysis of surface waters.* As has already been said, our attention in this work has been devoted mainly to the analysis of T-S curves relating to the principal oceanic structures; the upper layer of the ocean of approximately 100 m, subjected to considerable changes both of temperature and salinity, has somehow invariably eluded the attention of the authors who have investigated T-S relations. However, the study of the T-S relations of the surface waters – a kind of micro-T-S analysis – promises much that is interesting. Recently, studies have appeared devoted to "refined" T-S analysis and which have lead to new results. Thus, we may mention the works of Okuda (1962), Bary (1963), Bonchik (1967); they contain a T-S analysis of the surface waters of a region near the coast of Brazil, of the coastal waters of Great Britain and of the waters of the southern Baltic, respectively. In the work of Stommel and Fedorov (1966), the small-scale structure of the waters in the region of the islands of Timor and Mindanao is also considered by methods of T-S analysis. Finally, T-S analysis of waters in regions of interaction of sea and river waters may be of great interest (the author's attention was drawn to this aspect by A.M. Muromtsev). In these regions the stability of the saline composition is upset, and the equation of state deviates from the "oceanic" equation; the study of the deviations from the normal course of interaction of the mixing waters which occur in such cases is of great interest.

(4) *T-S analysis of the formation of bottom waters.* The methods of T-S analysis are highly effective for the study of processes of formation of bottom

waters of the ocean; however, there are few studies on this topic. Fofonoff (1956) shed light on the question of the effect of contraction on mixing on the formation of the Antarctic Bottom Water mass in the Weddell Sea. In the work of Bolin and Stommel (1961), the question of the origin of the Antarctic Bottom Waters and the Deep Waters of the Pacific Ocean is considered on the basis of a solution of systems of equations close to systems [30.7] and [30.9]. Finally, in an article by Worthington and Wright (1967), T-S analysis is carried out for cold waters in the entire North Atlantic, with a temperature of less than 4°. As a result of their investigation, the existence of five main sources of deep and bottom waters was ascertained, and a balance of the circulation of the cold waters of the North Atlantic was also drawn up. This balance was calculated using analysis (the values of potential temperature were utilized), the dynamic method and direct measurements of deep currents.

(5) *T-S analysis of convection processes*. The T-S diagram represents an effective instrument for the analysis of the processes of thermohaline convection, which occurs in the ocean on the destruction of stable stratification. Thus, Zubov's classical method for calculating vertical winter circulation is based on the use of the T-S diagram. A T-S diagram representing the course of vertical winter circulation, according to Zubov, is given in the author's monograph (Mamayev, 1963, fig. 27). Some simple problems of thermohaline convection according to Stommel (1961) and their representation in the T-S plane were considered in Section 46. Some other interesting conclusions, based on logical modelling of the process of convection on the T-S diagram, as well as on laboratory experiments, are also the work of the same author (Stommel, 1962a; Turner and Stommel, 1964). Another example of the use of the T-S diagram for the study of the process of convection may be observed in the work of Neumann and McGill (1961), relating to the waters of the Red Sea, the Gulf of Aden and the adjacent part of the Indian Ocean.

(6) *Development of the methods of T-S analysis*. The appearance in 1955 of the methods of statistical T-S analysis (Montgomery, 1955) showed that the "classical" method of analysis of individual T-S curves was not the only one; from this work of Montgomery a separate branch of theory arose, suitable for the analysis of waters interacting in any situation and lending itself to further improvement. Thus, the set of methods of T-S analysis considered in the present work is not finished, and new methods and procedures for the analysis of the interaction and transformation of waters must therefore be sought. As an example of some particular methods of analysis, which did not enter into the context of the present work, we may mention the so-called "equivalent-thickness method" of Jacobsen (1943; Defant, 1961), which makes it possible to determine from the straight line of mixing the thicknesses

which mixing water masses would have if inverse stratification of the waters occurred, and the "frequency method" of Rochford (1963), which supplements the "core method" of Wüst and makes it possible to clarify the position and form of the extreme of the intermediate water mass, by bringing into the analysis histograms of salinity distribution.

It is clear that the discovery of different new methods and procedures of T-S analysis, appropriate for the study of a particular situation arising in the interaction of water masses, is quite possible. The improvement of methods of instrumental observations in the sea should contribute substantially to the development of new methods and procedures.

We have quoted several examples of problems relating to the field of "pure" T-S analysis; it goes without saying that the points listed do not exhaust all the paths of its development. Without dwelling on other possible problems of "pure" T-S analysis, let us note in conclusion two interrelated problems in which, apart from the usual spatial and temporal parameters for T-S analysis (time, depth, etc.), there appear, as it were, "outside" parameters or numerical indicators. Such parameters may be, for example, dissolved oxygen content, nutrient elements and other hydrochemical indicators, or such indicators as plankton content. The appearance of additional parameters means that, on the one hand, they too may be plotted on the T-S diagram, and that, on the other hand, other characteristic diagrams may be plotted in addition to the T-S diagram. Some of the interrelated problems mentioned are as follows:

(7) *The comparison of the T-S diagram with other characteristic diagrams.* If dissolved oxygen content is considered as an "additional" parameter, then, apart from the T-S diagram, it is interesting to consider the following characteristic diagrams: σ_T–O_2, T–O_2 diagram, diagram and S–O_2 diagram. In the oceanographic literature there is a small number of studies in which such additional nomograms are considered and used for the analysis of waters. As an example, let us mention the work of Rochford (1966a), where σ_T–O_2 diagrams are used for the analysis of the origin of the oxygen minimum in the Indian Ocean. It is interesting that salinity is considered as a parameter "inside" the diagram in these diagrams of Rochford. Of no less interest to analysis may be diagrams where other indicators are used as an additional parameter, say, nitrate content (see, for example, Rochford, 1966b).

(8) *The development of T-S-P analysis.* Recently a distinctive branch of T-S analysis has arisen in which plankton content (P) in the water layers (along the vertical) appears as an additional parameter on the T-S diagram. T-S-P diagrams for the eastern regions of the North Atlantic and for the region to the south of New Zealand were constructed by Bary (1959; see also Bary, 1963). There are also only a few studies in this direction. The cycle

of studies of Yashnov (1966) is of substantial interest for the subsequent development of *T-S* analysis. *T-S* diagrams may also be used for other hydrobiological indicators; as a curious example, let us note the work of Kingsman (1964), which gives a *T-S* diagram of the tolerance of reef coral in the Indian Ocean. The boundaries of the optimal temperature and salinity region are plotted on this diagram, as well as the boundaries of the various admissible limits within which the growth of reef coral is possible.

Such, in the opinion of the author, are the principal aspects of the further development of the thermohaline analysis of the natural waters of the World Ocean and its seas.

APPENDIX

TABLE A1

Vertical stratification of the standard ocean
($T = 0°$C, $S = 35°/oo$)
(Defant, 1961)

Pressure (dbar)	Geom. depth (m)	Dynamic depth (dyn.m)	Spec. volume ($10^5 \alpha$)	Density (σ_t)	Dynamic depth (dyn.m)	Pressure (dbar)
0	0	0	97.264	28.23	0	0
100	99.24	97.242	97.219	28.61	100	102.837
200	198.45	194.438	97.174	29.12	200	205.724
300	297.60	291.590	97.129	29.64	300	308.659
400	396.71	388.696	97.084	30.03	400	411.643
500	495.78	485.758	97.040	30.50	500	514.677
600	594.80	582.776	96.995	31.02	600	617.758
700	693.77	679.749	96.951	31.45	700	720.889
800	792.69	776.678	96.901	31.92	800	824.068
900	890.57	873.564	96.863	32.41	900	927.296
1000	984.41	970.404	96.819	32.85	1000	1030.572
1500	1482.97	1453.955	96.602	35.17	1500	1547.696
2000	1975.43	1936.429	96.388	37.47	2000	2065.967
2500	2465.96	2417.836	96.177	39.75	2500	2585.445
3000	2956.20	2898.204	95.970	41.99	3000	3106.094
3500	3445.55	3377.544	95.766	44.21	3500	3627.903
4000	3932.89	3855.873	95.566	46.40	4000	4150.862
5000	4904.57	4809.556	95.173	50.72	5000	5200.185
6000	5873.38	5759.368	94.791	54.95	6000	6253.981
7000	6836.43	6705.421	94.421	59.08	7000	7312.174
8000	7796.89	7647.817	94.060	63.15	8000	8374.688

Table A2 represents the coordinates of the isosteres on the T-S diagram (Fig. 1), in multiples of whole values of v_T, and the coordinates of isotherms on the v_T-S diagram (Fig. 3), in multiples of whole values of a degree. In the first case, the coordinates of each of the v_T lines are represented by whole values of temperature (extreme left column) and by the corresponding fractional values of salinity. In the second case, the coordinates of each of the isotherms are represented by whole values of v_T (heading of the table) and by the corresponding fractional values of salinity. The precision of the table makes it possible to plot large-scale graphs (for example, a T-S diagram in the scale: $1°/oo - 50$ cm, $1° - 5$ or 10 cm). The isolines of tenths (and hundredths) of values of v_T and T are plotted with a sufficient degree of accuracy between the isolines of whole values of these quantities by means of linear interpolation (Mamayev, 1954; Callaway, 1951).

TABLE A2

Function $S = f(v_T, T)$

$T(°C)$ \ v_T	101	100	99	98	97	96	95
30	4.418	5.761	7.108	8.459	9.813	11.170	12.530
29	4.011	5.351	6.695	8.042	9.393	10.747	12.104
28	–	4.954	6.296	7.641	8.989	10.341	11.695
27	–	4.572	5.910	7.252	8.597	9.946	11.301
26	–	4.200	5.536	6.875	8.218	9.564	10.914
25	–	3.844	5.177	6.513	7.853	9.197	10.544
24	–	–	4.832	6.165	7.502	8.843	10.187
23	–	–	4.500	5.831	7.166	8.504	9.846
22	–	–	4.184	5.511	6.842	8.177	9.515
21	–	–	3.882	5.206	6.534	7.866	9.201
20	–	–	–	4.915	6.240	7.569	8.901
19	–	–	–	4.639	5.961	7.286	8.615
18	–	–	–	4.378	5.696	7.018	8.344
17	–	–	–	4.133	5.447	6.766	8.088
16	–	–	–	3.903	5.214	6.529	7.848
15	–	–	–	–	4.996	6.308	7.623
14	–	–	–	–	4.795	6.102	7.413
13	–	–	–	–	4.608	5.912	7.220
12	–	–	–	–	4.438	5.738	7.042
11	–	–	–	–	4.286	5.582	6.882
10	–	–	–	–	4.151	5.443	6.738
9	–	–	–	–	4.034	5.321	6.612
8	–	–	–	–	3.933	5.216	6.503
7	–	–	–	–	3.851	5.129	6.411
6	–	–	–	–	3.788	5.061	6.339
5	–	–	–	–	3.743	5.011	6.283
4	–	–	–	–	3.717	4.981	6.248
3	–	–	–	–	3.712	4.970	6.232
2	–	–	–	–	3.726	4.979	6.236
1	–	–	–	–	3.760	5.007	6.258
0	–	–	–	–	3.815	5.056	6.301
−1	–	–	–	–	3.892	5.127	6.367
−2	–	–	–	–	–	–	–

TABLE A2 (continued)

$T(^{\circ}C)$ / vT	94	93	92	91	90	89	88
30	13.892	15.257	16.625	17.996	19.370	20.746	22.125
29	13.465	14.829	16.195	17.564	18.935	20.309	21.685
28	13.053	14.413	15.777	17.194	18.514	19.886	21.260
27	12.657	14.015	15.376	16.740	18.106	19.475	20.846
26	12.267	13.623	14.982	16.344	17.708	19.075	20.445
25	11.894	13.248	14.605	15.964	17.326	18.690	20.057
24	11.534	12.885	14.239	15.596	16.956	18.319	19.684
23	11.190	12.538	13.889	15.243	16.600	17.960	19.323
22	10.857	12.202	13.551	14.903	16.258	17.615	18.975
21	10.540	11.882	13.227	14.576	15.927	17.281	18.638
20	10.237	11.576	12.918	14.263	15.612	16.964	18.318
19	9.948	11.284	12.623	13.966	15.311	16.660	18.011
18	9.674	11.006	12.342	13.682	15.024	16.369	17.717
17	9.414	10.743	12.076	13.412	14.751	16.093	17.438
16	9.169	10.495	11.824	13.157	14.493	15.832	17.174
15	8.941	10.263	11.589	12.918	14.250	15.585	16.924
14	8.728	10.046	11.368	12.694	14.023	15.355	16.690
13	8.531	9.846	11.165	12.487	13.812	15.140	16.471
12	8.350	9.661	10.976	12.294	13.615	14.939	16.267
11	8.185	9.492	10.803	12.117	13.434	14.754	16.077
10	8.037	9.340	10.646	11.955	13.268	14.584	15.903
9	7.906	9.204	10.506	11.812	13.121	14.433	15.748
8	7.793	9.087	10.384	11.685	12.989	14.297	15.608
7	7.697	8.987	10.280	11.576	12.876	14.179	15.485
6	7.620	8.905	10.193	11.485	12.780	14.078	15.379
5	7.559	8.839	10.122	11.409	12.700	13.994	15.291
4	7.519	8.794	10.073	11.355	12.640	13.929	15.221
3	7.498	8.768	10.041	11.318	12.598	13.882	15.169
2	7.497	8.761	10.028	11.300	12.575	13.853	15.136
1	7.513	8.772	10.035	11.301	12.571	13.844	15.121
0	7.554	8.806	10.062	11.323	12.587	13.855	15.126
−1	7.611	8.859	10.110	11.365	12.623	13.885	15.150
−2	−	−	−	−	−	−	−

TABLE A2 (continued)

$T(°C)$ \ vT	87	86	85	84	83	82	81
30	23.505	24.887	26.271	27.656	29.043	30.431	31.821
29	23.064	24.444	25.826	27.210	28.596	29.983	31.372
28	22.636	24.015	25.395	26.778	28.162	29.548	30.935
27	22.220	23.596	24.975	26.356	27.739	29.123	30.509
26	21.818	23.193	24.570	25.949	27.329	28.712	30.096
25	21.427	22.800	24.175	25.552	26.931	28.311	29.693
24	21.052	22.422	23.794	25.169	26.545	27.922	29.302
23	20.688	22.056	23.426	24.798	26.172	27.548	28.926
22	20.337	21.702	23.069	24.439	25.811	27.184	28.560
21	19.999	21.361	22.726	24.093	25.463	26.835	28.208
20	19.675	21.035	22.397	23.762	25.129	26.498	27.869
19	19.365	20.722	22.081	23.443	24.807	26.174	27.542
18	19.068	20.421	21.778	23.137	24.498	25.861	27.226
17	18.786	20.137	21.490	22.846	24.204	25.564	26.927
16	18.519	19.866	21.216	22.569	23.924	25.282	26.641
15	18.266	19.611	20.958	22.307	23.659	25.012	26.368
14	18.027	19.367	20.710	22.056	23.405	24.757	26.110
13	17.805	19.142	20.481	21.824	23.169	24.516	25.865
12	17.597	18.930	20.266	21.604	22.945	24.288	25.634
11	17.403	18.732	20.064	21.398	22.736	24.076	25.419
10	17.255	18.550	19.879	21.210	22.544	23.880	25.218
9	17.066	18.387	19.712	21.039	22.368	23.699	25.033
8	16.922	18.239	19.558	20.881	22.206	23.534	24.864
7	16.795	18.108	19.423	20.741	22.062	23.385	24.711
6	16.684	17.992	19.303	20.616	21.933	23.253	24.574
5	16.591	17.894	19.200	20.509	21.821	23.136	24.453
4	16.516	17.814	19.116	20.420	21.727	23.037	24.348
3	16.459	17.752	19.049	20.348	21.650	22.955	24.262
2	16.421	17.709	19.000	20.294	21.591	22.891	24.193
1	16.401	17.683	18.969	20.258	21.550	22.845	24.142
0	16.400	17.678	18.959	20.243	21.528	22.817	24.109
−1	16.419	17.691	18.966	20.244	21.525	22.809	24.095
−2	–	–	18.995	20.267	21.542	22.820	24.101

TABLE A2 (continued)

$T(°C)$ vT	80	79	78	77	76	75	74
30	33.212	34.604	35.997	37.391	38.786	40.181	–
29	32.763	34.154	35.546	36.939	38.333	39.727	41.122
28	32.323	33.713	35.104	36.496	37.889	39.282	40.676
27	31.896	33.285	34.675	36.066	37.457	38.849	40.242
26	31.481	32.867	34.255	35.644	37.035	38.427	39.819
25	31.077	32.462	33.849	35.237	36.626	38.016	39.406
24	30.684	32.068	33.453	34.839	36.227	37.616	39.005
23	30.306	31.687	33.070	34.454	35.839	37.226	38.614
22	29.938	31.317	32.698	34.080	35.464	36.848	38.234
21	29.583	30.960	32.339	33.719	35.101	36.483	37.867
20	29.242	30.616	31.992	33.370	34.749	36.129	37.510
19	28.912	30.284	31.657	33.032	34.409	35.787	37.166
18	28.594	29.964	31.335	32.708	34.082	35.457	36.834
17	28.292	29.659	31.027	32.397	33.768	35.141	36.515
16	28.002	29.365	30.730	32.097	32.466	34.837	36.208
15	27.726	29.086	30.448	31.813	33.179	34.546	35.914
14	27.464	28.821	30.180	31.541	32.904	34.268	35.633
13	27.217	28.571	29.927	31.284	32.643	34.003	35.365
12	26.982	28.332	29.684	31.038	32.394	33.751	35.110
11	26.764	28.111	29.459	30.809	32.161	33.515	34.871
10	26.558	27.901	29.246	30.593	31.942	33.292	34.644
9	26.369	27.708	29.049	30.392	31.737	33.084	34.432
8	26.197	27.532	28.869	30.208	31.549	32.891	34.235
7	26.039	28.369	28.702	30.037	31.373	32.711	34.051
6	25.898	27.224	28.552	29.882	31.214	32.548	33.884
5	25.772	27.093	28.417	29.743	31.071	32.401	33.732
4	25.663	26.981	28.300	29.622	30.945	32.270	33.597
3	25.572	26.884	28.199	29.516	30.835	32.156	33.478
2	25.498	26.805	28.115	29.427	30.741	32.057	33.375
1	25.442	26.744	28.049	29.356	30.665	31.976	33.289
0	25.404	26.701	28.001	29.303	30.608	31.914	33.222
−1	25.384	26.676	27.970	29.267	30.566	31.867	33.170
−2	25.385	26.671	27.960	29.251	30.546	31.842	33.140

TABLE A2 (continued)

$T(°C)$ \ vT	73	72	71	70	69	68
30	–	–	–	–	–	–
29	–	–	–	–	–	–
28	–	–	–	–	–	–
27	–	–	–	–	–	–
26	41.212	–	–	–	–	–
25	40.797	–	–	–	–	–
24	40.395	–	–	–	–	–
23	40.003	–	–	–	–	–
22	39.620	41.007	–	–	–	–
21	39.251	40.636	–	–	–	–
20	38.892	40.275	–	–	–	–
19	38.546	39.927	41.308	–	–	–
18	38.212	39.590	40.969	–	–	–
17	37.890	39.266	40.643	–	–	–
16	37.580	38.953	40.328	–	–	–
15	37.284	38.654	40.025	–	–	–
14	36.999	38.367	39.736	41.106	–	–
13	36.728	38.093	39.459	40.825	–	–
12	36.470	37.831	39.193	40.557	–	–
11	36.227	37.584	38.943	40.303	–	–
10	35.997	37.351	38.706	40.062	–	–
9	35.781	37.131	38.483	39.836	41.190	–
8	35.580	36.926	38.274	39.623	40.973	–
7	35.393	36.736	38.080	39.425	40.771	–
6	35.221	36.560	37.900	39.242	40.584	–
5	35.065	36.400	37.736	39.073	40.412	–
4	34.926	36.256	37.588	38.921	40.255	–
3	34.803	36.129	37.456	38.784	40.114	–
2	34.695	36.017	37.340	38.664	39.989	41.315
1	34.605	35.922	37.241	38.561	39.880	41.202
0	34.532	35.843	37.157	38.472	39.788	41.105
−1	34.476	35.782	37.090	38.400	39.712	41.025
−2	34.439	35.741	37.045	38.350	39.656	40.964

TABLE A3

Function $\Delta_{ST} = f(\sigma T)$
(Sverdrup et al., 1942)

σT	0.00	0.01	0.02	0.03	0.04	0.05	0.06	0.07	0.08	0.09
23.0	488	487	486	485	484	483	482	481	480	479
23.1	478	477	476	475	474	473	472	472	471	470
23.2	469	468	467	466	465	464	463	462	461	460
23.3	459	458	457	456	455	454	453	452	452	451
23.4	450	449	448	447	446	445	444	443	442	441

TABLE A3 (continued)

σT	0.00	0.01	0.02	0.03	0.04	0.05	0.06	0.07	0.08	0.09
23.5	440	439	438	437	436	435	434	433	432	431
23.6	430	430	429	428	427	426	425	424	423	422
23.7	421	420	419	418	417	416	415	414	413	412
23.8	411	410	410	409	408	407	406	405	404	403
23.9	402	401	400	399	398	397	396	395	394	393
24.0	392	391	390	389	388	387	386	386	385	384
24.1	383	382	381	380	379	378	377	376	375	374
24.2	373	372	371	370	369	368	367	366	366	365
24.3	364	363	362	361	360	359	358	357	356	355
24.4	354	353	352	351	350	349	348	347	346	346
24.5	345	344	343	342	341	340	339	338	337	336
24.6	335	334	333	332	331	330	329	328	327	326
24.7	325	324	324	323	322	321	320	319	318	317
24.8	316	315	314	313	312	311	310	309	308	307
24.9	306	306	305	304	303	302	301	300	299	298
25.0	297	296	295	294	293	292	291	290	289	288
25.1	287	286	286	285	284	283	282	281	280	279
25.2	278	277	276	275	274	273	272	271	270	269
25.3	268	267	266	266	265	264	263	262	261	260
25.4	259	258	257	256	255	254	253	252	251	250
25.5	249	248	248	247	246	245	244	243	242	241
25.6	240	239	238	237	236	235	234	233	232	231
25.7	230	229	228	228	227	226	225	224	223	222
25.8	221	220	219	218	217	216	215	214	213	212
25.9	211	210	210	209	208	207	206	205	204	203
26.0	202	201	200	199	198	197	196	195	194	193
26.1	192	191	190	190	189	188	187	186	185	184
26.2	183	182	181	180	179	178	177	176	175	174
26.3	173	172	171	170	170	169	168	167	166	165
26.4	164	163	162	161	160	159	158	157	156	155
26.5	154	154	153	152	151	150	149	148	147	146
26.6	145	144	143	142	141	140	139	138	137	136
26.7	135	134	134	133	132	131	130	129	128	127
26.8	126	125	124	123	122	121	120	119	118	117
26.9	116	116	115	114	113	112	111	110	109	108
27.0	107	106	105	104	103	102	101	100	99	98
27.1	98	97	96	95	94	93	92	91	90	89
27.2	88	87	86	85	84	83	82	81	80	79
27.3	78	78	77	76	75	74	73	72	71	70
27.4	69	68	67	66	65	64	63	62	61	60
27.5	60	59	58	57	56	55	54	53	52	51
27.6	50	49	48	47	46	45	44	44	43	42
27.7	41	40	39	38	37	36	35	34	33	32
27.8	31	30	29	28	27	26	26	25	24	23
27.9	22	21	20	19	18	17	16	15	14	13
28.0	12	11	10	9	8	8	7	6	5	4

TABLE A4

Function $10^4(\partial v_T/\partial T) = f(S, T)$

$T(°C)$ \ $S(°/oo)$	0	5	10	15	20	25	30	35	40
-2	-1057	-843	-640	-447	-265	-92	71	225	371
0	-679	-483	-297	-121	47	-206	358	501	637
2	-326	-148	22	183	337	484	623	757	884
4	5	166	320	467	608	743	872	995	1113
6	314	461	601	735	863	986	1105	1218	1327
8	606	738	865	987	1104	1216	1325	1429	1529
10	882	1001	1116	1226	1332	1434	1533	1628	1720
12	1114	1251	1354	1454	1550	1642	1732	1818	1902
14	1392	1489	1582	1671	1758	1841	1922	2000	2075
16	1629	1716	1800	1880	1958	2033	2105	2174	2242
18	1856	1934	2009	2081	2151	2217	2282	2343	2403
20	2073	2144	2211	2276	2338	2397	2453	2507	2559
22	2281	2346	2407	2465	2520	2571	2620	2667	2711
24	2482	2542	2597	2649	2697	2742	2784	2823	2859
26	2676	2731	2782	2828	2871	2910	2945	2977	3005
28	2864	2916	2962	3004	3042	3075	3104	3129	3150
30	3046	3095	3139	3177	3210	3238	3261	3280	3292

TABLE A5

Function $-10^4(\partial v_T/\partial S) = f(S, T)$

$T(°C)$ \ $S(°/oo)$	0	5	10	15	20	25	30	35	40
-2	8235	8121	8020	7931	7853	7786	7729	7683	7646
0	8150	8042	7945	7859	7785	7721	7668	7625	7591
2	8074	7969	7875	7793	7722	7662	7611	7571	7540
4	8004	7903	7812	7733	7665	7607	7559	7521	7492
6	7941	7843	7755	7678	7612	7556	7510	7474	7448
8	7885	7788	7703	7628	7564	7510	7466	7432	7407
10	7833	7739	7655	7583	7520	7468	7425	7392	7369
12	7787	7695	7613	7541	7480	7430	7388	7357	7335
14	7746	7655	7574	7504	7445	7395	7355	7324	7303
16	7708	7619	7540	7471	7412	7364	7325	7295	7275
18	7675	7586	7508	7441	7384	7336	7298	7270	7250
20	7644	7557	7481	7414	7358	7312	7275	7247	7229
22	7617	7531	7455	7390	7335	7290	7254	7228	7210
24	7591	7507	7433	7369	7315	7271	7236	7211	7195
26	7568	7485	7413	7350	7298	7255	7222	7197	7182
28	7545	7465	7394	7333	7283	7241	7210	7187	7173
30	7524	7446	7377	7318	7269	7230	7200	7179	7167

TABLE A6

$\cot\varphi = (\partial S/\partial T)_\rho = f(S,T)$

$T(°C)$ \ $S(°/oo)$	0	5	10	15	20	25	30	35	40
−2	+0.129	+0.104	+0.080	+0.056	+0.034	+0.012	−0.009	−0.029	−0.049
0	+0.082	+0.060	+0.037	+0.015	−0.006	−0.027	−0.047	−0.006	−0.084
2	+0.040	+0.018	−0.003	−0.024	−0.044	−0.063	−0.082	−0.100	−0.117
4	−0.001	−0.021	−0.041	−0.061	−0.079	−0.098	−0.115	−0.132	−0.149
6	−0.039	−0.059	−0.077	−0.096	−0.113	−0.131	−0.147	−0.163	−0.178
8	−0.077	−0.095	−0.112	−0.129	−0.146	−0.162	−0.177	−0.192	−0.206
10	−0.113	−0.129	−0.146	−0.162	−0.177	−0.192	−0.206	−0.220	−0.233
12	−0.147	−0.163	−0.178	−0.193	−0.207	−0.211	−0.234	−0.247	−0.259
14	−0.180	−0.194	−0.209	−0.223	−0.236	−0.249	−0.261	−0.273	−0.284
16	−0.211	−0.225	−0.239	−0.252	−0.264	−0.276	−0.287	−0.298	−0.308
18	−0.242	−0.255	−0.268	−0.280	−0.291	−0.302	−0.313	−0.322	−0.331
20	−0.271	−0.284	−0.296	−0.307	−0.318	−0.328	−0.337	−0.346	−0.354
22	−0.300	−0.312	−0.323	−0.334	−0.344	−0.353	−0.361	−0.369	−0.376
24	−0.327	−0.338	−0.349	−0.360	−0.369	−0.377	−0.385	−0.392	−0.397
26	−0.354	−0.365	−0.375	−0.385	−0.393	−0.401	−0.408	−0.414	−0.418
28	−0.380	−0.391	−0.401	−0.410	−0.418	−0.425	−0.431	−0.435	−0.439
30	−0.406	−0.416	−0.426	−0.434	−0.442	−0.448	−0.453	−0.457	−0.459

TABLE A7

Function $10^4\left[\left(\dfrac{\partial\rho}{\partial S}\,ds + \dfrac{\partial\rho}{\partial T}\,dT\right) - (\beta\Delta S + \alpha\Delta T)\right] = f(S,T)$

$\Delta S = 1°/oo, \quad \Delta T = 1°C, \quad \beta = 8\cdot10^{-4}; \quad \alpha = -2\cdot10^{-4}$

$T(°C)$ \ $S(°/oo)$	0	5	10	15	20	25	30	35	40
−2	3.291	3.033	2.797	2.580	2.380	2.198	2.032	1.883	1.752
0	2.817	2.593	2.375	2.173	1.988	1.819	1.667	1.530	1.409
5	1.803	1.617	1.437	1.272	1.121	0.983	0.858	0.747	0.649
10	0.946	0.787	0.639	0.503	0.379	0.267	0.167	0.080	0.008
15	0.199	0.068	−0.055	−0.167	−0.268	−0.358	−0.437	−0.504	−0.560
20	−0.451	−0.565	−0.669	−0.762	−0.844	−0.915	−0.974	−1.022	−1.059
25	−1.025	−1.133	−1.224	−1.302	−1.368	−1.422	−1.463	−1.494	−1.510
30	−1.568	−1.655	−1.735	−1.802	−1.854	−1.893	−1.918	−1.928	

TABLE A8

Equation of state of water and sea water in Tumlirtz form
$[p + p_0(T,S)][\alpha - \alpha_0(T,S)] = \lambda(T,S)$
(numerical values of constants)

	α_0	p_0	λ
Eckart (1958)	0.6980	$5890 + 38T - 0.375T^2 + 3S$	$10^2(17.795 + 0.1125T - 0.000745T^2 - 0.0380S - 0.0001TS)$
Wilson and Bradley (1968)	0.70200	$5880.9 + 37.592T - 0.34395T^2 + 2.2524S$	$10^2(17.5273 + 0.1101T - 0.000639T^2 - 0.039986S - 0.000107TS)$
Fisher et al. (1969)	$0.6980 - 0.7435626 \cdot 10^{-3}\,T$ $+ 0.3704258 \cdot 10^{-4}T^2 - 0.6315724 \cdot 10^{-6}T^3$ $+ 0.9829576 \cdot 10^{-8}T^4 - 0.1197269 \cdot 10^{-9}T^5$ $+ 0.1005461 \cdot 10^{-11}T^6 - 0.5437898 \cdot 10^{-14}T^7$ $+ 0.169946 \cdot 10^{-16}T^8 - 0.2295063 \cdot 10^{-19}T^9$ $- (2.679 \cdot 10^{-4} + 2.02 \cdot 10^{-6}T - 6.0 \cdot 10^{-9}T^2)S$	$5918.499 + 58.05267T$ $-1.1253317T^2 + 6.6123869 \cdot 10^{-3}T^3$ $-1.4661625 \cdot 10^{-5}T^4 + (10.874$ $-4.1384 \cdot 10^{-2}T)S$	$1788.316 + 21.55053T$ $-0.4695911T^2 + 3.096363 \cdot 10^{-3}T^3$ $-0.7341182 \cdot 10^{-5}T^4 *$

* Here λ is the function of T only.

TABLE A9

Equation of state of sea water in polynomial form

(A) *Sea water at atmospheric pressure (Cox et al., 1970)*

$$\sigma_T(T,S) = \sum_{ij} a_{i,j} T^i S^j \qquad\qquad 0 \leqslant i,j \leqslant 3$$

$$= a_{0,0} + a_{1,0}T + a_{0,1}S + a_{2,0}T^2 + a_{1,1}ST + a_{0,2}S^2$$
$$+ a_{3,0}T^3 + a_{2,1}ST^2 + a_{1,2}S^2T + a_{0,3}S^3$$

The polynomial coefficients a_i, j

i	j	a_{ij}
0	0	$8.00969062 \cdot 10^{-2}$
	1	$7.97018644 \cdot 10^{-1}$
	2	$1.31710842 \cdot 10^{-4}$
	3	$-6.11831499 \cdot 10^{-3}$
1	0	$5.88194023 \cdot 10^{-2}$
	1	$-3.25310441 \cdot 10^{-3}$
	2	$2.87971530 \cdot 10^{-6}$
2	0	$-8.11465413 \cdot 10^{-3}$
	1	$3.89187483 \cdot 10^{-5}$
3	0	$4.76600414 \cdot 10^{-5}$

The formula is valid over the following ranges: $0 \leqslant T \leqslant 25°$; $9 \leqslant S \leqslant 41°/oo$

(B) *Sea water in situ (Crease, 1962)*

$$\alpha(p,T,S) = \sum_{i,j,k} a_{i,j,k} p^i T^j (S-35)^k \qquad\qquad 0 \leqslant i,j,k \leqslant 4$$

The polynomial coefficients a_i, j, k (expressed in floating decimals, exponent following the comma)

i	j	k	a_{ijk}	i	j	k	a_{ijk}
0	0	0	9.727149, -1	1	0	0	-4.42003, -5
		1	-7.6268, -4			1	1.5221, -7
		2	3.934, -7			2	-2.490, $-10*$
		3	-5.54, $-9*$			3	6.6 $-13*$
		4	2.55, $-11*$		1	0	2.5962, -7
	1	0	5.037, -5			1	-2.3956, $-9*$
		1	2.8892, -6			2	5.70, $-12*$
		2	-1.056, $-8*$		2	0	-4.8964, -9
		3	1.84, $-11*$			1	3.127, $-11*$
	2	0	6.4656, -6		3	0	2.556, -11
		1	-5.827, -8	2	0	0	6.7845, -9
		2	1.307, $-10*$			1	-3.348, -11
	3	0	-6.532, -8		1	0	-7.954, -11
		1	4.001, -10		2	0	1.35, -12
	4	0	6.201, -10	3	0	0	-0.297, -13
					1	0	6.52, -15
				4	0	0	1.076, -16

* The starred coefficients are relevant only to salinities below $30°/oo$.

The formula is valid over the following ranges:
(1) $p < 100 \text{ kg/cm}^2$, $5°/oo < S < 37°/oo$, freezing point $< T < 30°C$.
(2) $p < 500 \text{ kg/cm}^2$, $33°/oo < S < 37°/oo$, freezing point $< T < 30°C$.
(3) $p < 1000 \text{ kg/cm}^2$, $33°/oo < S < 37°/oo$, freezing point $< T < 6°C$.

TABLE A10

Approximation of the Knudsen–Ekman equation of state in situ

(A)   *By Bryan and Cox (1972)*

$$\sigma_T(T, S, Z) - \sigma_{T,0}(T_0, S_0, Z) = X_1(T-T_0) + X_2(S-S_0) + X_3(T-T_0)^2 \qquad [X.1]$$

Constants in the polynomial formula [X.1]

Z	$(\rho_0-1)10^{-3}$	T_0	S_0	X_1	X_2	X_3
0	24.458	13.50	32.600	−0.19494−00	0.77475−00	−0.49038−02
250	28.475	8.50	35.150	−0.15781−00	0.78318−00	−0.52669−02
500	29.797	6.00	34.900	−0.13728−00	0.78650−00	−0.55278−02
750	31.144	4.50	34.900	−0.12720−00	0.78807−00	−0.56610−02
1000	32.236	4.00	34.750	−0.12795−00	0.78710−00	−0.56274−02
1250	33.505	3.00	34.750	−0.12312+00	0.78763−00	−0.56972−02
1500	34.808	2.00	34.800	−0.11837+00	0.78822−00	−0.57761−02
1750	35.969	1.50	34.750	−0.11896+00	0.78751−00	−0.57631−02
2000	37.143	1.50	34.800	−0.12543−00	0.78560−00	−0.56422−02
2250	38.272	1.50	34.800	−0.13168−00	0.78368−00	−0.55239−02
2500	39.462	1.00	34.800	−0.13250−00	0.78300−00	−0.55116−02
2750	40.582	1.00	34.800	−0.13871−00	0.78109−00	−0.53946−02
3000	41.695	1.00	34.800	−0.14483−00	0.77920−00	−0.52793−02
3250	42.801	1.00	34.800	−0.15088−00	0.77733−00	−0.51654−02
3500	43.863	1.00	34.750	−0.15673−00	0.77544−00	−0.50557−02
3750	45.038	0.50	34.750	−0.15771−00	0.77475−00	−0.50466−02
4000	46.130	0.50	34.750	−0.16363−00	0.77292−00	−0.49360−02
4250	47.216	0.50	34.750	−0.16948−00	0.77110−00	−0.48268−02
4500	48.296	0.50	34.750	−0.17524−00	0.76930−00	−0.47193−02
4750	49.278	1.00	34.750	−0.18556−00	0.76641−00	−0.45102−02
5000	50.344	1.00	34.750	−0.19107−00	0.76467−00	−0.44074−02
5250	51.404	1.00	34.750	−0.19650−00	0.76295−00	−0.43061−02
5500	52.459	1.00	34.750	−0.20186−00	0.76126−00	−0.42068−02
5750	53.508	1.00	34.750	−0.20715−00	0.75958−00	−0.41138−02
6000	54.552	1.00	34.750	−0.21237−00	0.75792−00	−0.40172−02

(B)   *By Friedrich and Levitus (1972)*

Formula for the upper layer of the ocean ($0 < Z < 2$ km):

$$\sigma_T(T, S) = C_1 + C_2 T + C_3 S + C_4 T^2 + C_5 ST + C_6 T^3 + C_2 ST^2 \qquad [X.2]$$

Formula for the lower layer of the ocean ($Z \geqslant 2$ km):

$$\sigma_T(T, S) = C_1 + C_2 T + C_3 S + C_4 T^2 + C_5 ST \qquad [X.3]$$

Coefficients of the approximation, $C_N(Z) = \alpha_N + \beta_N Z + \gamma_N Z^2$, for the formula [X.2]

N	α_N	β_N	γ_N
1	$-7.2169 \cdot 10^{-2}$	$+5.1215 \cdot 10^{-0}$	$-5.012 \cdot 10^{-2}$
2	$+4.9762 \cdot 10^{-2}$	$-3.6349 \cdot 10^{-2}$	$+7.853 \cdot 10^{-4}$
3	$+8.0560 \cdot 10^{-1}$	$-8.5540 \cdot 10^{-3}$	$+1.070 \cdot 10^{-4}$
4	$-7.5911 \cdot 10^{-3}$	$+6.4295 \cdot 10^{-4}$	$-1.397 \cdot 10^{-5}$
5	$-3.0063 \cdot 10^{-3}$	$+1.9365 \cdot 10^{-4}$	$-3.899 \cdot 10^{-6}$
6	$+3.5187 \cdot 10^{-5}$	$-3.9740 \cdot 10^{-6}$	$-5.695 \cdot 10^{-8}$
7	$+3.7297 \cdot 10^{-5}$	$-2.8108 \cdot 10^{-6}$	$+1.147 \cdot 10^{-7}$

TABLE A10 (continued)

Coefficients of the approximation, $C_N(Z) = \alpha_N + \beta_N Z + \gamma_N Z^2$, for the formula [X.3]

N	α_N	β_N	γ_N
1	$-9.2163 \cdot 10^{-2}$	$+5.1140 \cdot 10^{-0}$	$-4.692 \cdot 10^{-2}$
2	$+4.3314 \cdot 10^{-2}$	$-3.5685 \cdot 10^{-2}$	$+7.689 \cdot 10^{-4}$
3	$+8.0640 \cdot 10^{-1}$	$-8.6826 \cdot 10^{-3}$	$+1.433 \cdot 10^{-4}$
4	$-6.2723 \cdot 10^{-3}$	$+5.1351 \cdot 10^{-4}$	$-1.246 \cdot 10^{-5}$
5	$-2.7762 \cdot 10^{-3}$	$+1.7792 \cdot 10^{-4}$	$-3.985 \cdot 10^{-6}$

TABLE A11

Polynomial expression for γ ("Density flux" function)
(Veronis, 1972)

$$\gamma = \sum_{i=1}^{6}\sum_{j=1}^{6} A_{ij}(\theta - 10°)^{i-1}(S - 35)^{j-1}$$

The polynomial coefficients A_{ij}, (k) meaning 10^k

j \ $i=$	1	2	3	4	5	6
1	5.089907	0.1529522	−0.2653792(−2)	0.9409070(−4)	−0.3334090(−5)	0.5986311(−7)
2	0.8424673	0.2309882(−1)	−0.8486828(−3)	0.5136553(−4)	−0.3040085(−5)	0.7860879(−7)
3	0.6951154(−1)	−0.1173089(−2)	0.6613862(−4)	−0.1241726(−5)	−0.7876663(−6)	0.4358532(−7)
4	−0.1263383(−2)	−0.1542035(−4)	−0.2640413(−5)	−0.3038523(−5)	0.1563445(−6)	0.4513363(−8)
5	0.2008858(−4)	−0.1205016(−4)	−0.9072095(−5)	0.6932661(−6)	0.1004874(−6)	−0.4025061(−8)
6	−0.1178271(−3)	0.7019903(−6)	0.4057835(−5)	0.1928200(−6)	−0.2404088(−7)	0.6693087(−9)

The formula is valid over the ranges $0 \leqslant \theta \leqslant 30°C$, $33 \leqslant S \leqslant 37°/oo$ and for the ratio of temperature–salinity scales on the T-S diagram, $1°C/1°/oo$, equal to 1/5.

TABLE A12

Specific heat c_p of sea water at atmospheric pressure (J g^{-1} °C^{-1})
(Cox and Smith, 1959)

$S(°/oo)$ \ $T(°C)$	−2	−1	0	1	2	5	10	15	20	25	30
0	—	—	4.217	4.214	4.210	4.202	4.192	4.186	4.182	4.179	4.178
5	—	—	4.179	4.176	4.174	4.168	4.161	4.157	4.154	4.153	4.152
10	—	—	4.142	4.140	4.138	4.135	4.130	4.128	4.126	4.126	4.126
15	—	—	4.107	4.106	4.105	4.103	4.100	4.099	4.098	4.099	4.100
20	—	4.075	4.074	4.074	4.073	4.072	4.071	4.071	4.071	4.072	4.074
25	—	4.043	4.043	4.043	4.042	4.042	4.042	4.043	4.045	4.046	4.048
30	—	4.013	4.013	4.013	4.013	4.014	4.015	4.016	4.018	4.020	4.023
32	—	4.001	4.002	4.002	4.002	4.003	4.004	4.006	4.008	4.010	4.013
34	—	3.990	3.990	3.991	3.991	3.992	3.993	3.995	3.998	4.000	4.003
35	3.984	3.984	3.985	3.985	3.985	3.986	3.988	3.990	3.993	3.995	3.999
36	3.979	3.979	3.979	3.980	3.980	3.981	3.983	3.985	3.988	3.991	3.994
38	3.968	3.968	3.968	3.968	3.969	3.970	3.972	3.975	3.978	3.981	3.985
40	3.957	3.957	3.957	3.958	3.958	3.959	3.962	3.965	3.968	3.972	3.976

REFERENCES

Alekin, O.A., 1966, *Chemistry of the Ocean.* Gidrometeoizdat, Leningrad (in Russian).

Amagat, E.H., 1893. Mémoires sur l'élasticité et la dilabilité des fluides jusqu'aux très hautes pressions. *Ann. Chem. Phys.,* 6 (29).

Anosov, V.Ya. and Pogodin, S.A., 1947. *Fundamental Principles of Physico-Chemical Analysis.* Izd. Akad. Nauk S.S.S.R., Moscow–Leningrad (in Russian).

Aramanovich, I.G. and Levin, V.I., 1964. *Equations of Mathematical Physics.* Nauka, Moscow (in Russian).

Baranov, G.I., 1961. On the possibility of isolating the components of the heat reserve of the sea. *Probl. Arktiki Antarktiki,* 9 (in Russian)

Baranov, G.I., 1965. Method of isolating the components of the heat reserve of a sea of infinite depth. *Probl. Arktiki Antarktiki,* 19 (in Russian).

Bark, L.S., Ganson, P.P. and Meister, N.A., 1961. *Tables of the Speed of Sound in Sea Water.* Computing Center of the Academy of Sciences of the U.S.S.R., Moscow (in Russian).

Bary, B. McK., 1959. Biogeographic boundaries. The use of temperature–salinity–plankton diagrams. *Int. Oceanogr. Congr., Washington, D.C.*

Bary, B. McK., 1963. Temperature, salinity and plankton in the eastern North Atlantic and coastal waters of Britain, 1957. I. The characterization and distribution of surface waters. *J. Fish. Res. Board Can.,* 20 (3).

Batuner, L.M. and Pozin, M.E., 1963. *Mathematical Methods in Chemical Engineering.* Goskhimizdat, Leningrad (in Russian).

Bazarov, I.P., 1961. *Thermodynamics.* Fizmatgiz, Moscow (in Russian).

Bein, W., Hirsehorn, H. and Möller, L., 1935. Konstantenbestimmungen der Meerwasser und Ergebnisse über Wasserkörper. *Berlin Univ., Inst. Meeresk., Veröff. N.F.A. Geogr., Natürwiss. Reihe,* 23.

Belinskii, V.A., 1948. *Dynamic Meteorology.* Gostekhizdat, Moscow–Leningrad (in Russian).

Bellas, F.W., 1961. Comparison of a familiar formula for speed of sound in sea water with the Kuwahara-Mackenzie formula. *J. Acoust. Soc. Am.,* 33 (2).

Beyer, R.T., 1954. Formulas for sound velocity in sea water. *J. Mar. Res.,* 13 (1).

Bialek, E.L., 1964. Adoption of new sound speed equations by U.S. Naval Oceanographic Office. *Int. Hydrogr. Rev.,* 41 (2).

Bjerknes, V. and Sandström, J., 1910. Dynamic meteorology and hydrography. I. Statics. *Carnegie Inst., Publ.* 88, Washington, D.C.

Bjerknes, V. and Sandström, J., 1912. *Statik der Atmosphäre und der Hydrosphäre.* Braunschweig.

Bjerknes, V., Bjerknes, J., Solberg, H. and Bergeron, T., 1933. *Physikalische Hydrodynamik.* Springer, Berlin.

Blokh, L.C., 1952. *Triangular System of Coordinates.* Metallurgizdat, Moscow (in Russian).

Bolin, B. and Stommel, H., 1961. On the abyssal circulation of the World Ocean. IV. Origin and rate of circulation of deep ocean water as determined by the aid of tracers. *Deep-Sea Res.,* 8 (2).

Bonchik, I., 1967. Water masses of the southern Baltic and their principal properties. *Okeanologiya,* 7(2) (in Russian).

Bradshaw, A. and Schleicher, K.E., 1965. The effect of pressure on the electrical conductance of sea water. *Deep-Sea Res.,* 12.

Bradshaw, A. and Schleicher, K.E., 1970. Direct measurement of thermal expansion of sea water under pressure. *Deep-Sea Res.,* 17.

Bromley, L.A., 1968. Heat capacity of sea water solutions. Partial and apparent values for salts and water. *J. Chem. Eng. Data,* 13 (1).

Bromley, L.A., Desaussure, V.A., Clipp, J.C. and Wright, J.S., 1967. Heat capacities of sea water solutions at salinities of 1 to 12% and temperatures of 2° to 80°C. *J. Chem. Eng. Data,* 12 (2).

Bromley, L.A., Diamond, A.E., Salami, E. and Wilkins, D.G., 1970. Heat capacities and enthalpies of sea salt solutions to 200°C. *J. Chem. Eng. Data,* 15 (2).

Bronshtein, I.N. and Semendiaev, K.A., 1953. *Handbook on Mathematics.* Gostekhteorizdat, Moscow (in Russian).

Bruevich, S.V., 1961. On the basic saline composition of ocean water and the calculation of salinity according to Knudsen. *Tr. Inst. Okeanol. Akad. Nauk S.S.S.R.*, 47 (in Russian).

Bryan, K. and Cox, M.D., 1972. An approximate equation of state for numerical models of ocean circulation. *J. Phys. Oceanogr.*, 2 (4).

Bryden, H.L., 1973. New polynomials for thermal expansion, adiabatic temperature gradient and potential temperature of sea water. *Deep-Sea Res.*, 20 (4).

Bubnov, V.A., 1960. On the effect of the process of contraction on mixing on the dynamics of the waters of the subarctic frontal zone of the Pacific Ocean. Collection *Probl. Okeanol.* – Moscow State Univ. Pub., (in Russian).

Bubnov, V.A., 1967. Contraction on mixing and the transformation of the Mediterranean waters in the Atlantic Ocean. *Vestn. Mosk. Univ., Ser. Geogr.*, 1967 (4) (in Russian).

Bubnov, V.A., 1968. On the limits of distribution of intermediate water masses in the northern part of the Atlantic Ocean. *Okeanologiya*, 8 (3) (in Russian).

Bubnov, V.A., 1971. Structure and dynamics of the Mediterranean waters in the Atlantic Ocean. Chapter IV in: Baranov, E.I., Bubnov, V.A., Bulatov, R.P. and Privalova, I.B., *Study of the Circulation and Transport of the Waters of the Atlantic Ocean. Okeanol. Issled.*, 22 – Nauka, Moscow (in Russian).

Bulgakov, N.P., 1960. On the phenomenon of contraction on mixing of waters. *Izv. Akad. Nauk S.S.S.R., Ser. Geofiz.*, 1960 (2) (in Russian).

Burkov, V.A., Godanov, K.T., Gamutilov, A.E. and Shirei, V.A., 1957. Methods of hydrological work in the open sea. *Tr. Inst. Okeanol. Akad. Nauk S.S.S.R.*, 24 (in Russian).

Caldwell, D.R. and Tucker, B.E., 1970. Determination of thermal expansion of sea-water by observing onset of convection. *Deep-Sea Res.*, 17.

Callaway, E.B., 1951. Graphical determination of specific volume anomaly and current. *Trans. Am. Geophys. Union*, 32 (5).

Carritt, D.E. and Carpenter, J.H., 1959. The composition of sea-water and the salinity–chlorinity–density problem. In: *Physical and Chemical Properties of Sea Water. Nat. Acad. Sci., Nat. Res. Counc., Publ.* 600.

Chambadal, P., 1963. *Évolution et Applications du Concept d'Entropie*, Dunod, Paris.

Chappius, P., 1907. Dilatation de l'eau. *Trav. Mém. Bur. Int. Mes.*, 13, D1–D40.

Cochrane, J.D., 1956. The frequency distribution of surface water characteristics in the Pacific Ocean. *Deep-Sea Res.*, 4 (1).

Cochrane, J.D., 1958. The frequency distribution of water characteristics in the Pacific Ocean. *Deep-Sea Res.*, 5 (2).

Connors, D.N., 1970. On the enthalpy of sea water. *Limnol. Oceanogr.*, 15 (4).

Connors, D.N. and Weyl, P.K., 1968. The partial equivalent conductances of salts in sea water and the density–conductance relationship. *Limnol. Oceanogr.*, 13 (1).

Cox, R.A., 1958. The thermostat salinity meter. *Nat. Inst. Oceanogr., Intern. Rep.* C2.

Cox, R.A. and Smith, N.D., 1959. The specific heat of sea water. *Proc. R. Soc. Lond., Ser. A*, 252, no. 1268.

Cox, R.A., McCartney, M.J. and Culkin, F., 1968. Pure water for relative density standard. *Deep-Sea Res.*, 15.

Cox, R.A., McCartney, M.J. and Culkin, F., 1970. The specific gravity–salinity–temperature relationship in natural sea water. *Deep-Sea Res.*, 17.

Craig, H., 1960. The thermodynamics of sea water. *Proc. Nat. Acad. Sci.*, 46.

Crease, J., 1962. The specific volume of sea water under pressure as determined by recent measurements of sound velocity. *Deep-Sea Res.*, 9.

Crease, J., 1971. Determination of the density of seawater. *Nature*, 223 (5318).

Crease, J., Catton, D. and Cox, R.A., 1962. Tables of adiabatic cooling of sea-water. *Deep-Sea Res.*, 9.

Defant, A., 1929. Stabile Lagerung ozeanischer Wasserkörpern und dazu gehörige Stromsysteme. *Berlin Univ., Inst. Meeresk., N.F., Ser. A*, 19.

Defant, A., 1954. Turbulenz und Vermischung im Meer. *Dtsch. Hydrol. Z.*, 7 (1/2).

Defant, A., 1961. *Physical Oceanography*, I. Pergamon, Oxford.

Defant, A. and Wüst, G., 1930. Die Mischung von Wasserkörpern im System *s=f (t)*. *Rapp. Procès Verb. Réun. Cons. Perm. Int. Explor. Mer,* 62.

Del Grosso, V.A., 1952. The velocity of sound in sea water of zero depth. *U.S. Nav. Res. Lab. Rep.* 4002, Washington.

Del Grosso, V.A., 1959. Speed of sound in pure water and sea water. In: *Physical and Chemical Properties of Sea Water. Nat. Acad. Sci., Res. Counc., Publ.* 600.

Dietrich, G., 1950. Kontinentale Einflüsse auf Temperatur und Salzgehalt des Ozeanwassers. *Dtsch. Hydr. Z.,* 3 (1/2).

Dietrich, G., 1964. *General Oceanography.* Wiley, N.Y.

Dobrovol'skii, A.D., 1961. On the determination of water masses. *Okeanologiya,* 1 (1) (in Russian).

Dobrovol'skii, A.D., 1962. *On the Water Masses of the Northwestern Part of the Pacific.* Pishchepromizdat, Moscow (in Russian).

Dobrovol'skii, A.D., Leont'eva, V.V. and Kuksa, V.I., 1960. Towards a description of the structures and water masses of the western and central parts of the Pacific Ocean. *Tr. Inst. Okeanol. Akad. Nauk S.S.S.R.,* 40 (in Russian).

Dorsey, N.E., 1940. Properties of ordinary water-substance. *Am. Chem. Soc., Monogr. Ser.,* 81 – Reinhold, N.Y.

Dorsey, N.E., 1968. Properties of ordinary water-substance. *Am. Chem. Soc., Monogr. Ser.* (facsimile of 1940 edition) – Hafner, New York, N.Y.

Dubrovin, B.I., 1965. Volumetric statistical *T-S* analysis of the water masses of the Arabian and Red Seas. *Okeanologiya,* 5 (4) (in Russian).

Duedall, I.W. and Weyl, P.K., 1965. Apparatus for determining the partial equivalent volumes of salts in aqueous solutions. *Rev. Sci. Instr.,* 36 (4).

Duedall, I.W. and Weyl, P.K., 1967. The partial equivalent volumes of salts in seawater. *Limnol. Oceanogr.,* 12 (1).

Eckart, C., 1958. Properties of water. II. The equation of state of water and sea water at low temperatures and pressures. *Am. J. Sci.,* 256 (4).

Eckart, C., 1959. On the need of a revision of the equation of state of sea water. In: *Physical and Chemical Properties of Sea Water. Nat. Acad. Sci., Nat. Res. Counc., Publ.* 600.

Eckart, C., 1960. *Hydrodynamics of Oceans and Atmospheres.* Pergamon, Oxford.

Eckart, C., 1962. The equations of motion of sea-water. In: M.N. Hill (General Editor), *The Sea. Ideas and Observations on Progress in the Study of the Seas.* 1. *Physical Oceanography.* Interscience, New York–London.

Efimov, N.V., 1964. *Quadratic Forms and Matrices.* Nauka, Moscow (in Russian).

Ekman, V.W., 1908. Die Zusammendruckbarkeit des Meerwassers nebst einigen Werten für Wasser und Quecksilber. *Cons. Perm. Int. Explor. Mer,* Publ. Circonstance, 43, Copenhagen.

Ekman, V.W., 1910. Tables for seawater under pressure. *Cons. Perm. Int. Explor. Mer,* Publ. Circonstance, 49, Copenhagen.

Ekman, V.W., 1914. Der adiabatische Temperaturgradient im Meere. *Ann. Hydr. Mar. Meteorol.,* 42 (6).

Eucken, A., 1948, Zur Struktur des flüssigen Wassers. *Angew. Chem.,* 60.

Everett, D.H., 1960. *An Introduction to the Study of Chemical Thermodynamics.* Longmans, London.

Fabrikant, N.Ya., 1949. *Aerodynamics.* Gostekhizdat, Moscow–Leningrad (in Russian).

Fikhtengolts, G.M., 1956. *Principles of Mathematical Analysis.* Gostekhizdat, Moscow (in Russian).

Fisher, F.H., Williams, R.B. and Dial Jr., O.E., 1970. Analytic equation of state for water and sea water. Fifth report of the Joint Panel on Oceanographic Tables and Standards. *Unesco Techn. Pap. Mar. Sci.,* 14 – Unesco, Paris.

Fofonoff, N.P., 1956. Some properties of sea water influencing the formation of Antarctic Bottom Water. *Deep-Sea Res.,* 4 (1).

Fofonoff, N.P., 1959. Interpretation of oceanographic measurements – thermodynamics. In: *Physical and Chemical Properties of Sea Water. Nat. Acad. Sci., Nat. Res. Counc., Publ.* 600.

Fofonoff, N.P., 1961. Energy transformations in the sea. *Fish. Res. Board Can., Manuscr. Rep. Ser. (Oceanogr. Limnol.)*, 109.

Fofonoff, N.P., 1962. Physical properties of sea water. In: M.N. Hill (General Editor), *The Sea. Ideas and Observations on Progress in the Study of the Seas. 1. Physical Oceanography*. Interscience, New York—London.

Fofonoff, N.P. and Froese, C., 1958. Tables of physical properties of sea water. *Fish. Res. Board Can., Manuscr. Rep. Ser. (Oceanogr. Limnol.)*, 24.

Forch, C., Knudsen, M. and Sörensen, S.P.L., 1902. Berichte über die Konstantenbestimmungen zur Aufstellung der hydrographischen Tabellen. *K. Dan. Vidensk. Selsk. Skr.*, 6 — *Raekke, Naturvidensk. Math. Afd.*, 12 (1).

Friedrich, H. and Levitus, S., 1972. An approximation to the equation of state for sea water, suitable for numerical ocean models. *J. Phys. Oceanogr.*, 2 (4).

Fuglister, F.C., 1960. *Atlantic Ocean Atlas of Temperature and Salinity Profiles and Data from the International Geophysical Year of 1957—1958*. Woods Hole Oceanogr. Inst., Woods Hole, Mass.

Ganson, P.P., 1958. On formulae for the computation of the speed of propagation of sound in sea water. *Meteorol. Gidrol.*, 1958 (4) (in Russian).

Garner, D.M., 1967. Oceanic sound channels around New Zealand. *N.Z. J. Mar. Freshwater Res.*, 1 (1).

General Report of the Participation of Japan in the International Indian Ocean Expedition, V-4, Physical Oceanography 1966. *Res. Oceanogr. Works Jap.*, 8 (2).

Gibbs, J., 1950. On the equilibrium of heterogeneous substances. In: J. Gibbs, *Thermodynamic Works*. Gostekhizdat, Moscow (in Russian).

Greenspan, M. and Tschiegg, C.E., 1959. Tables of speed of sound in water. *J. Acoust. Soc. Am.*, 31 (1).

Gröber, H. and Erk, S., 1955. *Die Grundgesetze der Wärmeübertragung*. Springer, Berlin—Göttingen—Heidelberg, 3rd ed.

Guggenheim, E.A., 1950. *Thermodynamics. An Advanced Treatment for Chemists and Physicists*. North-Holland, Amsterdam.

Haase, R., 1963. *Thermodynamik der Irreversiblen Prozesse*. Dr. Dietrich Steinkopff Verlag, Darmstadt.

Hansen, W., 1965. Ein einfaches Model der Zirkulation im Meere. *Dtsch. Hydr. Z.*, 9 (2).

Hansen, H.J., 1904. Experimental determination of the relation between the freezing point of sea water and the specific gravity at 0°C. *Medd. Komm. Havunders., Ser. Hydrogr.*, 1 (2): 1—10.

Heck, H.H. and Service, I.H., 1924. Velocity of sound in sea water. *U.S. Coast Geod. Surv.*, Spec. Publ., 108.

Helland-Hansen, B., 1912. The ocean waters, 1. *Int. Rev. Ges. Hydrobiol. Hydrogr.*, Leipzig, 5.

Helland-Hansen, B., 1918. Nogen hydrografiske metoder. *Førh. Skand. Naturforskeres*, 16, Kristiania.

Helland-Hansen, B., 1930. *Physical Oceanography and Meteorology*. Report on the Scientific Results of the 'Michael Sars' North Atlantic Deep-Sea Expedition 1910, Bergen, 1.

Helland-Hansen, B. and Nansen, F., 1926. The eastern North Atlantic. *Geofys. Publ. (Oslo)*, 4 (2).

Hesselberg, Th. and Sverdrup, H.U., 1915a. Beitrag zur Berechnung der Druck- und Massenverteilung im Meere. *Bergens Museums Årbok 1914—1915*, 14, Bergen.

Hesselberg, Th. and Sverdrup, H.U., 1915b. Die Stabilitätsverhältnisse des Seewassers bei vertikalen Verschiebungen. *Bergens Museums Årbok 1914—1915*, 15, Bergen.

Hirano, T., 1957. The oceanographic study on the subarctic region of the northwest Pacific Ocean. II. On the formation of the subarctic water systems. *Bull. Tokai Reg. Fish. Res. Lab., Tokyo*, 1957 (15).

Holser, W.T. and Kennedy, C.C., 1958. Properties of water. IV. Pressure—volume—temperature relations of water in the range 100—400°C and 100—1400 mbars. *Am. J. Sci.*, 256 (10).

Horne, R.A., 1969. *Marine Chemistry. The structure of Water and the Chemistry of the Hydrosphere*. Wiley—Interscience, New York, N.Y.

Horne, R.A., 1970. Sea water. *Adv. Hydrosci.*, 6 (107).

Horne, R.A., 1972. Structure of sea water and its role in chemical mass transport between the sea and the atmosphere. *J. Geophys. Res.*, 77 (27).

Horton, H., 1953. *Fundamentals of Sonar*. U.S. Naval Inst., Annapolis, Md.

Institute of Oceanology, U.S.S.R., 1968. *The Pacific Ocean. 2. Hydrology of the Pacific Ocean.* Nauka, Moscow (in Russian).

International Oceanographic Tables, 1966. Nat. Inst. Oceanogr. Great Britain and Unesco.

Iselin, C.O.D., 1936. A study of the circulation of the western North Atlantic. *Pap. Phys. Oceanogr. Meteorol.,* 4 (4).

Istoshin, Yu.V., 1961. The region of distribution of the 18-degree waters of the Sargasso Sea. *Okeanologiya,* 1 (4) (in Russian).

Ivanov, A.V., 1943. The development of the theory of *T-S* curves. *Probl. Arktiki* (in Russian).

Ivanov, A.V., 1944. A graphic method for determining the coefficient of mixing, based on the use of *T-S* curves. *Probl. Arktiki* (in Russian).

Ivanov, A.V., 1946. On the theory of *T-S* curves. *Izv. Akad. Nauk S.S.S.R., Ser. Geogr. Geofiz.,* 10 (6) (in Russian).

Ivanov, A.V., 1949. Determination of the percentage composition of a mixture of water masses of the sea. *Tr. Inst. Okeanol. Akad. Nauk S.S.S.R.,* 4 (in Russian).

Ivanov-Frantskevich, G.N., 1953. The vertical stability of water layers as an important oceanographic characteristic. *Tr. Inst. Okeanol. Akad. Nauk S.S.S.R.,* 7 (in Russian).

Ivanov-Frantskevich, G.N., 1956. On the question of the vertical stability of water layers. *Tr. Inst. Okeanol. Akad. Nauk S.S.S.R.,* 19 (in Russian).

Ivanov-Frantskevich, G.N., 1963. Meeting of the Joint Panel on the Equation of State of Sea Water (Paris, May 23–25, 1962). *Okeanologiya,* 3 (6) (in Russian).

Jacobsen, J.P., 1927. Eine graphische Methode zur Bestimmung des Vermischungskoeffizienten im Meere. *Gerlands Beitr. Geophys.,* 16 (4).

Jacobsen, J.P., 1929. Contribution to the hydrography of the North Atlantic. *The danish Dana Exped., 1920–1922,* 1 (3), Copenhagen.

Jacobsen, J.P., 1943. The Atlantic current through the Faroer–Shetland Channel and its influence on the hydrographic conditions in the northern part of the North Sea, the Norwegian Sea and the Barentz Sea. *Rapp. Verb., Cons. Int. Explor. Mer,* 112.

Jacobsen, J.P. and Knudsen, M., 1940. Urnormal 1937 or primary standard sea water 1937. *Int. Assoc. Phys. Oceanogr., Publ. Sci.,* 7.

Jaeger, W. and Steinwehr, 1915. *Sitzungsber. Berl. Akad. Wiss.*

Kalle, K., 1941. Fluchtentafeln zur Bestimmung der horizontalen Schallgeschwindigkeit (V m/sec) aus Salzgehalt (S ‰) und Temperatur (t °C) des Meerwassers. *Ann. Hydr. Meteorol.,* 69 (4).

Kalle, K., 1942. Das anomale Verhalten des Wassers in Bezug auf einige für die Meereskunde wichtige physikalische Eigenschaften. *Ann. Hydr. Mar. Meteorol.,* 70 (5).

Karavaeva, V.I. and Radzikhovskaia, M.A., 1965. Volumes of the main water masses of the northern part of the Pacific Ocean. *Okeanologiya,* 5 (2) (in Russian).

Kell, G.S. and Whalley, E., 1965. The *PVT* properties of water. I. Liquid water in the temperature range 0 to 150°C and at pressures up to 1 kb. *Phil. Trans. Roy. Soc. Lond., Ser. A,* 258.

Kennedy, G.C., 1957. Properties of water. I. Pressure–volume–temperature relations in steam to 1000°C and 100 bars pressure. *Am. J. Sci.,* 255.

Kennedy, G.C., Knight, W.L. and Holser, W.T., 1958. Properties of water. II. Specific volume of liquid water to 100°C and 1400 bars. *Am. J. Sci.,* 256 (8).

Khanaichenko, N.K., 1947. On types of transformation of water masses. *Meteorol. Gidrol.,* 1947 (1) (in Russian).

Kin'diushev, V.I., 1965. On contraction on mixing of waters in the process of transformation of the Mediterranean water mass in the Atlantic Ocean. *Okeanologiya,* 5 (4) (in Russian).

Kingsman, D.J.J., 1964. Reef coral tolerance of high temperatures and salinities. *Nature,* 202 (4939).

Knudsen, M., 1901. *Hydrographical Tables.* Copenhagen.

Koizumi, M., 1955. Researches on the variations of oceanographic conditions in the region of the Ocean Weather Station "Extra" in the North Pacific Ocean. I. *Pap. Meteorol., Geophys.,* 6 (2).

Kosarev, A.N., 1966. On the water masses of the equatorial part of the Atlantic Ocean. *Okeanologiya,* 6 (3) (in Russian).

Kremling, K., 1972. Comparison of specific gravity in natural sea water from hydrographical tables and measurements by a new density instrument. *Deep-Sea Res.*, 19 (5).

Kusunoki, K., 1962. Hydrography of the Arctic Ocean with special reference to the Beaufort Sea. *Contrib. Inst. Low Temp. Sci., Ser. A*, 17.

Kuwahara, S., 1939. Velocity of sound in sea water and the calculation of the velocity for use in sonic sounding. *Int. Hydrogr. Rev.*, 16 (2).

Lacombe, H., 1965. *Cours d'Océanographie Physique. (Théories de la Circulation Générale. Houles et Vagues).* Gauthiers-Villars, Paris, 392.

La Fond, E.C., 1957. Sea water density at four stations on the East Coast of India. *Ind. J. Meteorol. Geophys.*, 8 (2).

Lamb, H., 1932. *Hydrodynamics.* Cambridge Univ. Press, N.Y., 6th ed.

Landolt, H. and Börnstein, R., 1952. *Zahlenwerte und Funktionen aus Physik, Chemie, Astronomie, Geophysik und Technik. III. Astronomie und Geophysik.*, 6th ed.

Leont'eva, V.V. and Radzikhovskaia, M.A., 1963. On identifying hydrological structures and water masses in the ocean. *Tr. Inst. Okeanol. Akad. Nauk S.S.S.R.*, 16 (in Russian).

Li, Yuan-Hui, 1967. Equation of state of water and sea-water. *J. Geophys. Res.*, 72 (10).

Lineikin, P.S., 1955. On the determination of the thickness of the baroclinic layer of the sea. *Dokl. Akad. Nauk S.S.S.R.*, 101 (3) (in Russian).

Lineikin, P.S., 1957. *Basic Questions of the Dynamic Theory of the Baroclinic Layer of the Sea.* Gidrometeoizdat, Leningrad (in Russian).

Lineikin, P.S., 1961. Wind and thermohaline circulation in the ocean. *Dokl. Akad. Nauk S.S.S.R.*, 138 (6) (in Russian).

Lineikin, P.S., 1962. On zero surface and the deep-water currents of the northern part of the Atlantic Ocen. *Izv. Akad. Nauk S.S.S.R., Ser. Geofiz.*, 6 (in Russian).

Lyman, J., 1959. Chemical considerations. In: *Physical and Chemical Properties of Sea Water. Nat. Acad. Sci., Nat. Res. Counc. Publ. 600.*

Lyman, J., 1969. Redefinition of salinity and chlorinity. *Limnol. Oceanogr.*, 14 (6).

Lyman, J. and Fleming, R.H., 1940. Composition of sea-water. *J. Mar. Res.*, 3 (2).

Mackenzie, K.V., 1960. Formulas for the computation of sound speed in sea water. *J. Acoust. Soc. Am.*, 32 (1).

Mamayev, O.I., 1954. Precise auxiliary table for the construction of specific volume graphs of sea water. *Vestn. Mosk. Univ.*, 12 (in Russian).

Mamayev, O.I., 1958. On the effect of stratification on vertical turbulent mixing in the sea. *Izv. Akad. Nauk S.S.S.R., Ser. Geofiz.*, 7 (in Russian).

Mamayev, O.I., 1960a. On the water masses of the North Atlantic and their interaction. *Tr. Morsk. Gidrofiz. Inst., Akad. Nauk S.S.S.R.*, 19 (in Russian).

Mamayev, O.I., 1960b. On a particularity of the transformation of the intermediate water masses of the World Ocean. *Tr. Morsk. Gidrofiz. Inst., Akad. Nauk S.S.S.R.*, 19 (in Russian).

Mamayev, O.I., 1962a. *T-S* analysis of moving ocean water masses of finite depth. *Okeanologiya*, 2 (2) (in Russian).

Mamayev, O.I., 1962b. *T-S* curves and the vertical stability of ocean waters. *Dokl. Akad. Nauk S.S.S.R.*, 146 (2) (in Russian).

Mamayev, O.I., 1963. *Oceanographic Analysis in the α-S-T-p System.* Moscow State Univ., Moscow (in Russian).

Mamayev, O.I., 1964a. A simplified relationship between the density, temperature and salinity of sea water. *Izv. Akad. Nauk S.S.S.R., Ser. Geofiz.*, 2 (in Russian).

Mamayev, O.I., 1964b. On a description of oceanic *T-S* curves. *Izv. Akad. Nauk S.S.S.R., Ser. Geofiz.*, 4 (in Russian).

Mamayev, O.I., 1964c. On particularities of the distribution of salinity and temperature, connected with the Atlantic equatorial sub-surface counter-current. *Vestn. Mosk. Univ., Ser. Geogr.*, 5 (3) (in Russian).

Mamayev, O.I., 1964d. On the question of the growth and decay of an ice sheet on the sea. *Okeanologiya*, 4 (5) (in Russian).

Mamayev, O.I., 1965a. On methods of analysis of the water masses of the ocean. *Vestn. Mosk. Univ., Ser. Geogr.,* 4 (in Russian).

Mamayev, O.I., 1965b. On calculating the line integral along a *T-S* curve. *Okeanologiya,* 5 (5) (in Russian).

Mamayev, O.I., 1966a. The latitudinal transformation of the water masses of the ocean in the light of the theory of *T-S* curves. *Okeanologiya,* 6 (3) (in Russian).

Mamayev, O.I., 1966b. On the development of the theory of *T-S* curves for an ocean of semi-infinite depth. *Tr. Morsk. Gidrofiz. Inst., Akad. Nauk S.S.S.R.,* 37 (in Russian).

Mamayev, O.I., 1967. Some aspects of the equation of state of sea water. *Int. Union Geodesy and Geophysics, General Assembly. Abstr. Pap.,* 5 – Int. Assoc. Phys. Oceanogr., Berne.

Mamayev, O.I., 1968. Some aspects of the equation of state of sea water. *Izv. Akad. Nauk. S.S.S.R., Fiz. Atmos. Okeana,* 4 (6) (in Russian).

Mamayev, O.I., 1969a. Generalized *T-S* diagrams of the water masses of the World Ocean. *Okeanologiya,* 9 (1) (in Russian).

Mamayev, O.I., 1969b. On the mixing of four water masses of the ocean. *Vestn. Mosk. Univ. Ser. Geogr.,* 5 (3) (in Russian)

Mamayev, O.I., 1970. On the contraction on mixing of four water masses of the ocean. *Okeanologiya,* 10 (3) (in Russian).

Mamayev, O.I., 1973a. Water masses of the South-East Pacific. In: *Oceanography of the South Pacific, 1972.* Papers presented at an International Symposium held in Wellington, New Zealand 9–15 Febr. 1972, compiled by Ronald Frazer, New Zealand Nat. Comm. Unesco, Wellington.

Mamayev, O.I., 1973b. Comments on the paper by G. Veronis "Properties of sea water defined by temperature, salinity and pressure". *J. Mar. Res.,* 31 (1).

Masuzawa, J., 1964a. Flux and water characteristics of the Pacific North Equatorial Current. *Stud. Oceanogr., Tokyo.*

Masuzawa, J., 1964b. A typical hydrographic section of the Kuroshio Extension. *Oceanogr. Mag.,* 16.

Masuzawa, J., 1965. Water characteristics of the Kuroshio. *Oceanogr. Mag.,* 17 (1–2).

Matthews, D.J., 1927. *Tables of the Velocity of Sound in Pure Water and Sea Water for Use in Echo Sounding and Ranging.* Hydrogr. Dep. Admirality, London.

Matthews, D.J., 1939. *Tables of the Velocity of Sound in Pure Water and Sea Water.* Hydrogr. Dept. Admiralty, London, 2nd ed.

Matthews, D.J., 1944. *Tables of the Velocity of Sound in Pure Water and Sea Water.* Hydrogr. Dept. Admiralty, London, 3rd ed.

McLellan, H.J., 1957. On the distinctness and origin of the slope water off the Scotian Shelf and its easterly flow south of Grand Banks. *J. Fish. Res. Board Can.,* 14 (2).

McLellan, H.J., 1965. *Elements of Physical Oceanography.* Pergamon, London.

Ménaché, M., 1971. Vérification, par analyse isotopique, de la validité de la méthode de Cox, McCartney et Culkin tendant à l'obtention d'un étalon de masse volumique. *Deep-Sea Res.,* 18.

Miller, A.R., 1950. A study of mixing processes of the edge of the continental shelf. *J. Mar. Res.,* 9 (2).

Miller, A.R. and Stanley, R.J., 1963. Volumetric *T-S* diagrams for the Mediterranean Sea. *Rapp. Proces-Verb. Réun. C.I.E.S.M.,* 17 (3).

Miller, A.R., Tchernia, P. and Charnock, H., 1970. Mediterranean Sea atlas of temperature, salinity, oxygen. Profiles and data from cruises of R.V. "Atlantis" and R.V. "Chain". *Woods Hole Oceanogr. Inst. Atlas Ser.,* 3, Woods Hole, Mass.

Millero, F.J., 1969. The partial molal volumes of ions in sea water. *Limnol. Oceanogr.,* 14 (3).

Milne-Thomson, L.M., 1960. *Theoretical Hydrodynamics.* MacMillan, London–New York.

Montgomery, R.B., 1937. A suggested method for representing gradient flow in isotropic surfaces. *Bull. Am. Met. Soc.,* 18.

Montgomery, R.B., 1938. Circulation in upper layers of the southern North Atlantic deduced with the use of isotropic analysis. *Pap. Phys. Oceanogr. Meterorol.,* 6 (2).

Montgomery, R.B., 1950. The Taylor diagram (temperature against vapour pressure) for air mixtures. *Arch. Met. Geophys. Bioklim., Ser. A,* 2.

Montgomery, R.B., 1955. Characteristics of surface water at Weather Ship I. *Pap. Mar. Biol. Oceanogr., Deep-Sea Res.,* Suppl. 3.

Montgomery, R.B., 1958. Water characteristics of the Atlantic Ocean and of the World Ocean. *Deep-Sea Res.*, 5 (2).

Montgomery, R.B. and Wooster, W.S., 1954. Thermosteric anomaly and the analysis of serial oceanographic data. *Deep-Sea Res.*, 2 (1).

Mosby, H., 1938. Svalbard waters. *Geofysk. Publ., Oslo*, 12 (4).

Nazarov, V.S., 1963. The quantity of ice-flows of the World Ocean and their changes. *Okeanologiya*, 8 (2) (in Russian).

Nefed'ev, B.P., 1961. Deep water masses of the seas of Indonesia. *Okeanologiya*, 1 (1) (in Russian).

Neumann, A.C. and McGill, D.A., 1961. Circulation of the Red Sea in early summer. *Deep-Sea Res.*, 8 (3/4).

Neumann, G. and Pierson Jr., W.J., 1966. *Principles of Physical Oceanography*. Prentice-Hall, Englewood Cliffs, N.J.

Newton, M.S. and Kennedy, G.C., 1965. An experimental study of *P-V-T-S* relations of sea water. *J. Mar. Res.*, 23 (2).

Oceanological Tables for the Caspian Sea, 1949. U.S.S.R. Nav. Hydrogr. Off. (in Russian).

Oceanological Tables for the Caspian, Aral and Azov Seas, 1964. Gidrometeoizdat, Moscow (in Russian).

Okada, M., 1938. Use of *TS*-diagram for determining the rates of mixing and of replacement of sea water in the coastal regions. I. *Bull. Jap. Soc. Sci. Fish.*, 6 (5).

Okubo, A., 1951. On the heat of mixing of sea water. *Oceanogr. Rep. Centr. Meteorol. Obs.*, 2 (1) (in Japanese).

Okuda, T., 1962. Physical and chemical oceanography over the continental shelf between Cabo Frio and Vitoria (Central Brazil). *J. Oceanogr. Soc. Jap.*, 20th. Anniversary Vol.

Osborne, N.S., Stimson, H.F. and Ginnings D.C., 1939. Measurements of heat capacity and heat of vaporization of water in the range 0° to 100°C. *J. Res. Nat. Bur. Stand.*, 23.

Parr, A.E., 1938. Isopycnic analysis of current flow by means of identifying properties. *J. Mar. Res.*, 1 (2).

Phillips, O.M., 1966. *The Dynamics of the Upper Ocean*. Cambridge Univ. Press.

Pingree, R.D., 1972. Mixing in the deep stratified ocean. *Deep-Sea Res.*, 19.

Pollak, M.I., 1954. Static stability parameters in oceanography. *J. Mar. Res.*, 13 (1).

Pollak, M.I., 1958. Frequency distribution of potential temperatures and salinities in the Indian Ocean. *Deep-Sea Res.*, 5 (2).

Pollak, M.I., 1961. Units for specific volume of sea water. *J. Mar. Res.*, 19 (1).

Polosin, A.S., 1967. On calculating the speed of sound in sea water. *Vestn. Mosk. Univ., Ser. Geogr.*, 3 (in Russian).

Proudman, J., 1953. *Dynamical Oceanography*. Methuen, London – Wiley, New York, N.Y.

Radzikhovskaia, M.A., 1965. Volumes of the principal water masses of the southern part of the Pacific Ocean. *Okeanologiya*, 5 (in Russian).

Rama Sastry, A.A., 1963. Surface water characteristics in the Bay of Bengal off Madras. *Ind. J. Meteorol. Geophys.*, 14 (4).

Rattray, M., Love, C.M. and Heggarty, D.E., 1962. Distribution of physical properties below the level of seasonal influence in the Eastern North Pacific Ocean. *J. Geophys. Res.*, 67 (3).

Reid, R.O., 1959. Influence of some errors in the equation of state in observations on geostrophic currents. In: *Physical and Chemical Properties of Sea-Water. Nat. Acad. Sci., Nat. Res. Counc., Publ. 600.*

Robinson, A.R. and Stommel, H., 1959. The oceanic thermocline and the associated thermohaline circulation. *Tellus*, 11 (3).

Robinson, A.R. and Stommel, H., 1965. The oceanic thermocline and thermohaline circulation. In: *Problems of Ocean Circulation.* Mir, Moscow (in Russian).

Robinson, M., 1960. Statistical evidence indicating no long-term climatic change in the deep waters of the North and South Pacific Oceans. *J. Geophys. Res.*, 65 (7).

Rochford, D.J., 1963. Mixing trajectories of intermediate depth waters of the South East Indian Ocean as determined by a salinity frequency method. *Aust. J. Mar. Fresh-water Res.*, 14 (1).

Rochford, D. J., 1966a. Source regions of oxygen maxima in intermediate depths of the Arabian Sea. *Aust. J. Mar. Fresh-water Res.*, 17 (1).

Rochford, D.J., 1966b. Some hydrological features of the Eastern Arafura Sea and the Gulf of Carpentaria in August 1964. *Aust. J. Mar. Fresh-water Res.*, 17 (1).

Rochford, D.J., 1966c. Distribution of Banda Intermediate Water in the Indian Ocean. *Aust. J. Mar. Fresh-water Res.*, 17 (1).

Sage, B.H., 1966. *Thermodynamics of Multicomponent Systems.* Reinhold, New York, N.Y.

Sarkisian, A.S., 1966. *Theoretical Principles and Calculation of Ocean Currents.* Gidrometeoizdat, Leningrad (in Russian).

Schleicher, K.E. and Bradshaw, A., 1956. A conductivity bridge for measurement of the salinity of sea water. *J. Cons. Int. Explor. Mer*, 22 (1).

Scholl, G., 1914. Adiabatische Temperaturänderung in grossen Meerestiefen. *Ann. Hydr. Mar. Meteorol.*, 42 (6).

Schumacher, A., 1924. Hydrographische Bemerkungen und Hilfsmittel zur akustischen Tiefenmessung. *Ann. Hydr. Mar. Meteorol.*, 52 (4).

Shtokman, V.B., 1938. Some characteristic features of the horizontal mixing of water masses in the system $\theta = f(S)$. *Izv. Akad. Nauk S.S.S.R.*, 8 (in Russian).

Shtokman, V.B., 1939. On the turbulent diffusion of Atlantic waters in the northwestern part of the Kara Sea. *Probl. Arktiki*, 5 (in Russian).

Shtokman, V.B., 1943a. Principles of the theory of θ-S curves as a method for the study of the mixing and transformation of water masses. *Probl. Arktiki*, 1 (in Russian).

Shtokman, V.B., 1943b. On the water masses of the central part of the Arctic Ocean. *Probl. Arktiki*, 2 (in Russian).

Shtokman, V.B., 1944. Geometrical properties of θ-S curves in the mixing of three water masses in an infinite sea. *Izv. Akad. Nauk S.S.S.R.*, 13 (8) (in Russian).

Shtokman, V.B. (Stockmann, W.B.), 1946. A theory of T-S curves as a method for studying the mixing of water masses in the sea. *J. Mar. Res.*, 6 (1).

Shtokman, V.B., 1962. Some observations on O.I. Mamayev's article "T-S analysis of moving ocean water masses of finite depth". *Okeanologiya*, 2 (5) (in Russian).

Shuleikin, V.V., 1953. *Physics of the Sea.* Akad. Nauk S.S.S.R., 3rd ed. – 4th ed. in 1968 (in Russian).

Shumilov, A.V., 1964. On one particular case of the theory of T-S curves. *Okeanologiya*, 4 (3) (in Russian).

Stepanov, V.N., 1965. Main types of structure of World Ocean waters. *Okeanologiya*, 5 (5) (in Russian).

Stommel, H., 1947. Note on the use of the T-S correlation for dynamic height anomaly computations. *J. Mar. Res.*, 6 (2).

Stommel, H., 1961. Thermohaline convection with two stable regimes of flow. *Tellus*, 13 (2).

Stommel, H., 1962a. Examples of mixing and of self-stimulating convection on a T-S diagram. *Okeanologiya*, 2 (2) (in Russian).

Stommel, H., 1962b. On the cause of the temperature–salinity curve in the ocean. *Proc. Nat. Acad. Sci.*, 48 (5).

Stommel, H., 1966. *The Gulf Stream. A Physical and Dynamical Description.* Univ. Calif. Press, Berkeley – Cambridge Univ. Press, London, 2nd ed.

Stommel, H. and Fedorov, K.N., 1966. Small-scale structure in temperature and salinity near Timor and Mindanao. *Tellus*, 19.

Stupochenko, E.V., 1956. The equation of state. In: *Great Soviet Encyclopedia*, 44, 2nd ed. (in Russian).

Sturges, W., 1965. Water characteristics of the Caribbean Sea. *J. Mar. Res.*, 23 (2).

Sturges, W., 1970. On the thermal expansion of sea water. *Deep-Sea Res.*, 17 (3).

Sund, O., 1926. Graphical calculation of specific volume and dynamic depth. *J. Cons.*, 1.

Sverdrup. H.U., 1933. Vereinfachtes Verfahren zur Berechnung der Druck- und Massenverteilung im Meere. *Geofysisk. Publ.*, 10.

Sverdrup, H.U. and Fleming, R.H., 1941. The waters off the coast of southern California, March to July 1937. *Bull. Scripps. Inst. Oceanogr., Tech. Ser.*, 4 (10).

Sverdrup, H.U., Johnson, M.W. and Fleming, R.H., 1942. *The Oceans, their Physics, Chemistry and General Biology.* Prentice-Hall, New York, N.Y.

Tables of Sound Speed in Sea Water, 1962. U.S. Nav. Oceanogr. Off.

Tait, P.G., 1888. Report on some of the physical properties of fresh water and sea water. *Rep. Sci. Res. Voy. H.M.S. "Challenger", Phys. Chem.,* 2.

Takano, K., 1955. An example of thermal convective current. *Res. Oceanogr. Works Jap.,* 2 (1).

Thiesen, M., 1897. *Z. Intrumentenk.,* 17.

Thompson, E.F., 1939. A rapid method for the determination of dynamic heights (or depths) at successive lowerings at an anchor station. *J. Mar. Res.,* 2 (2).

Thompson, T.G., 1932. The physical properties of sea water. *Bull. Nat. Res. Counc.,* 85.

Thompson, T.G. and Wirth, H.E., 1931. The specific gravity of sea water of zero degree in relation to the chlorinity. *J. Cons.,* 6.

Thomsen, H., 1935. Entstehung und Verbreitung einiger charakteristischen Wassermasses in dem Indischen und südlichen Pazifischen Ozean. *Ann. Hydr. Mar. Meteorol.,* 63 (8).

Thoulet, J. and Chevaillier, A., 1889. Sur la chaleur spécifique de l'eau de mer à divers degrés de dilution et de concentration. *Compt. Rend. Acad. Sci. Paris,* 108.

Tibby, R.B., 1941. The water masses off the west coast of North America, *J. Mar. Res.,* 4 (2).

Tilton, L.W. and Taylor, J.K., 1937. Accurate representation of the refractivity and density of distilled water as a function of temperature. *J. Res. Nat. Bur. Stand.,* 18.

Timofeev, V.T., 1960. *Water Masses of the Arctic Basin.* Gidrometeoizdat, Moscow (in Russian).

Timofeev, V.T. and Panov, V.V., 1962. *Indirect Methods of Identifying and Analysing Water Masses.* Gidrometeoizdat, Leningrad (in Russian).

Tiuriakov, B.I., 1964. On the zoning of the North Atlantic according to the principle of identity of structure of water masses. In: *Investigations of the Northern Part of the Atlantic Ocean,* 3. Leningrad Gidrometeorol. Inst. (in Russian).

Tiuriakov, B.I. and Zakharchenko, N.E., 1965. Zoning of the southern half of the North Atlantic according to the principle of identity of structure of water masses. In: *Investigations of the Northern Part of the Atlantic Ocean,* 4. Leningrad Gidrometeorol. Inst. (in Russian).

Tumlirz, O., 1909. Die Zustandgleichung der Flüssigkeiten bei hohem Drucke. *Sitzungber. Akad. Wiss. Wien, Math., Naturwiss. Kl.,* 118a.

Turner, J.S. and Stommel, H., 1964. A new case of convection in the presence of combined vertical salinity and temperature gradients. *Proc. Nat. Acad. Sci.,* 52 (1).

Unesco, 1962. *Report of the Joint Panel on the Equation of State of Sea Water.* Ref. NS/9/114 B, Paris.

Vasiliev, A.S., 1968. On the application of models of density, temperature and salinity in the theory of ocean currents. In: *Problems of the Theory of Wind-Driven and Thermohaline Currents.* Morsk. Gidrofiz. Inst., Akad. Nauk Ukr. S.S.S.R., Sevatopol (in Russian).

Veronis, G., 1972. On properties of seawater defined by temperature, salinity and pressure. *J. Mar. Res.,* 30 (2).

Weyl, P.K., 1970. On the annual temperature–salinity variation of the ocean surface. *J. Geophys. Res.,* 75 (12).

Wilson, W.D., 1959. Speed of sound in distilled water as a function of temperature and pressure. *J. Acoust. Soc. Am.,* 31 (8).

Wilson, W.D., 1960a. Speed of sound in sea water as a function of temperature, pressure and salinity. *J. Acoust. Soc. Am.,* 32 (6).

Wilson, W.D., 1960b. Equation for the speed of sound in sea water. *J. Acoust. Soc. Am.,* 32 (10).

Wilson, W.D., 1962. Extrapolation of the equation for the speed of sound in sea water. *J. Acoust. Soc. Am.,* 34 (6).

Wilson, W.D. and Bradley, D., 1968. Specific volume of sea water as a function of temperature, pressure and salinity. *Deep-Sea Res.,* 15.

Wooster, W.S., Lee, A.J. and Dietrich, G., 1969. Redefinition of salinity. *Limnol. Oceanogr.*, 14 (3).
Worthington, L.V., 1959. The 18° water in the Sargasso Sea. *Deep-Sea Res.*, 5.
Worthington, L.V. and Wright, W.R., 1967. On the deep water masses and circulation of the North Atlantic. *Int. Un. Geodes. Geophys., XIVth General Assembly, Abstr. Paps.,* 5 – *Int. Assoc. Phys. Oceanogr., Berne.*
Wright, W.R. and Worthington, L.V., 1970. *The Water Masses of the North Atlantic Ocean. A Volumetric Census of Temperature and Salinity. Serial Atlas of the Marine Environment,* folio 19. Am. Geograph. Soc., New York.
Wüst, G., 1935. Schichtung und Zirkulation des Atlantischen Ozeans. Die Stratosphäre. *Dtsch. Atl. Exped. "Meteor" 1925–1927, Wiss. Ergeb.,* 6 (1).
Wüst, G., Brogmus, W. and Noodt, E., 1954. Die zonale Verteilung von Salzgehalt, Niederschlag, Verdünstung, Temperatur und Dichte an der Oberflache der Ozeane. *Kieler Meeresforsch.,* 10 (2).
Wyrtki, K., 1963. The horizontal and vertical field of motion in the Peru Current. *Bull. Sci. Inst. Oceanogr.,* 8.
Wyrtki, K., 1967. Circulation and water masses in the Eastern Equatorial Pacific Ocean. *Int. I. Oceanol. Limnol.,* 1 (2).
Wyrtki, K., 1971. *Oceanographic Atlas of the International Indian Ocean Expedition.* Nat. Sci. Found., Washington, D.C.

Yashnov, V.A., 1961. Water masses and plankton. *Zool. J.,* 40 (9); 1963. *Zool. J.,* 42 (7); 1965. *Okeanologiya,* 5 (5); 1966. *Okeanologiya,* 6 (3) (in Russian).
Yasui, M., Yasuoka, T., Tanioka, K. and Shiota, O., 1967. Oceanographic studies of the Japan Sea. I. Water characteristics. *Oceanogr. Mag.,* 19 (2).

Zubov, N.N., 1929. Calculation of elements of sea current based on the data of hydrological sections. *Zap. Gidrogr.,* 68 (in Russian).
Zubov, N.N., 1938. *Sea Waters and Ice.* Gidrometeoizdat, Moscow (in Russian).
Zubov, N.N., 1945. *Arctic Ice.* Izd. Glavsevmorputi, Moscow (in Russian).
Zubov, N.N., 1947. *Dynamical Oceanology.* Gidrometeoizdat, Moscow–Leningrad (in Russian).
Zubov, N.N., 1957a. *Oceanological Tables.* Gidrometeoizdat, Leningrad (in Russian).
Zubov, N.N., 1957b. *Contraction on Mixing of Sea Waters of Different Temperatures and Salinities.* Gidrometeoizdat, Leningrad (in Russian).
Zubov, N.N. and Chigirin, N.I., 1940. *Oceanological Tables.* Gidrometeoizdat, Moscow (in Russian).
Zubov, N.N. and Mamayev, O.I., 1956. *Dynamical Method of Calculation of the Elements of Sea Currents.* Gidrometeoizdat, Leningrad (in Russian).
Zubov, N.N. and Sabinin, K.D., 1958. *Calculation of Contraction on Mixing of Sea Waters.* Gidrometeoizdat, Moscow (in Russian).

SELECTED LIST OF RECENT PUBLICATIONS*

Bezrukov, Yu.F., 1973. Contraction on mixing of water masses of the East China Sea. *Izv. Tikhook. Nauchno-Issled. Inst. Rybnogo Khoz. i Okeanogr. (TINRO)*, 87, Vladivostok (in Russian).
Triangles and quadrangles of contraction on mixing are used to study the effects of contraction on mixing processes over the continental shelf in the East China Sea.
(Relevant to Sections 22, 55–57).

Carmack, E.C., 1974. A quantitative characterization of water masses in the Weddell Sea during summer. *Deep-Sea Res.*, 20.
Statistical θ-S diagrams for the oceanic domain and shelf domain of the Weddell Sea are presented for the following characteristics in bivariate classes: volume, mean depth, oxygen content and silicate concentration.
(Relevant to Section 58).

Carmack, E. and Aagaard, K., 1973. On the deep water of the Greenland Sea. *Deep-Sea Res.*, 20 (8).
Volumetric T-S diagrams for the Upper and Deep Waters of the Greenland Sea are presented and analysed. Also the possible influence of the double-diffusive process on the formation of bottom waters is discussed.
(Relevant to Sections 41 and 58).

Fujino, K., Lewis, E.L. and Perkin, R.G., 1974. The freezing point of seawater at pressures up to 100 bars. *J. Geophys. Res.*, 79 (12).
The following new formulae for the freezing point of seawater are derived:

$$\tau(S,0) = -0.036 - 0.0499S - 0.000112\, S^2\ (17.7 \leqslant S \leqslant 35^\circ/\text{oo})$$

$$\tau(S,p) = \tau(S,0) + 0.00759\, p\ (1 < p < 100\ \text{bars})$$

(Relevant to Section 5).

Gill, A.E., 1973. Circulation and bottom water production in the Weddell Sea. *Deep-Sea Res.*, 20 (2).
The T-S diagrams with isopycnals *appropriate to the depth concerned* (of the type shown in Fig.4 of this book) are utilized in analyzing stabilities and vertical movements of the water layers in the Weddell Sea.
(Relevant to Sections 5, 25).

Kamenkovich, V.M., 1973. *Fundamentals of Ocean Dynamics*. Gidrometeoizdat, Leningrad (in Russian).
The monograph contains chapters on thermodynamics of equilibrium states as well as irreversible processes as applied to the sea water as a two-component system.
(Relevant to Chapter 3).

Mamayev, O.I., 1974. Some theoretical considerations on the equation of state of sea water. (Appendix 2 in: Sixth Report of the Joint Panel on Oceanographic Tables and Standards–Kiel, 24–26 January 1973). *Unesco Technical Papers in Marine Science*, 16, Unesco, Paris.
Provides additional discussion of the hypothesis formulated by the author (see Section 33) that the density $\rho(S,T)$ of sea water under constant pressure is a harmonic function of S and T. Knudsen's model of sea water in fact satisfies the Poisson equation:

$$\frac{\partial^2 \rho}{\partial S^2} + \frac{\partial^2 \rho}{\partial T^2} = \delta(S,T)$$

---

* Added in proof.

equation [33.14], in which the origin of function δ is unknown. The solution of Dirichlet problem and Neuman's problem for the S-T plane is discussed.
(Relevant to Sections 33 and 38).

Mamayev, O.L., 1974. On the problem of conjugated function on the oceanographic T-S diagram. *Vestn. Mosk. Univ., Ser. Geogr.*, 2 (in Russian).
Density flux function $\gamma(S,T)$ is calculated by means of formula [33.17] where the value fo the density of sea water $\rho(S,T)$ is taken according to approximation of Friedrich and Levitus formula [10.5]. The table of values of $\gamma(T,S)$ is presented for the ranges $-2 \leqslant T \leqslant 30°C$ and $30 \leqslant S \leqslant 40°/oo$. Ratio of scales is taken as $\Delta T/\Delta S = 1°/1°/oo = 1$.
(Relevant to Sections 33 and 38).

Monin, A.S., 1973. The hydrothermodynamics of the ocean. *Izv. Acad. Sci., USSR, Atmos. Oceanic Phys.*, 9 (10).
The problem of determination of thermodynamic parameters of sea water is discussed, together with the questions of slow adiabatic motions in the ocean, its stability and energy.
(Relevant to Chapter 3).

Pingree, R.D., 1973. A component of Labrador Sea Water in the Bay of Biscay. *Limnol. Oceanogr.*, 18 (5).
The T-S characteristics are considered along a neutral (or, otherwise, appropriate to the depth concerned – see Fig.4 of this book).isopycnal surface on the T-S diagram, corresponding approximately to the depth 1900 m. Thus the hypothesis of large-scale mixing along such surfaces is tested. Salinity minimum at the above depth in the Bay of Biscay region might originate is such case from the Labrador Sea.
(Relevant to Sections 5, 25).

Turner, J.S. *Buoyancy Effects in Fluids, – Cambridge Monographs on Mechnics and Applied Mathematics.* Cambridge, at the University Press, 1973.
The monograph describes recent developments in the study of mixing and turbulence in stratified fluids. Chapter 8 "Double-diffusive convection" is devoted to the study of thermohaline convection observed in presence of different (molecular) diffusivities of heat and salts.
(Relevant to Section 41).

Wang, Dong-Ping and Millero, F.J., 1973. Precise representation of the P-V-T properties of water and seawater determined from sound speeds. *J. Geophys. Res.*, 78 (30)
Equation of the state of water and seawater is presented in the form of the second-degree secant modulus equation:

$$(v_0 - v_p)/p = D/(B + A_1 p + A_2 p^2)$$

where v_0 and v_p are specific volumes of water or sea water at atmospheric pressure and applied pressure, correspondingly. The results are based on Knudsen's (1901) equations for v_0 and sound speed measurements of Wilson (1959, 1960b). D is specific volume of pure water at $p = 0$, empirical constants B, A_1 and A_2 are functions of temperature and salinity.
(Relevant to Sections 3, 4, 7, 8).

INDEX

Numbers in italics refer to tables